INTRODUCTION TO THE CHARACTERIZATION OF RESIDUAL STRESS BY NEUTRON DIFFRACTION

T0179369

INTRODUCTION TO THE CHARACTERIZATION OF RESIDUAL STRESS BY NEUTRON DIFFRACTION

Michael T. Hutchings

Department of Materials Engineering
The Open University
Milton Keynes, United Kingdom

Philip J. Withers

School of Materials
University of Manchester
Manchester, United Kingdom

Thomas M. Holden

Northern Stress Technologies
Deep River, Ontario, Canada

Torben Lorentzen

DanStir ApS
Danish Stir Welding Technology
Brondby, Denmark

CRC Press
Taylor & Francis Group
Boca Raton London New York

CRC Press is an imprint of the
Taylor & Francis Group, an **informa** business
A TAYLOR & FRANCIS BOOK

CRC Press
Taylor & Francis Group
6000 Broken Sound Parkway NW, Suite 300
Boca Raton, FL 33487-2742

First issued in paperback 2019

ISBN-13: 978-0-415-31000-0 (hbk)
ISBN-13: 978-0-367-39326-7 (pbk)
Library of Congress Card Number 2004054549

Library of Congress Cataloging-in-Publication Data

Introduction to characterization of residual stress by neutron diffraction / M.T. Hutchings [et al.].
     p. cm.
Includes bibliographical references and index.
ISBN 0-415-31000-8 (alk. paper)
1. Neutron radiography. 2. Residual stresses—Measurement. I. Hutchings, Michael T.

TA417.25.I59 2004
620.1'123'0287--dc22

    2004054549

Visit the Taylor & Francis Web site at
http://www.taylorandfrancis.com

and the CRC Press Web site at
http://www.crcpress.com

Preface

After nearly a quarter-century of active research, residual stress measurement by neutron diffraction has come of age. There are now instruments where engineering measurements can be made at nearly all neutron facilities, the first generation of dedicated instruments designed to measure strain in engineering components has been commissioned, and industrial standards for the reliable and repeatable measurement of strain have been drafted. This book provides an introduction to the techniques used, and a handbook for all those interested in undertaking neutron strain measurements. It is written in particular for engineers and materials scientists who have little prior knowledge of neutron scattering techniques.

Many books outlining the principles and application of neutron diffraction have been published, but few thus far have considered the application of neutron diffraction to problems arising in engineering. Moreover, essentially none aims specifically at residual stress measurement by neutron diffraction. The principal objectives of the present volume are to correct that omission and provide the first textbook devoted entirely to the treatment of the neutron diffraction technique for stress measurement. In order to put the technique in context, brief reference will be made to complementary methods, where appropriate, particularly to the rapidly developing techniques using high-intensity synchrotron x-ray diffraction.

Rather than being a diverse collection of articles of relevance to the technique, this book is therefore written as a coherent tutorial guide to prepare students, engineers, and other newcomers to the field for their first neutron diffraction experiment. It will also serve as a reference book for those with more experience in the field. Following the introduction to the subject given in Chapter 1, Chapter 2 focuses on the basic physics of neutron diffraction, especially the particular features of neutron radiation that make it useful as a tool for probing strains and stresses inside solids. After reading this chapter, researchers will be well equipped to understand the physical principles underlying how neutrons are scattered. They will possess the basic groundwork necessary in order to appreciate the extent and limitations of what can be derived from the data, and thus take full advantage of the opportunities that the technique provides. Building on this knowledge, an introduction to the instrumentation and techniques available are given in Chapter 3. This chapter covers the types of sources available, basic elements of the various types of instrument, and a description of the different detector arrangements employed in practice. Through this chapter the reader will acquire a broad insight into the various instrument designs available, and will be helped to understand the benefits and drawbacks of various instrumental configurations, and thereby assisted in identifying the optimum experimental conditions for a given application. At this point the focus turns toward the collection and analysis of data. In Chapter 4, the measurement strategy for the determination of lattice strain is discussed. In Chapter 5, a framework is presented for the reliable interpretation of strain data and the derivation of stress from strain data, in terms of the micromechanics of polycrystalline deformation. Only by taking into account the mechanics of materials at the scale at which the experimental technique probes the strains are we able to reliably interpret and analyze the data obtained. Examples of strategies for modeling the micromechanics are supported by experimental data showing how neutron diffraction may be used in the evaluation of modeling schemes, and how the technique offers opportunities for their refinement. While Chapter 5 focuses on the fundamental aspects of data interpretation, we turn to more practical examples of how the technique may be applied in Chapter 6. By considering selected examples, the reader is introduced to the various scientific and engineering challenges involved in collecting and utilizing the results for the optimization of materials processing, product manufacture, and engineering performance.

It is our hope that this volume will serve as an appetizer for people joining the field as well as a reference book for established experts, and that it will be a useful guide for undertaking successful experiments and data analysis as well as serve as an inspiration for novel applications in the future.

Acknowledgments

Torben Lorentzen had the original concept for an edited book based around a series of separately authored chapters. However, as the book developed, it became apparent that it would be better to blend these into a single integrated co-authored volume. Consequently, the original chapters have been brought together and greatly expanded and the style and nomenclature standardized, by Michael Hutchings and Philip Withers. TL acknowledges Director, Dr. Techn. A. N. Neergaard and Wife's Foundation, and PJW, a Royal Society-Wolfson Merit Award, for respective financial support in undertaking the preparation of this book. TL, MTH, and PJW are grateful to former head of the Risø Materials Research Department, Dr. Techn. Niels Hansen, for his support through the years. The help of Sue Edwards in collating and processing the figures is also acknowledged.

We would like to thank colleagues from all the neutron sources worldwide who shared their expertise and enthusiasm with us in applying neutron diffraction to strain measurement, most notably Andrew Allen and Colin Windsor from their time at Harwell; Andrew Taylor, Mark Daymond, and Mike Johnson at ISIS (Rutherford Appleton Laboratory, Oxfordshire, United Kingdom); Gerald Dolling, Richard Holt, Judy Pang, Brian Powell, Ron Rogge, John Root, and Varley Sears at Chalk River Laboratories (Ontario, Canada); Dorte Juul Jensen (Risø), Mark Bourke, Robert Von Dreele, and Joyce Roberts at Los Alamos National Laboratory (Los Alamos, New Mexico); Hans Priesmeyer (GKSS, Geesthacht), Aaron Krawitz at the University of Missouri Research Reactor (Columbia, Missouri). We also thank Gernot Kostorz; Lyndon Edwards, Michael Fitzpatrick, George and Peter Webster, and most important by, all the doctoral students, post-docs, and instrument scientists with whom we have burned the midnight oil at facilities around the world.

Michael Hutchings

Phil Withers

Tom Holden

Torben Lorentzen

Contents

1

Introduction

In this chapter a brief historical overview of the development of the subject is provided, and the features that make neutron strain measurement a unique tool are examined. While it is not the aim of this book to discuss the nature and origins of residual stress in detail, the latter sections of this chapter illustrate briefly how residual stress can arise, and how they may affect mechanical properties.

1.1 Residual Stress: Friend or Foe?

Residual stresses are self-equilibrating stresses within a stationary solid body when no external forces are applied. As such they are not usually immediately apparent, can be difficult to measure, are hard to predict, and can give rise to unexpected failure if not accounted for. On the other hand, if the origins are fully understood, residual stress can be deliberately introduced into a component in order to lengthen its service life. In view of this, residual stresses have caught the interest of engineers and scientists for many decades. Indeed, it is very rare to encounter engineered components that do not contain residual stresses, caused either by fabrication, joining, assembly, heat or surface treatment, service use, or more probably, by a complex combination of all of these.

In many cases, the stresses are not life-limiting and need not be considered at all. In others, the stresses must be monitored, modified, introduced, and managed intelligently to maximize life. Often, in the past, the question of residual stress has only become an issue because of premature failure of a component, or after other observations had indicated that the structure did not perform as previously expected. In many cases, the influence of unquantified residual stresses has provided a handy catch-all explanation to account for mechanical behavior when all other explanations failed. This situation arose largely because only rarely could residual stress be quantified experimentally. As a result, it remained a feature that engineers knew were present but tried to ignore.

For many years, the calculation of residual stresses and their effects, often termed the "assessment" of residual stress, was limited by the lack of numerical algorithms and computers. Even today calculation of residual stress is rarely found to play a role in the design of structures, and consequently, assumptions are made that often result in an unnecessarily conservative and expensive design. This is no longer due to the lack of numerical tools or powerful computers, since modern computational and analytical approaches to the assessment of residual stress are becoming quite reliable, but rather is more the result of the lack of data characterizing fully the process or processes by which residual stresses are generated. In addition, insufficient knowledge of the constitutive parameters for describing the behavior of materials under the complex deformation history that they experience, during manufacturing and/or in service, often prevents reliable residual stress analysis. As a consequence, the study of residual stress has tended to be essentially a research topic away from the engineering shop floor. This is a great pity because of the potential impact on the structural integrity of engineering components that knowledge of the residual stresses present would provide, helping engineers in the design of structures and in the optimization of manufacturing processes. However, the availability of a suite of standardized destructive and nondestructive experimental tools for stress measurement combined with a drive for finer safety margins is leading to a greater appreciation of residual stresses and their potentially detrimental and beneficial effects. In this respect, diffraction-based techniques are a useful component of an engineer's toolbox since they are able to quantify the distances between atomic planes very accurately, and thereby provide a nondestructive probe for elastic strain and thus stress characterization. In particular, the penetration of thermal neutron radiation makes neutron diffraction an increasingly important technique in assisting engineering design and the advancement of engineering materials.

1.2 Historical Development of Stress Measurement by Diffraction

As discussed in Section 2.3, the basic concepts of diffraction rest on the fundamental relation formulated by W.H. Bragg in 1913 [1,2], namely a relationship between the wavelength of the radiation, λ, the distance between selected lattice planes hkl in a crystalline material, d_{hkl}, and the angle, $2\theta^B_{hkl}$, at which the radiation is scattered coherently and elastically by the correctly oriented crystal lattice planes hkl (see Sections 2.3.1 and 2.32):

$$2d_{hkl}\sin\theta^B_{hkl} = \lambda \tag{1.1}$$

In a sense, the crystal lattice, which constitutes a natural building block of crystalline solids, is adopted as a natural and ever-present *atomic plane strain gauge* embedded in each crystallite or grain. Although the technique does not probe the deformation with atomic spatial resolution, it does probe the average deformation of the lattice planes in a certain sampled volume, as discussed in Sections 3.5.2 and 3.6.1.

Detailed descriptions of the scattering of waves by regular arrangements of atoms can be found in textbooks on solid state physics [3], but Equation (1.1) is, in its simplicity, the essential basis for diffraction-based techniques for strain and hence stress characterization. The average elastic lattice strain in the sampled volume of the component under examination is given in terms of the difference in its lattice plane spacing d_{hkl} relative to that of a strain-free or stress-free reference, d_{hkl}^0. For measurements made on a continuous source diffractometer, this difference manifests itself in a difference or "shift" as it is often termed, in the measured Bragg diffraction angle θ^B of the stressed sample from that measured of the reference sample, as given in Equation (3.8).

1.2.1 Laboratory X-Ray Diffraction

As x-ray diffraction predates and is closely related to the use of neutron diffraction for residual stress measurement, it is useful first to review briefly this technique.

About 20 years after the discovery of x-ray radiation in 1896 [4], the work by Bragg stimulated scientists to pursue the task of residual strain determination using this novel technique. It was not until the 1920s that the topic was addressed in a number of articles, primarily originating in Germany [5–7], as retold in a comprehensive review of its historical development [8]. Since then, laboratory-based x-ray techniques have been used extensively in both scientific and engineering fields, and many important topics have been addressed over the years [9]. However, the major shortcoming of laboratory-based x-ray techniques for stress measurement remains the lack of capability to penetrate deeply into typical crystalline materials used in the engineering industry. Such materials will be termed "engineering" or "industrial" materials in this book, and components and materials under investigation will be referred to as *samples*. The penetration path length is, to some extent, adjustable by appropriate selection of specific x-ray producing targets, and hence the x-ray energies and wavelengths, but in general the technique is limited to penetrations of a few tens of microns as summarized in Table 1.1. Hence, nondestructive laboratory x-ray measurements have been limited to studies of near-surface effects or otherwise thin structures. Only by successively removing surface layers, either by etching, polishing, or gentle machining, can the technique be adapted to probe stresses farther below the surface of a material or component, and correction must then be made to the measured data to compensate for the effect of material removed.

TABLE 1.1

Comparison of Various Properties Important for Diffraction Strain Measurement of the Electron, X-ray Photon, and Neutron

	Laboratory X-rays			Hard X-rays			Electrons			Neutrons		
Energy (E) (Target Used)	6.40 keV (Fe)	8.04 keV (Cu)	17.4 keV (Mo)	35 keV	80 keV	250 keV	100 keV	200 keV	500 keV	1 meV	10 meV	100 meV
Relativistic mass (in m_e)	0.012	0.016	0.034	0.068	0.16	0.48	1.2	1.4	2.0	1839	1839	1839
Wavelength (Å)	1.94	1.54	0.71	0.35	0.15	0.05	0.037	0.025	0.014	9.0	2.9	0.9
Velocity (m/s)	3×10^8	3×10^8	3×10^8	3×10^8	3×10^8	3×10^8	1.65×10^8	2.1×10^8	2.6×10^8	437	1390	4370
Temperature (K) $=E/k_B$	0.74×10^8	0.93×10^8	2.0×10^8	4.1×10^8	9.2×10^8	29×10^8	11×10^8	23×10^8	57×10^8	12	116	1160
Attenuation length in Fe	18 μm	4 μm	34 μm	0.24 mm	2.18 mm	10.5 mm	~100 nm	~100 nm		~0.8cm		
Typical gauge volume	$1 \times 5 \times 0.01$(deep) mm³			$50 \times 50 \times 1000$ μm³			$5 \times 5 \times 100$(thick) nm³			$1 \times 1 \times 1$ mm³		

Note: Relativistic masses are in units of the rest mass of the electron. The neutron attenuation length cited is for a neutron wavelength of 1.8 Å (25 meV).

1.2.2 Synchrotron X-Ray Diffraction

Today many of the limitations of the laboratory-based x-ray techniques have been overcome by the rapid introduction of third-generation synchrotron sources [10]; Figure 1.1 illustrates their historical development. These sources provide access to higher x-ray energies, or hard x-rays as they are commonly known. At these high energies where there are no absorption edges, the attenuation length, defined as the path length over which the intensity falls to e^{-1} or 36.8% of the incident intensity, increases markedly with increasing energy (Table 1.1). This combined with the relatively very high x-ray intensities that they produce leads to path lengths of centimeters in steel and even tens of centimeters in aluminum. As a probe of samples important to engineering and materials science, synchrotron sources now offer the opportunity to study phenomena within most samples. The main advantages, however, are the high intensity and the high collimation of the beam that allow data acquisition rates on the order of seconds if not milliseconds, and the definition of millimeter- to micron-size sampled gauge dimensions. For certain engineering and materials science problems, such as the study of thin films or coatings, hard x-rays may be the optimum choice, although the grain size of many materials sets a lower limit on the sampled volume that can be used with powder diffraction techniques. However, for many engineering problems requiring residual stress measurement, such an extremely high data acquisition rate is not necessary and submillimeter-size spatial resolution is not essential. Employing a lower spatial resolution combined with a higher level of penetration reaching to several centimeters into materials like steel, nickel, or titanium is often necessary. This is not readily achievable even with the hard x-rays available today. Furthermore, a general consequence of using high x-ray energies is that they lead to very small (<10°) scattering angles, resulting in extended diamond-shaped instrumental gauge volumes (see Section 3.5.2). These can lead to impracticably high path lengths required to measure strain in more than one or two directions, even for simple component geometries. Since the determination of stress requires strain measurement in a number of directions, this can pose severe problems if stress rather than strain is the object of measurement.

1.2.3 Neutron Diffraction

Within the last two decades neutron diffraction has been developed as an alternative and complementary technique to the conventional x-ray method. Neutron diffraction has been an important tool in solid-state physics research ever since the pioneering investigations by Nobel Prize winners Brockhouse and Schull. An historical perspective was given in the lectures delivered on the occasion of the presentation of the 1994 Nobel Prize in physics [11,12]. However, not until the early 1980s was an experimental setup for the measurement of residual strain reported. The first papers appeared in 1981 by Allen et al. [13], Pintschovius et al. [14], and Krawitz et al. [15], soon to be

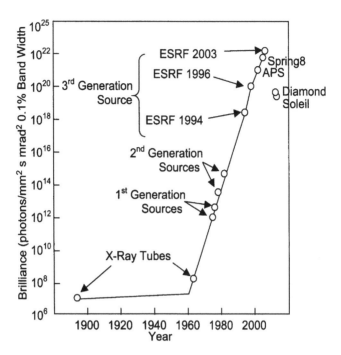

FIGURE 1.1
Historical overview of the development of the brightness of x-ray sources adapted from [10].

followed by further contributions [16–22]. A comprehensive review was given in 1985 [23]. The first conference paying some attention to this emerging technique was the 28th Army Sagamore conference in the United States in 1981 (see, for example, [14], [15]). A NATO Advanced Research Workshop, titled "Measurement of Residual and Applied Stress Using Neutron Diffraction," held in Oxford in 1991, reviewed the progress and emerging questions at that date [24]. Since then, the most popular public fora are the International and European Conferences on Residual Stresses — the ICRS and ECRS conference series, respectively. A recent book edited by Fitzpatrick and Lodini has reviewed the use of neutrons and x-rays for stress management. The preeminent journals reporting neutron strain measurements and analysis are *Acta Materialia, Materials Science and Engineering A, Journal of Neutron Research, Journal of Strain Analysis, Materials Science and Technology, Journal of Applied Crystallography,* and *Journal of Applied Physics.*

1.3   Special Characteristics of Neutron Strain Measurement

Neutron diffraction is an experimental technique with immense potential for the characterization of residual stress via probing the interior of solids. It brings the opportunity to acquire otherwise inaccessible information on the state of strain within the bulk of a structure, and it has been attributed to be "the engineer's dream come true" [25]. The main advantages of the technique are:

- Penetration power of the order of centimeters in most engineering materials
- Nondestructive and can be used to monitor the evolution of residual stress in realistic environments and loading conditions
- Provides a spatial resolution that is readily adjustable and can be adequate for resolving strain gradients in engineering components
- Can be used to infer bulk macroscopic engineering stresses, average phase-specific stresses, and intergranular stresses.

The benefits of these advantages are elaborated in the remainder of this section.

The high neutron penetration power, which is about three orders of magnitude higher than that for conventional laboratory-based x-rays for most materials (Table 1.1 and Figure 2.2), provides access to the interior of solids. For engineering applications, it can therefore provide access to residual elastic strain profiles extending centimeters into structural components. Moreover, the penetrating power in principle enables a free choice of the strain measurement direction. The technique has proved to be a valuable engineering tool used in product design and development, process optimization, and for postfailure evaluation. Common applications, some of which are discussed in detail in Chapter 6, include measurement of stresses due to welding, plastically deformed structures such as autofrettaged tubes and cold expanded holes, shrink-fit assemblies, automotive components such as crank shafts, centrifuges and sheet metal–formed products, and surface treatments such as nitriding and peening.

Through the nature of the diffraction process, which focuses on specific lattice plane spacings of a subset of crystallites, or grains, with specific orientations relative to the scattering geometry, the technique provides unique insights for both fundamental and applied materials science studies. This grain-selective character facilitates the separation of the strain response of different phases in a multiphase material, provided that the phases are distinguishable in a crystallographic sense. A typical example of a multiphase engineering material is duplex stainless steel, which consists of a 50/50 mixture of a face-centered cubic austenitic phase and a body-centered cubic ferritic phase. Neutron diffraction facilitates the investigation of the stress/strain response of the two phases separately, and thereby can provide valuable insight into, for example, the mechanisms of low-cycle fatigue [26,27]. Indeed, many engineering materials comprise two or more phases, whether by design or as a result of their production process.

Composite materials are another class of multi-phase materials for which important information has been provided by neutron diffraction as described fully in Section 6.3. In this case, the properties of one phase, the matrix, are modified by the addition of another, often called the reinforcement. This reinforcing phase may have various morphologies ranging from small sub-micron-sized particles to continuous fibers or layers having dimensions comparable to the structure itself. Whatever the composition, provided that the phases are crystalline, the neutron diffraction technique gives access to the strain response of the various constituents separately. A common example is the Al/SiC family of metal matrix composite materials, the micromechanics of which has been widely studied by neutron diffraction [28–31].

Sensitivity to the crystallographic character of the phases also makes the technique an attractive probe for the investigation of these composite systems where one phase may be actuated through imposed variations in some external field, the so-called "smart" composites. An example is a system where one phase possesses piezoelectric, magnetostrictive, or shape-memory properties, the last group being an example of using a phase transformation. While phase transformations have been studied by diffraction for a long time, it is only in recent years that neutron diffraction has been brought into play in studying their micromechanical aspects. By providing access to the interior of solids, the technique is especially attractive for studies of the micromechanical effects associated with ongoing phase transformations of embedded phases. As an example of this, neutron diffraction has been used to monitor the ongoing phase transformation of embedded NiTi fibers, and this transformation was related directly to observations of the macroscopic response of the composite [32,33]. In this case, the flexibility of the neutron technique with regard to various sample environments is also important due to the neutron's easy penetration of the walls of cells and furnace heat shields and the consequent relative ease with which *in situ* studies can be undertaken.

The selective nature of the technique also adds a new dimension to fundamental investigations into the micromechanics of monolithic materials at the grain level. Through its sensitivity to specific lattice plane spacing and crystallite orientation, the technique facilitates the probing of the strain response of selected families of grain orientations in polycrystalline aggregates as discussed in Sections 5.5, 5.6, and 6.4. While the typical spatial resolution of the neutron technique does not usually facilitate measurements at the individual grain scale, it does allow monitoring of the volume average strain response of families of grains with identical plane orientation within a specified gauge volume, typically a few cubic millimeters. Such investigations now provide valuable insights enhancing our understanding of polycrystal deformation, and are stimulating the refinement of existing modeling techniques as described in Sections 5.5 and 5.6.

Without doubt, the use of neutron diffraction for the characterization of residual stress has now matured to a state where it is used widely in materials research and in related aspects of education. Moreover, it is rapidly being accepted and approved by the engineering industry as a readily available tool. This is leading to its adoption for publicly funded research and development, as well as to direct use by industry for wholly private-sector funded stress measurement. This commercial exploitation of the technique is being accelerated by the establishment of global standards for reliable and repeatable measurement of stress by neutron diffraction. Indeed, at the time this book went to press, a draft standard, "Standard Test Method for Determining Residual Stresses by Neutron Diffraction" (Technology Trends Assessment, ISO VAMAS 2001) [34] had been prepared. Where possible, this book adopts the conventions and approaches recommended by this draft standard.

Residual stress characterization using neutron diffraction is, however, a flux-limited application of a neutron source. As a result the development of the method has depended on the availability of sources with sufficient thermal neutron flux. Further progress and prospects for new applications in the fields of materials science and engineering rely heavily on future developments of new sources with enhanced thermal neutron flux. The advancement of the flux of thermal neutron sources since the discovery of the neutron is shown in Figure 1.2, indicating that fluxes now reach 10^{15} neutrons/cm^2 for some of today's most brilliant sources [35]. The trend line, showing prospects for a further three orders of magnitude with the next generation of pulsed sources such as the Spallation Neutron Source (SNS) at Oak Ridge, Tennessee, and plans for a possible European Spallation Source (ESS), indicates a potential giant leap forward with respect to industrial applications and the exploitation of neutron beams for measuring residual stresses.

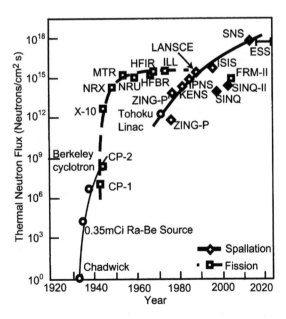

FIGURE 1.2
Historical overview of the development of the intensity of neutron sources (updated from Carpenter and Yelon [35]). Spallation sources are characterized by their peak thermal flux.

1.4   Nature and Origin of Residual Stress

While this book is focused on the measurement of residual strain and hence stress, it is perhaps appropriate at this early stage to examine their nature and origins, if only briefly. Residual stresses arise because of shape misfits, sometimes called "eigenstrains," between the unstressed shapes of different parts, different regions, or different phases of the component. Shape misfits can arise in many ways, as illustrated in Figure 1.3. Often these regions span large distances, such as misfits caused by the plastic bending of a bar, or from sharp thermal gradients such as those caused during welding or heat treatment operations. On the other hand, they can arise on a much shorter scale, such as between the matrix and reinforcement of a composite material, due to differences in properties between the phases. These short-range stresses add to any long-range macrostresses.

Residual stresses may be categorized in several ways:

- By cause, such as thermal or elastic misfits
- By the scale (a) of the size of the misfitting region (see Figure 1.3), which determines the scale over which the stress varies rapidly, l_0; or (b) over which the stress self-equilibrates, L_0
- According to the method by which they are measured.

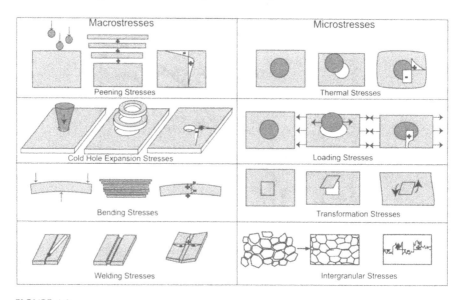

FIGURE 1.3
Residual stresses arise from misfits, either between different regions of a material or between different phases within the material. Examples of different types of residual macro- and micro-residual stress are illustrated schematically. In each case, the process is indicated on the left, the misfit in the center, and the resulting stress pattern on the right.

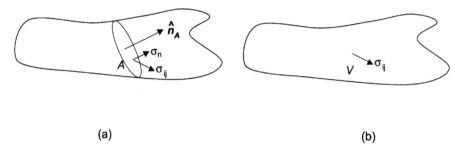

(a)   (b)

FIGURE 1.4
Schematic showing that (a) the normal stress, σ_n must average to zero over any cross-sectional plane, A, within the body, and (b) a stress component, σ_{ij}, must average to zero over the sample volume, V.

All categories are reviewed here from a length-scale perspective. The stresses are commonly divided into three classes, or types, by the length scales over which they vary and over which they self-equilibrate, namely, Type I, Type II, and Type III stresses. They are often categorized as macrostresses (Type I) and microstresses (Types II and III).

1.4.1 Macrostresses and Microstresses

Since residual stress, $\boldsymbol{\sigma}$, is self-equilibrating it must balance across any cross-section, A, within the body, as illustrated in Figure 1.4a. As a result, it must obey the relation

$$0 = \int_A \sigma_n dA = \int_A \boldsymbol{\sigma} \cdot \hat{n}_A dA \qquad (1.2)$$

where $\hat{n}_A$ is a unit vector normal to the area A. In fact, any component of the stress σ_{ij} must average to zero over the whole volume, V, of the body (Figure 1.4b).

$$0 = \int_V \sigma_{ij} dV \qquad (1.3)$$

As noted above, residual stress arises from shape misfits. The residual stress arising from a single misfitting region will fall off with distance from it. In fact, the range of the residual stress field tends to scale with the extent of the misfitting region, l_0, such that beyond $2l_0$ to $3l_0$ the residual stress field caused by it is small. As a consequence, the stress caused by a misfit will often essentially average to zero over a volume, V_0, including the misfit, which is much smaller than the sample volume, V, in Equation (1.3). The distance normal to a stress component over which it averages to zero is denoted L_0. It should be noted that the characteristic lengths, l_0 and L_0, may

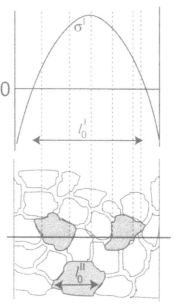

FIGURE 1.5

Type I stress varies on a length scale l_0^I, which is a considerable fraction of the sample size. In this example, the misfit has been introduced by quenching rapidly, setting up long-range misfits. Type I stresses are considered to be continuous even across grain and phase boundaries.

vary with direction in the sample, and in order to satisfy Equation (1.2), it is the variation perpendicular to the direction of the component of stress that is of concern. [36]

Type I stresses self-equilibrate over a length L_0^I, which is comparable to the macroscopic dimension of the structure or component in question. These stresses are typically a consequence of macroscopic misfits generated, for example, by macroscopic plastic deformation or quenching of a hot sample. They are often referred to as *macrostresses*. They are assumed to be continuous from grain to grain, and indeed, even from phase to phase as illustrated in Figure 1.5. Provided that the spatial extent of the sampled gauge volume V_v is much smaller than L_0^I, these stresses generally give rise to detectable diffraction peak shifts.

Type II stresses self-equilibrate over a length scale of L_0^{II}, comparable to that of the grain structure. They arise from misfits having a characteristic length scale, l_0^{II}, comparable to the grain size of polycrystalline solids, usually a few tens of microns. Type II stresses are discontinuous from grain to grain as shown in Figure 1.6. Low-level Type II stresses nearly always exist in polycrystalline materials, simply because elastic — and for symmetries lower than cubic, thermal properties of differently oriented neighboring grains are different. Because of the propensity for large property mismatches between the phases,

Type II Microstress

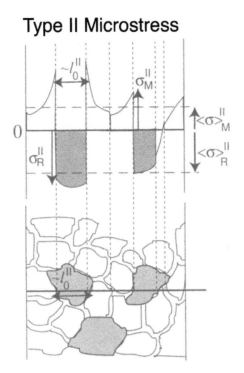

FIGURE 1.6
Type II stress varies on a length scale l_0^{II}, which is of the order of the grain size. The grain-to-grain and phase-to-phase misfits are shown schematically by separating the ill-fitting grains. In this example, the major Type II misfit is caused by differential thermal contraction, which on average, generates tensile stresses ($\langle \sigma \rangle_M^{II}$) in the matrix (white) and compressive average phase stresses ($\langle \sigma \rangle_R^{II}$) in the reinforcement (grey). Elastic mismatches between grains, or phases, in combination with macrostresses will also generate Type II stress.

more significant grain-scale stresses occur in multiphase materials or when phase transformations take place. Note that despite the common name *intergranular stresses* given to some of these stresses, they have nothing to do with stresses directly associated with the grain boundary region.

Type III stresses self-equilibrate over a length scale L_0^{III}, smaller than the characteristic length scale of the microstructure; that is, the grain size or the fiber/particle spacing for composite materials. These could be stresses varying within a specific grain, such as due to grain subdivision into cell structures. In this case, their origin is misfits, such as crystal defects, with a scale shorter than the grain scale shown in Figure 1.7. The Type III category typically includes stresses due to coherency at interfaces and dislocation stress fields.

Types II and III stresses are also sometimes referred to collectively as *microstresses* and can vary in the manner shown in Figure 1.8. The sampled gauge of neutron diffraction is usually too coarse to resolve these, so that they generally give rise to diffraction peak broadening.

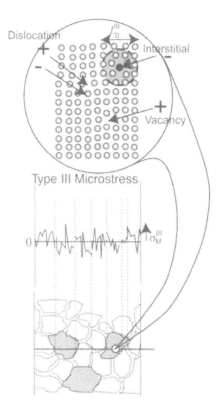

FIGURE 1.7
Type III stress varies on a subgrain length scale l_0^{III}, which is much less than the grain size and equilibrate over a length scale L_0^{III}. In this example, the misfits are introduced by dislocations, vacancies, and interstitials.

1.4.2   Measurement of Macrostresses and Microstresses

Clearly, considerable care must be taken when selecting the most appropriate residual stress measurement technique to gain insight into a particular mechanical problem. First, it is important to ask what components of the stress tensor and types of stress are required by the materials designer to improve performance, or for the engineer to make structural integrity assessment. For example, the designer of composite materials might be interested in the development of Type II volume-averaged phase stresses in order to learn about the transfer of load from the matrix to the reinforcement. In contrast, in the life assessment of monolithic metallic components, Types II and III are often unimportant and attention is focused on Type I macrostresses. According to the problem under consideration, this might, for example, be the hydrostatic tensile component responsible for creep cavitation, or the Mode I crack opening stress component affecting crack growth. It may therefore happen that an unexpected behavior of a piece of equipment or

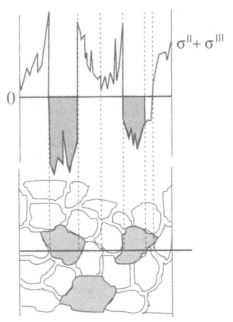

FIGURE 1.8
Type II and Type III stresses are often grouped together and termed microstresses as shown here.

plant may not in fact be due to an incorrect measurement of stress, but to the measurement of the wrong stress contribution caused by an inappropriate choice of technique. A few of the more common destructive and nondestructive stress measurement techniques are summarized in Chapter 6 and by Withers and Bhadeshia [37].

1.4.2.1 Diffraction Techniques

No measurement technique can measure the strain at a single point in a component sample. Diffraction techniques measure a strain averaged over a sampled gauge volume, V_v, usually defined by apertures or collimators as described in Section 3.6.1. Typical sizes of the sampled gauge volume can range from 0.2 to 1000 mm^3.

The manner in which neutron diffraction samples the different types of strain can be understood by comparing the size of the sampled gauge volume, V_v, with the characteristic volume, $V_0^j \sim L_0^{j3}$ [38], over which a given type of stress, j, associated with the stress-causing misfit averages to zero. If the sampled gauge volume is greater than the characteristic volume, then the corresponding strain will not be measured since it averages to zero. In general, neutron diffraction can only record a peak shift for strains and stresses for which the characteristic volume is larger than the sampled gauge dimension, that is, $V_0^j > V_v$. However, since the sampled gauge volume is usually much larger than the grain size, as indeed it must be for true powder diffraction, it is also much larger than the characteristic volumes for Type II and III microstresses. Thus, microstresses, for which V_0^I and V_0^{II} are less than V_v, give rise only to peak broadening. However, because diffraction is selective with respect to phase and to grain orientation, net peak shifts can in fact be recorded for a given reflection when that phase, or family of diffracting grains, is strained differently from the average, even when $V_0^j < V_v$. Of course in such cases, the average over *all* phases or grain orientations within V_0 would indeed be zero.

A useful concept is the local strain, and hence associated stress, averaged over the sampled gauge volume. In practice, this is determined either from a family of planes {hkl} contributing to the diffracted intensity from a monolithic material, or for a given phase from a multi-phase material. The average over all suitably oriented planes within the sampled volume will be implicitly assumed when we write $\underline{\varepsilon}^{hkl}$ and $\underline{\sigma}^{hkl}$ for the strain and stress determined from the {hkl} lattice planes. In the case of a multi-phase material, the *average* stress, derived from the *average* strain as measured by diffraction, will be represented by $\overline{\underline{\sigma}}_i$ for phase i. This average stress is the superposition of the macrostress, $\underline{\sigma}^I$ (or $\underline{\sigma}^{Macro}$) (as in Figure 1.5) and the *mean* phase, $\langle\underline{\sigma}\rangle_i^{II}$, or grain family, $\langle\underline{\sigma}\rangle_{hkl}^{II}$, Type II stress.

$$\overline{\underline{\sigma}}_i = \underline{\sigma}^I + \langle\underline{\sigma}\rangle_i^{II} \tag{1.4}$$

In this way the phase-independent macrostress and the phase-dependent microstresses can be measured separately, as discussed in Section 6.3.3. By studying more than one grain family, or more than one phase, it is possible to obtain information about the bulk engineering macrostress variation, as well as the extent of interphase or intergranular stressing.

1.4.2.2 Other Measurement Techniques

When considering other techniques for measurement of stress, as for diffraction methods it is useful to consider the characteristic volume V_0^j over which a given type of stress, j, averages to zero. If the measurement volume sampled by the destructive technique V_v is greater than V_0^j, then no stress will be recorded. For example, most material-removal techniques, such as hole drilling and layer removal, remove macroscopically sized regions of volume so that $V_v \gg V_0^{II}, V_0^{III}$, so that only the macrostress is recorded. Destructive methods are not phase- or grain-family selective, and so phase-specific average stresses cannot normally be determined using these methods.

1.5   Effects of Residual Stress

Although it is not the aim of this book to discuss the effects of residual stress on performance of engineering components in detail, it is important to examine briefly why residual stress needs to be known so that the appropriate types of stress as defined in the previous section can be measured.

1.5.1   Static Loading

The static loading performance of brittle materials can be improved markedly by the intelligent use of beneficial residual stresses. Common examples include thermally toughened glass and prestressed concrete. In the former, rapid cooling of the glass from elevated temperature generates compressive surface stress that is counterbalanced by tensile stress in the interior. The surface compressive stress, typically of ~100 MPa, means that any surface flaws that would otherwise cause failure at very low levels of applied tensile stress experience in-plane compression. While the interior experiences counterbalancing tensile stress this region is largely defect-free, and so the inherent strength of the glass is sufficient to prevent failure. Of course, once a crack penetrates to the interior of the material that is under tension, it can grow rapidly and catastrophically to give the characteristic shattered mosaic-like pattern. In fact, it is possible to assess the original residual stress from the average dimensions of the shattered pieces, since the greater the original elastic energy the smaller the pieces. Like glass, concrete is also brittle and hence has low defect tolerance and an associated low tensile strength. Concrete can nevertheless be used in tension, as in cantilever beams, after it has been prestressed in compression. Considerable applied tensile stress can then be tolerated before the superposition of the applied and residual stresses lead to a net tension of the concrete.

For plastically deformable materials, the residual and applied stresses can only be added together directly until the yield strength is reached. In this respect, residual stress may accelerate or delay the onset of plastic deformation and can be important in situations where failure is by plastic collapse. However, the effect of residual stress on static ductile fracture is often small because the residual *elastic* misfit strains are usually small compared with the *plastic* strains generated prior to failure. As a result, they are obliterated long before failure occurs. Indeed, uniform plastic prestraining is an accepted means of residual stress leveling [39].

Neutron diffraction has proved to be an important tool in the study of the mechanical behavior of composite materials under static loading. Here the phase selectivity is invaluable, allowing the experimenter to separate the Type II stress behavior of the various phases in order to optimize composite performance [40]. Issues such as reinforcement cracking, de-bonding, matrix yielding, and the effects of thermal residual stresses have all been elucidated by neutron diffraction measurements of Type II average phase stresses. Some examples are given in Chapter 6.

1.5.2 Fatigue Loading

Most components do not fail immediately under static loads, but fail at lower stresses by cyclic fatigue processes. The effects of residual stress on fatigue lifetimes are covered in detail in, for example, Bouchard [41]. There are many types of fatigue processes, including cyclic fatigue, fretting fatigue, thermo-mechanical fatigue, and stress corrosion fatigue. Here we will consider only cyclic tension–tension fatigue (σ_{max} and $\sigma_{min} \geq 0$), as in Figure 1.10, but many of the principles are transferable to understanding other fatigue processes. Fatigue is generally divided into two types [42]: low-cycle fatigue (LCF) and high-cycle fatigue (HCF). Furthermore, it depends on whether the component is already cracked, in which case crack propagation is life controlling, or uncracked, in which case crack initiation may comprise a significant part of fatigue life.

LCF includes fatigue at stresses above the general yield point and involves less than 10,000 cycles to failure. This usually occurs under imposed plastic strain-controlled cycling, such as the thermal cycling of constrained parts. Generally, mean stress, and therefore residual stress, has a negligible effect on LCF life, as the original residual stresses tend to be obliterated within the first or second cycle by the large amplitude of the oscillating strains.

HCF involves larger numbers of cycles below general yield, is generally stress controlled, and includes rotating or vibrating systems such as axles. HCF is sensitive to residual stress levels. At least two important parameters govern HCF life: the mean stress $\sigma_{mean} = (\sigma_{max} + \sigma_{min})/2$, and the cyclic stress amplitude, $\Delta\sigma = (\sigma_{max} - \sigma_{min})$. For uncracked samples, Basquin's Law states that for zero mean stress the number of cycles to failure, N_f, can be expressed in the form

$$\Delta\sigma \, (N_f)^n = C \tag{1.5}$$

where n and C are empirical constants [42]. A macroresidual stress will superimpose a mean stress, as a constant offset, on any fatigue loading. Two empirical relationships have been developed to quantify the detrimental effect of mean stress on fatigue life. These are the linear Goodman relation proposed in 1899 and the Gerber parabolic relation proposed in 1874 [43], shown in Figure 1.9. From these it can be seen that because a tensile residual stress increases the mean stress, the stress amplitude must be reduced accordingly if lifetime is to be unaffected. On the other hand, in combination with Equation (1.5) they quantify the extent to which compressive residual stress may extend fatigue life.

Large engineering structures usually contain cracks, or defects or stress concentrators, from which cracks can readily propagate. In such

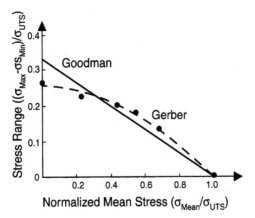

FIGURE 1.9
In practice, it is observed that the fatigue life under HCF is sensitive to the mean stress as well as the stress amplitude. In order to estimate a specific fatigue life, N_f cycles for a previously uncracked sample, two empirical relations have been developed by Goodman (continuous line) and Gerber (dashed line) describing how the amplitude of the fluctuating stress must be reduced as the mean stress increases. The stresses are normalized by the ultimate tensile stress, °UTS. The data points correspond to Wohler's data for Krupp steel, after Sendeckyj [43].

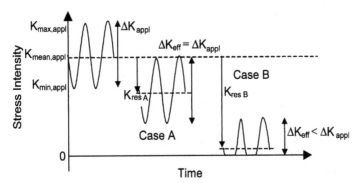

FIGURE 1.10
Graph showing the way that the effective stress intensity range ΔK_{eff} resulting from an applied fatigue cycle ΔK_{appl} is modified by a compressive residual stress. Both levels of compressive residual stress (A and B) act to reduce the tensile mean stress $K_{mean, eff}$. In contrast to Case A, in Case B the compressive residual stress (equivalent to $K_{res\,B}$) is sufficient to close the crack for part of the cycle, reducing the stress intensity range experienced at the crack tip, and thereby retarding the steady-state crack growth rate described in Equation (1.7).

circumstances, one needs to know how many cycles (i.e., how much time) it will take for one of these cracks to grow to such an extent that catastrophic fracture can occur. Because the stress felt at the crack-tip is related to the current crack length as well as the loading geometry, the loading of the crack tip is usually expressed in terms of the "stress intensity factor," K, rather than the applied stress σ, where

$$K = Y\sigma\sqrt{\pi a} \qquad (1.6)$$

Here, a is the length of an edge crack or half the length of an internal crack, and Y is a geometrical constant approximately equal to 1. The applied mean stress intensity factor K_{mean}, and the applied stress intensity range ΔK are important (Figure (1.10)). Residual stress can raise or lower the mean stress experienced over a fatigue cycle. Free surfaces are often a preferred site for the initiation of a fatigue crack, which means that considerable advantage can be gained by engineering a compressive in-plane stress in the near-surface region, such as by peening, auto-frettage, cold-hole expansion, or case hardening. During fatigue crack growth, the "near threshold," that is, ΔK just large enough to propagate a fatigue crack, and "high-growth" rate, large ΔK regimes are strongly affected by mean stress. In the intermediate steady-state "Paris" regime, the crack growth rate da/dN is given by [42]

$$\frac{da}{dN} = A(\Delta K)^m \tag{1.7}$$

where A and m are constants and is largely insensitive to mean stress. As shown in Figure 1.10, imposed macro residual stress alters the mean stress, but does not necessarily affect ΔK. However, if the residual stress causes the crack to close over part of the fatigue cycle, the stress intensity range experienced by the crack decreases, slowing the crack growth rate. Thus, unless a change in the mean stress brings about crack closure, residual stress has little effect on crack growth rates in the Paris regime [44].

On the whole, unless the crack-tip plastic zone is smaller than V_0^{II}, Type II and Type III stresses tend to be relieved by crack-tip plasticity, so that only Type I stress need be considered from a fatigue viewpoint. As a result, it is important to be able to derive an accurate measure of the engineering macrostress when making residual stress measurements by neutron diffraction, although this is not always easy because one must make sure to account for the possibility of intergranular stresses, discussed in Chapter 5. This is not true for short crack growth, where the crack is comparable in length to the scale of the microstructure. In this regime, crack volume is comparable to V_0^{II} and is microstructurally Type II stress dependent. Furthermore, Type II stresses can be important when considering the fatigue of composite materials for which crack bridging, crack deflection, and microcracking can all be important crack-retarding mechanisms. Another example where Type II stresses may become important is in stress corrosion cracking, which can often be intergranular in nature.

These examples illustrate that the effects of residual stress on structural integrity must be taken seriously where component life and safety are important. They influence crack initiation, crack growth and fracture processes. For accurate assessments of component life and structural integrity, it is therefore important to characterize the residual stress field and quantify its effect on degradation processes. Neutron diffraction provides an excellent nondestructive means of doing this, and recently great progress has been made in standardizing measurement techniques to arrive at reliable stress values for incorporation into lifing assessments. Indeed, at the present time, more sophisticated methods are required for the inclusion of measured residual stress into lifing assessments in order to exploit the recent gains in experimental measurement, and thereby to maximize economic advantage and safe component operating lifetimes.

References

1. W.H. Bragg, The reflection of x-rays by crystals, *Proc. R. Soc.*, 88, 428–438, 1913.
2. W.H. Bragg, The reflection of x-rays by crystals, *Proc. R. Soc.*, 89, 246–248, 248–277, 1913.
3. C. Kittel, *Introduction to Solid State Physics*, 6th ed., John Wiley & Sons, New York, 1986.
4. W.C. Röntgen, Üeber eine neue art von strahlen (vorlaufige mittheilung), *Sitzungsberichte der Wurzburger Physikalisches Medizines Gesellschaft*, 137–147, 1896.
5. H.H. Lester and R.H. Aborn, Behaviour under stress of iron crystal in steel, *Army Ordnance*, 6, 120–127, 200–207, 283–287, 364–369, 1925.
6. A.E. van Arkel, Über die verformung des kristallgitters von metallen durch mechanische bearbeitung, *Physica B*, 5, 208–212, 1925.
7. A.F. Joffe and M.V. Kirpitcheva, Röntgenograms of strained crystals, *Phil. Mag.*, 43, 204–206, 1922.
8. V. Hauk, *Structural and Residual Stress Analysis by Non-Destructive Methods*, Elsevier Science, Oxford, 1997.
9. B.D. Cullity, *Elements of X-Ray Diffraction*, 2nd ed., Addison-Wesley, Reading, MA, 1978.
10. European Synchronition Radiation Facility, Grenoble, France, Highlights 95–96, 2000, available at: www.esrf.fr, accessed on 9 Sept. 2004.
11. C.G. Schull and E.O. Wollan, Early development of neutron scattering, *Rev. Mod. Phys.*, 67, 753–757, 1995.
12. B.N. Brockhouse, Slow neutron spectroscopy and the grand atlas of the physical world, *Rev. Mod. Phys.*, 67, 735–751, 1995.
13. A.J. Allen, C. Andreani, M.T. Hutchings, and C.G. Windsor, Measurement of internal stress within bulk materials using neutron diffraction, *NDT Int.*, 15, 249–254, 1981.
14. L. Pintschovius, V. Jung, E. Macherauch, R. Schäfer, O. Vöhringer, Determination of residual stress distributions in the interior of technical parts by means of neutron diffraction, in *Residual Stress and Stress Relaxation*, E. Kula, and V. Weiss, Eds., Proceedings of the 28th Army Materials Research Conference, July 1981, Plenum Press, New York, 1981, 467–482.
15. A.D. Krawitz, J.E. Brune, M.J. Schmank, E. Kula, and V. Weiss, Eds.,. Measurement of stress in the interior of solids with neutrons in Residual Stress and Stress Relaxation, in Proceedings of the 28th Army Materials Research Conference, July 1981, Plenum Press, New York, 1981, 139–155.
16. M.J. Schmank and A.D. Krawitz, Measurement of stress gradient through the bulk of an aluminum alloy using neutrons, *Metall. Trans. A*, 13, 1069–1075, 1982.

17. L. Pintschovius and V. Jung, Residual stress measurements by means of neutron diffraction. *Mater. Sci. Eng.*, 61, 43–50, 1983.

18. L. Pintschovius, B. Scholtes, and R. Schröder. Determination of residual stresses in quenched steel cylinders by means of neutron diffraction, in *Microstructural Characterization of Materials by Non-Destructive Techniques*, Proceedings of the 5th Risø International Symposium on Materials Science, N.H. Andersen et al., Eds., Risø National Laboratory, Roskilde, 1984, 419–424.

19. S.R. MacEwen, J.Faber Jr., and P.L. Turner, The use of time-of-flight neutron diffraction to study grain interaction stresses, *Acta Metall.*, 31, 657–676, 1983.

20. B.L. Allen and W. Lord. Finite element modeling of pulsed eddy current phenomena, in *Tenth Annual Review of Progress in Quantitative Nondestructive Evaluation*, 1983. University of California, Santa Cruz, California, D.O. Thompson and D.E. Chimenti, Eds., Plenum Press, New York, 1983, 561–568.

21. T.M. Holden, B.M. Powel, G. Dolling, and S.R. MacEwen. Internal strain measurements in over rolled cold-worked Zr2.5 wt% Nb pressure tubes by neutron diffraction, in *Microstructural Characterization of Materials by Non-Destructive Techniques*, Proceedings of the 5th Risø International Symposium on Materials Science, N.H. Andersen et al., Eds., Risø National Laboratory, Roskilde, 1984, 291–294.

22. R.A. Holt, G. Dolling, B.M. Powel, T.M. Holden, and J.E. Winegar. Assessment of residual stresses in Incoly-800 tubing using x-ray and neutron diffraction, in *Microstructural Characterization of Materials by Non-Destructive Techniques*, Proceedings of the 5th Risø International Symposium on Materials Science, N.H. Andersen et al., Eds., Risø National Laboratory, Roskilde, 1984, 295–300.

23. A.J. Allen, M.T. Hutchings, C.G. Windsor, and C. Andreani, Neutron diffraction methods for the study of residual stress fields, *Adv. Physics*, 34, 1985, 445–473.

24. M.T. Hutchings and A.D. Krawitz, Eds., *Measurement of Residual and Applied Stress Using Neutron Diffraction*, NATO ASI Series E: Applied Sciences, vol. 216, Kluwer, Dordrecht, 1992.

25. M.T. Hutchings, Neutron diffraction measurements of residual stress fields — the engineer's dream come true? *Neutron News*, 3, 14–19, 1992.

26. A.J. Allen, M. Bourke, W.I.F. David, S. Dawes, M.T. Hutchings, A.D. Krawitz, and C.G. Windsor. Effect of elastic anisotropy on the lattice strains in polycrystalline metals and composites measured by neutron diffraction, in Proceedings of 2nd International Conference on Residual Stresses, C. Beck, S. Denis, and A. Simon, Eds., Elsevier Applied Science, London, 1989, 78–83.

27. M.T. Jensen, P. Brøndsted, B.S. Johansen, T. Lorentzen, and O.B. Pedersen. Phase coupling and low cycle fatigue in a duplex stainless steel, in ICSMA-10, *Fundamental Physical Aspects of the Strength of Crystalline Materials*, H. Oikawa et al., Eds., Japan Institute of Metals, Sendai, 1994, 489–492.

28. P.J. Withers, D.J. Jensen, H. Lilholt, and W.M. Stobbs. The evaluation of internal stresses in a short fibre metal matrix composite by neutron diffraction, in Proceedings ICCM VI/ECCM2, 1987, F.L. Matthews et al., Eds., Elsevier, London, 1987, 255–264.

29. P.J. Withers, *The Development of the Eshelby Model and Its Application to Metal Matrix Composites*, Ph.D. thesis, University of Cambridge, 1988.

30. G.L. Povirk, M.G. Stout, M. Bourke, J.A. Goldstone, A.C. Lawson, M. Lovato, S.R. MacEwen, S.R. Nutt, and A. Needleman, Thermally and mechanically induced residual strains in Al-SiC composites, *Acta Metall.*, 40, 2391–2412, 1992.

31. T. Lorentzen and A.P. Clarke, Thermo-mechanical induced residual strains in Al/SiCp metal matrix composites, *Composites Sci. Technol.*, 58, 345–353, 1998.

32. D.C. Dunand, D. Mari, M.A.M. Bourke, and J.A. Roberts, NiTi and NiTi-TiC composites, 4 and Neutron diffraction study of twinning and shape memory recovery, *Met. Mater. Trans. A*, 27, 2820–2836, 1996.

33. W.D. Armstrong, T. Lorentzen, P. Brøndsted, and P.H. Larsen, An experimental and modeling investigation of the external strain, internal stress and fiber phase transformation behaviour of a NiTi actuated aluminum metal matrix composite, *Acta Metall.*, 46, 3455–3466, 1998.

34. G.A. Webster, Ed., Polycrystalline materials — Determinations of residual stresses by neutron diffraction, ISO/TTA3 Technology Trends Assessment, Geneva, 2001.

35. J.M. Carpenter and W.B. Yelon, Neutron sources, in *Methods of Experimental Physics*, vol. 23A *Neutron Scattering*, K. Sköld and D.L. Price, Eds., Academic Press, New York, 1986, pp. 99–196.

36. P.J. Bouchard and P.J. Withers, The appropriateness of residual stress length scales in structural integrity, *J. Neutron Res.*, 12, 81–91, 2004.

37. P.J. Withers and H.K.D.H. Bhadeshia, Residual stress. I. Measurement techniques, *Mater. Sci. Technol.*, 17, 355–365, 2001.

38. P.J. Withers, Residual stresses: measurement by diffraction, in *Encyclopedia of Materials: Science & Technology, IV: Structural Phenomena*, K.H.J. Buschow et al., Eds., Elsevier, Oxford, 2001, 8158–8170.

39. M.B. Prime and M.R. Hill, Residual stress, stress relief and inhomogeneity in Al plate, *Scripta Mat.*, 46, 77–82, 2002.

40. T.W. Clyne and P.J. Withers, *An Introduction to Metal Matrix Composites*, Cambridge Solid State Series, Cambridge University Press, Cambridge, 1993.

41. P.J. Bouchard, Residual stresses in lifing and structural integrity assessment, in *Encyclopedia of Materials: Science and Technology, IV: Structural Phenomena*, K.H.J. Buschow et al., Eds., Elsevier, Oxford, 2001, 8134–8142.

42. M.F. Ashby and D.R.H. Jones, *Engineering Materials: An Introduction to Their Properties and Applications*, Pergamon, Oxford, 1980.

43. G.P. Sendeckyj, Constant life diagrams — a historical review, *Int. J. Fatigue*, 23, 347–353, 2001.

44. P.J. Withers, Residual stress: definition, in *Encyclopedia of Materials: Science and Technology, IV: Structural Phenomena*, K.H.J. Buschow et al., Eds., Elsevier, Oxford, 2001, 8110–8113.

45. M.E. Fitzpatrick and A. Lodini, Eds., *Analysis of Residual Stress by Diffraction using Neutron and Synchrotron Radiation*, Taylor & Francis, London, 2003.

2

Fundamentals of Neutron Diffraction

A neutron beam has a unique character which makes it a powerful nonde-structive probe of materials. Neutron scattering is a large and important field, but here we concentrate only on aspects that relate directly to the strain measurement technique. In this chapter, the concepts of diffraction are intro-duced, and the formulation of the scattering and absorption cross-sections described. In each section, formal expressions are introduced, and if there are important experimental consequences of the theory, reference is made to the following chapters in which they are discussed in more detail.

2.1 Introduction

The widespread use of both thermal neutrons and x-rays for investigating the arrangement of atoms in solids stems from the fact that they both have wavelengths of the same order as the separations of the atoms in crystalline materials. In a scattering process, a neutron may be described by its wave vector k, of magnitude $2\pi/\lambda$ and directed along its velocity path. Because of the wave nature of matter, the de Broglie wavelength of the neutron, λ, is related to the momentum, p, of the particle, by

$$p = m_n v = hk/2\pi = \hbar k \qquad (2.1)$$

where m_n and v are the mass and velocity of the neutron, and h is Planck's constant. The energy of the neutron is $E = 1/2\, m_n v^2 = h\nu$ where ν is the frequency of the radiation. These relations are discussed further in Section 2.3.

Table 2.1 gives a summary of the common units used in neutron scatter-ing and the conversion between them. Note from the table that a neutron of energy 1 meV has an equivalent temperature of 11.60 K, a wavelength of 9.045 Å and a velocity of 437 m/s. It can be seen from Table 2.1 that a wavelength of 1.8 Å corresponds to a neutron velocity of 2198 m/s. This

TABLE 2.1

Conversion Factors Between Neutron Energy (in meV), Temperature T (=E/k_B) (in K), Wavelength (in Å), Wave Vector (in Å^{-1}), and Velocity (m/s)

	E (meV)	T (K)	λ (Å)	k (Å)$^{-1}$	v (m/s)
E (meV)	—	0.0862T	81.80/λ^2	2.0721k^2	5.227 × 10^{-6} v^2
T (K)	11.60E	—	949.3/λ^2	24.05k^2	60.66 × 10^{-6}v^2
λ (Å)	9.045/√E	30.81/√T	—	6.2832/k	3.956 × 10^3/v
k (Å)$^{-1}$	0.695√E	0.2039√T	6.2832/λ	—	1.588 × 10^{-3}v
v (m/s)	437.4√E	128.4√T	3.956 × 10^3/λ	0.630 × 10^3k	—

Note: 1 meV = 0.2418 THz.

wavelength is frequently chosen as a standard reference value. This typical neutron velocity is somewhat faster than the speed of sound in air, and can easily be measured from the flight time over a known distance — the basis of all pulsed neutron techniques (Section 3.3.2).

By comparison, in the case of the x-ray photon, the momentum is $p = h/\lambda$, and the energy $E = h\nu$. The wavelength is given by

$$\lambda = c/\nu = 12.398/ E, \qquad (2.2)$$

where c is the speed of light and ν the frequency of the radiation. Here the numerator is for λ in Å and E in keV. Note that for wavelengths useful for diffraction, thermalized neutron energies are orders of magnitude less than the corresponding energies of x-rays or electrons, as seen from Table 1.1.

Many textbooks have been written on the theory and applications of neutron scattering over the 40 years that intense beams of neutrons have been available, including those by Bacon [1], Egelstaff [2], Lovesey [3], Squires [4], Windsor [5], Sköld and Price [6], and Sears [7]. Windsor [5] concentrates on pulsed neutrons, while Sears's [7] theoretical treatment is especially thorough. A recent review of neutron diffraction by Von Dreele [8] covers many of the topics treated here. All these primary references contain complete descriptions of the technique, and have been drawn upon in selecting aspects that are relevant to the accurate measurement of strain by neutron diffraction.

An historical treatment of diffraction was given in Chapter 1, with the applications of x-ray diffraction for the measurement of strain dating back almost as long as the concept of Bragg diffraction itself. To the present day, strain measurement by x-ray diffraction is more widely practiced than that by neutron diffraction. The textbook by Cullity [9] on x-ray diffraction remains a classic in the field for its comprehensive treatment and simplicity. The textbook by Noyan and Cohen [10] gives an excellent treatment of the topic of stress measurement, principally by x-ray diffraction.

2.2 Scattering and Absorption of Neutrons and X-Rays by Atoms

2.2.1 Neutron Scattering

As charge-neutral particles, neutrons are scattered by a nuclear interaction with the nuclei of atoms. The strength of this nuclear scattering is described by the magnitude of an experimentally determined nuclear scattering length, b, which has order of magnitude $\sim 10^{-12}$ cm. The scattering event is described in terms of a cross-section, defined as the rate of occurrence of the event per incident flux Φ_0. The flux Φ_0 is the number of neutrons incident on the atom per unit area per unit time, usually expressed as number/cm² s. The cross-section has the dimensions of area $[L]^2$ with order of magnitude $\sim 10^{-24}$ cm², and the unit 10^{-24}cm² is called a "barn." In some sense, it represents the area of each nucleus as seen by the neutron. If we consider the event as neutron scattering by a single stationary atom, of one isotope, and no nuclear spin, the scattering is isotropic since the wavelength of the neutron is very much larger than the size of the nucleus. We can define a total cross-section σ as

$$\sigma = (\text{No. of neutrons scattered into } 4\pi \text{ solid angle/sec})/\Phi_0 \quad (2.3a)$$

In this case,

$$\sigma = 4\pi b^2 \quad (2.3b)$$

It is more helpful when considering diffraction to define a differential cross-section $d\sigma/d\Omega$ per atom, where the event is the scattering of a neutron by the atom into a solid angle $d\Omega$. This will give the detection count rate of a 100%-efficient detector accepting the solid angle $d\Omega$. In this simple case,

$$\frac{d\sigma}{d\Omega} = b^2 \quad (2.4)$$

As the neutron possesses a spin of $S = 1/2$ and a magnetic moment, it can also be magnetically scattered by the dipolar interaction with any moment the atom might possess through magnetic electrons in unfilled shells. The atoms in ferritic iron and nickel carry a moment, for example. If absolute scattering intensities need be considered, this interaction must be also taken into account through a magnetic scattering cross-section. However, for strain measurement we are usually not concerned with absolute intensities of scattering, and we shall largely ignore this magnetism as it normally does not affect the lattice spacing of engineering materials. The magnetic scattering intensity is usually less than that from the nuclear scattering in most industrial materials. It is confined to the low-order, low-angle, diffraction peaks since it is governed by a magnetic form factor, analogous to the x-ray form factor (see Section 2.2.2), arising from the spatial distribution of the magnetic moment of the order of an atomic radius.

It is often appropriate to consider a macroscopic, rather than the atomic, cross-section, particularly when considering attenuation of a neutron beam by a sample. This is designated by Σ or $d\Sigma/d\Omega$ for the total or differential macroscopic cross-section, respectively. It is defined as the scattering per unit volume of the material, and is related to the atomic cross-section by

$$\Sigma = \sigma N_a, \tag{2.5}$$

where N_a is the number of atoms per unit volume. The macroscopic cross-sections have the dimensions of $[L]^{-1}$, and the usual units are cm^{-1}.

Whereas the above relations for the scattering of a neutron from a single atomic nucleus are quite straightforward, in practice atoms often have several isotopes, some of which may have a nuclear spin, and the scattering length differs for each isotope and nuclear spin. Also, many materials contain more than one type of atom as a compound or as an alloy, and in our case of interest they are present as an ordered crystalline array. We will first discuss the effect of nuclear spin and isotopic constituency.

If the isotope has a nonzero nuclear spin, the value of b is different when the spin of the neutron is parallel or antiparallel to the nuclear spin. For an unpolarized neutron beam, the scattering length of an assembly of nuclei of an isotope, i, for which the nuclear spin is nonzero, may be written in terms of an average value $\bar{b}_i$. If there are several isotopes, then the average scattering length is

$$\bar{b} = \Sigma_i a_i^I \bar{b}_i, \tag{2.6}$$

where the sum is over the relative abundance a_i^I of each isotope i.

There are two important cross-sections for the atom, these involve the average $\bar{b}$ for an atom, and the mean square deviation of the b's from this average,

$$\overline{\left(b - \bar{b}\right)^2} = \overline{b^2} - (\bar{b})^2.$$

They are rigorously derived from the detailed theory of neutron scattering (not discussed here). The square of the average scattering length, $\bar{b}^2$, appears as the coefficient of the *coherent* scattering from an assembly of atoms of the same element, that is, when there are definite correlations in the positions of different atoms in a material such as when they are positioned on a lattice.

There is then a well-defined phase difference between the neutron waves scattered from each atom. In the simplest case of an assembly of fixed uncorrelated isolated atoms with a range of isotopes, some with nuclear spin, the *average* coherent total cross-section and differential cross-section per atom are

$$\sigma^{coh} = 4\pi\bar{b}^2 \text{ and } \frac{d\sigma^{coh}}{d\Omega} = \bar{b}^2 \qquad (2.7)$$

The mean square deviation of the random distribution of scattering lengths over spin and isotopes, $(\overline{b^2} - (\bar{b})^2)$, appears as the coefficient of the *incoherent* cross-section of an assembly of atoms. This incoherent cross-section represents the scattering from individual atoms, and is present if the positions of the atoms are correlated or not. Thus, for any assembly of atoms,

$$\sigma^{incoh} = 4\pi\left(\overline{b^2} - \bar{b}^2\right) \text{ and } \frac{d\sigma^{incoh}}{d\Omega} = \left(\overline{b^2} - \bar{b}^2\right) \qquad (2.8)$$

In general, coherent scattering tells us about the correlations in space and time between all the atoms, including each atom with itself, whereas incoherent scattering tells us about the self spatial and time correlations of each individual atom. We are mainly concerned with the coherent elastic scattering, which concerns the time-averaged spatial positions of the atoms.

As well as being scattered, a neutron may be absorbed by a nucleus to form a compound nucleus with emission of daughter products and γ-rays. The products may radioactively decay. The cross-section per atom for the absorption event is σ^{abs}. Absorption removes the neutron from the beam, and so contributes to the beam attenuation. In general, the absorption of neutrons by most nuclei is small, and for most industrially relevant materials neutrons penetrate on the order of tens of millimeters through the material (e.g., Table 2.2). However, there are some notable exceptions such as B, Li, Cd, and Gd, which are put to good use in the construction of neutron counters and shielding.

Thus, when a neutron approaches a single nucleus, four events are possible:

- The neutron passes by the atom unscattered if the separation is large compared to the nuclear interaction distance, which is the most likely outcome
- The neutron can be coherently scattered
- The neutron can be incoherently scattered
- The neutron can be absorbed by the nucleus.

TABLE 2.2

Comparison of Distances in Several Engineering Materials Over Which Beam Intensity of Neutrons, Synchrotron X-Rays at 3 Instruments, and Laboratory X-Rays, is Attenuated by 63.2%

Radiation	Energy (keV)	Wavelength (Å)	Attenuation Length l_μ (mm) for 63.2% Intensity Decrease				
			Al	Ti	Fe	Ni	Cu
Thermal neutrons	2.53×10^{-5}	1.8	96	17	8.3	4.8	10
ID 15 (ESRF) x-rays	150	0.08	39	14	7	5	5
ID 11 (ESRF) x-rays	49	0.25	10	1.9	0.7	0.5	0.4
16.3 SRS x-rays	30	0.41	3.3	0.46	0.16	0.11	0.10
Laboratory x-rays (Cu Kα)	8.05	1.54	0.076	0.011	0.004	0.023	0.021

ESRF, European Synchrotron Radiation Facility; SRS, Synchrotron Radiation Source, Daresbury Laboratory, UK.

2.2.1.1   Coherent Scattering Cross-Section

A major difference, usually advantageous, of neutron diffraction over x-ray diffraction methods arises from the fact that the average scattering length, $\bar{b}$, for each element varies essentially randomly with atomic number and with atomic mass. The average scattering length and coherent scattering cross-section for some common elements in engineering materials are listed in Table 2.3. The variation in scattering cross-section with atomic number is shown schematically in Figure 2.1. In contrast, the analogous x-ray scattering cross-sections increase monotonically with atomic number, Z. The neutron scattering length may have positive or negative values. Most nuclei have positive values, but a few, notably H, Mn, and Ti, have negative scattering lengths, and the sign must be included when calculating the intensity of scattering as discussed in Section 2.2.5. By taking appropriate fractions of two isotopes (e.g., titanium and niobium), it is possible to design alloys with zero average scattering length, that is, they do not coherently scatter neutrons at all. However, they will strongly scatter incoherently. Because of the irregularity of the scattering length according to atomic number, the scattering from light atoms such as hydrogen may be as strong as that from heavy atoms, such as uranium. Thus, the positions of light atoms in the presence of heavy atoms in a material can be determined using neutron diffraction.

2.2.1.2   Incoherent Scattering Cross-Section

The incoherent scattering cross-section for some common elements is given in Table 2.3. Incoherent scattering is almost isotropic and contributes to the background underneath the diffraction peaks discussed in Section 2.3. Its contribution to the background is usually small relative to diffraction peak

TABLE 2.3

Atomic Coherent Scattering Length, and Coherent, Incoherent, and Absorption Cross-section for 1.8 Å Neutrons, for Selected Elements

Element	$\bar{b}$ (10^{-12} cm)	σ^{coh} (barn)	σ^{incoh} (barn)	σ^{abs} (1.8 Å) (barn)	N_v (10^{22}) cm^{-3}	$l_{0.95}$ (mm)
H	−0.374	1.758	80.27	0.383	0.0054 (gas)	
D	0.667	5.592	2.05	0.0005	0.0054 (gas)	
B	0.530	3.54	1.70	767.0	13.04	0.3
C	0.665	5.550	0.001	0.004	9.527	47
N	0.936	11.01	0.5	1.9	0.0054 (gas)	
O	0.580	4.232	0.000	0.0002	0.0054 (gas)	
Al	0.345	1.495	0.008	0.231	6.024	287
Ti	−0.344	1.485	2.87	6.09	5.708	50
V	−0.038	0.018	5.07	5.08	7.222	41
Cr	0.364	1.66	1.83	3.05	8.316	55
Mn	−0.373	1.75	0.4	13.3	7.9	24
Fe	0.945	11.22	0.4	2.56	8.491	25
Ni	1.03	13.3	5.2	4.49	9.131	14
Cu	0.772	7.485	0.55	3.78	8.491	30
Zr	0.716	6.44	0.02	0.185	4.295	105
Nb	0.7054	6.253	0.0024	1.15	5.54	73
Mo	0.672	5.67	0.04	2.48	6.415	57
Cd	0.487	3.04	3.46	2520	4.634	0.27
Sn	0.623	4.87	0.022	0.626	2.92 (gray)	147
Gd	0.65	29.3	151	49700	3.026	0.03
Ta	0.691	6.00	0.01	20.6	5.543	20
W	0.486	2.97	1.63	18.3	6.322	21
Pb	0.941	11.115	0.003	0.171	3.299	81
U	0.842	8.903	0.005	7.57	4.794	38

Note: Macroscopic values are provided in Appendix 4. The number of atoms per unit volume, N_v, at normal temperature and pressure (NTP), and the distance in millimeters, $l_{0.95}$ over which a neutron beam is attenuated by 95% of its initial value ($l_{0.95} = 3l_\mu$) are also tabulated [12].

heights, partly because the incoherent scattering is isotropic, whereas all the coherent scattering occurs at specific well-defined scattering angles, and also because σ^{coh} is often greater than σ^{incoh}. However, the value of the spin-incoherent cross-section for vanadium is 5.07 barns, while that for hydrogen is the largest of all the elements with a value of 80.27 barns.

2.2.1.3 Absorption Cross-Section

The cross-sections for absorption of neutrons for a number of common elements, $\sigma^{abs}_{1.8}$, at the standard wavelength of 1.8 Å, are also tabulated in Table 2.3. The absorption cross-section is proportional to wavelength, so that at wavelength λ the cross-section, σ^{abs}_λ, is given by

$$\sigma^{abs}_\lambda = \sigma^{abs}_{1.8} \frac{\lambda}{1.8} \tag{2.9}$$

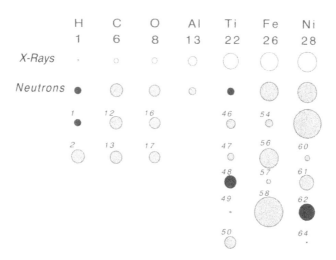

FIGURE 2.1

Illustrating the variation of coherent x-ray and neutron-scattering cross-sections for various elements and their isotopes. The x-ray cross-sections, which are not scaled relative to the neutron cross-sections, are represented by open circles whose radii are proportional to the atomic number Z given below the element. The neutron cross-sections are represented by shaded circles whose radii are proportional to the coherent scattering length. A black circle indicates a negative scattering length. The second row shows the average neutron scattering cross-section of each element as a shaded circle. Where several isotopes contribute, the individual cross-sections are labeled by the mass number A in italics (adapted from Bacon [1]).

2.2.1.4   Tabulation of Neutron Cross-Sections

The average coherent scattering length, incoherent scattering cross-section, and absorption cross-section for all the elements, as well as the values for the individual isotopes are available in many places, including Sears [11,12] and in the "Barn Book" [13]. Table 2.3 provides a list of these properties for a number of industrially relevant materials. A more complete list is given in Appendix 4, Table A.4.1.

2.2.2   X-Ray Scattering

It is useful to compare the above neutron cross-sections with corresponding cross-sections for x-ray diffraction. In contrast to the nuclear scattering of neutrons, x-rays are scattered by the electron cloud surrounding the nucleus. The atomic scattering factor, analogous to the scattering length for neutrons, increases systematically with atomic number Z. Since the distribution of electrons in space, of the order of the atomic radius, is comparable in dimension to the x-ray wavelength, interference between the wavelets scattered by the electrons around an atomic site gives rise to a reduction of the atomic scattering as the angle of scattering increases. This is expressed as a form factor. Although neutron scattering from a nucleus is isotropic, there is an analogous form factor for neutron magnetic scattering from the magnetic

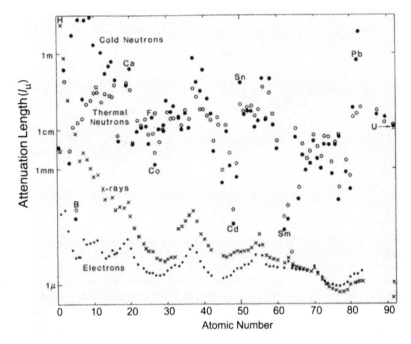

FIGURE 2.2
The penetration depth, or attenuation length l_μ, over which the intensity falls to $1/e$ of that incident, for cold neutrons (closed circles), thermal neutrons (open circles), x-rays (crosses), and electrons (dots), as a function of atomic number (adapted from Hutchings and Windsor [24]).

electrons as mentioned above. Generally, the atomic scattering factor for x-rays far exceeds the corresponding scattering length for neutrons, which reflects the much greater probability of an x-ray being scattered by an atom. However, because of x-ray dependence of the atomic scattering factor on atomic number, it is hard to distinguish between two elements with close atomic numbers, and to identify the scattering from light elements in the presence of heavy elements (Figures 2.1 and 2.2).

The absorption of characteristic laboratory x-rays of wavelength around 1 Å in solids is extremely high compared with neutrons, as illustrated in Table 2.2. This absorption is associated with electronic transitions between the electron levels in the material, such as the K-levels of the atom, and varies as $\lambda^3 Z^4$. An exchange of energy occurs between the x-rays and the electrons, and a photoelectron is emitted, accompanied by fluorescent radiation as the atom returns to its ground state. As a result, for x-rays from conventional laboratory sources, the penetration of the radiation into solids is in general measured in microns, whereas for neutrons the penetration is of the order of tens of millimeters (Tables 1.1 and 2.2). This means that neutrons provide a method of measuring lattice spacing, and hence strain, at depth in samples such as industrial components, while conventional x-rays provide the spacing and the strain at the surface. One advantage of laboratory-type x-ray

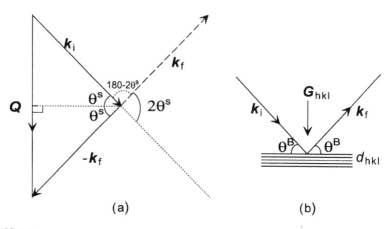

FIGURE 2.3
(a) Vector representation of the incident and scattered neutron wave vectors, and the scattering vector, Q, for elastic scattering. The scattering angle, $\phi^s = 2\theta^s$, is the angle between the incident and scattered beams. The direction of the scattering vector is along the bisector of the nonincluded angle, $(180 - 2\theta^s)°$, between the incident and scattered neutron beam. (b) Showing the incident and scattered neutron wavevectors in the Bragg reflection condition, with the *hkl* lattice planes and their normal vector G_{hkl}.

systems is that portable sources and detectors are available, and so measurement of near-surface strains and stresses can be made *in situ* on structures in the field. However, for short wavelength x-rays, the absorption becomes orders of magnitude smaller. This is of major significance for strain measurements at synchrotron sources where short-wavelength x-rays are copious [14–16]. Short wavelength x-rays penetrate much farther than laboratory x-rays, and permit strain measurement at depth (Table 2.2).

2.3   Neutron Diffraction from Crystalline Solids

The direction of a beam of neutrons incident on a sample or scattered from it, and hence the direction of its associated wave vector k, is determined by apertures or collimators placed in the beam path, as discussed in Section 3.2. If a neutron is scattered by a sample, there is, in general, an interchange of momentum and energy between them. If the incident and scattered neutrons are described by wave vectors k_i and k_f, as shown in Figure 2.3a, and energy E_i and E_f, respectively, the change in momentum and energy of the neutron is given by the conservation equations

$$\hbar Q = \hbar k_i - \hbar k_f \qquad (2.10)$$

and

$$hv = \hbar\omega = E_i - E_f \tag{2.11}$$

The quantity $Q = k_i - k_f$ is the wave vector transfer to the sample, called the *scattering vector*. As this book is concerned almost entirely with elastic scattering of the neutron from the sample, in which there is no energy change, we mainly consider the case where $E_i = E_f$ and so $v = \omega = 0$, and hence $|k_i| = |k_f|$. The incident and scattered neutron wave vectors and the scattering vector are shown in Figure 2.3a for this case. The angle, $2\theta^s$, between the incident and scattered wave vectors is called the *scattering angle*, often designated as ϕ^s, is also shown in Figure 2.3a. It follows from Equation (2.10) and Figure 2.3a that the magnitude of the scattering vector to the sample, $|Q|$, is

$$|Q| = 2k \sin\theta^s = 4\pi \sin\theta^s/\lambda \tag{2.12}$$

The direction of Q is along the bisector of the nonincluded angle, that is, $(180 - 2\theta)°$ between the incident and scattered neutron wave vectors.

2.3.1 Crystal Lattice

The crystalline structure of materials and conditions for diffraction are dealt with in detail in a number of standard textbooks (e.g., Kittel [17] and Ashcroft and Mermin [18]), or classic crystallography textbooks (e.g., Buerger [19]). A good introduction to diffraction is given by Krawitz [20]. Therefore, we outline only the main points here.

Every crystalline *structure* describing the regular position of atoms in a crystalline material may be considered in terms of a *lattice* of points in real space. Associated with each point is an array of atoms called a *basis*. Thus, a structure comprises a lattice with a basis. The lattice of points is described in terms of lattice vectors in real space, with every point on the lattice l given in terms of the fundamental translation vectors, or *crystal axes*, a, b, and c, which are chosen so that

$$l = n_1 a + n_2 b + n_3 c \tag{2.13}$$

where n_1, n_2, and n_3 are integers. The lattice points can thus be said to have translational symmetry. The magnitudes of a, b, c, namely a, b, c, are termed the lattice constants, which define the lattice together with the angles between these axes, α, β, and γ, where α is the angle between b and c and so on.

Associated with the lattice is the concept of a *unit cell*, a building block that by repetition through linear translation, covers all points on that lattice, that is, all the crystal. The smallest such unit cell is called the *primitive unit*

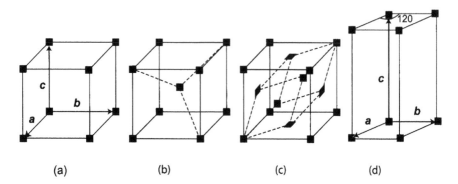

(a)   (b)   (c)   (d)

FIGURE 2.4
(a) Primitive cubic (e.g., Ni$_3$Al), (b) body-centered cubic (e.g., ferritic steel), (c) face-centered cubic (e.g., Al, Cu, Ni, austenitic steel), and (d) hexagonal (e.g., titanium) crystal structures showing the position of the lattice points. The solid lines indicate the conventional cubic nonprimitive cells for the bcc and fcc structures, while the lattice vectors describing the primitive unit cells are shown as dashed lines.

cell, usually chosen to have one lattice point at each corner; the lattice points will then lie on what is called a *primitive lattice.* Sometimes it is more convenient to use a larger building block. For cubic lattices, it is more intuitive to use three perpendicular translational vectors rather than the primitive ones, with the cubic cell referred to as a "conventional" unit cell. While the description of the crystal structure is the same, there will now be a larger basis of atoms associated with each conventional lattice point and unit cell.

The simplest monatomic cubic materials have a basis of just one atom situated at each lattice point of the primitive cell, but more complex structures and compounds have a basis of several atoms. In fact, in the case of the hexagonal close-packed structure, there must always be at least two atoms per lattice point. For example, for Ti there is a basis of two atoms per lattice point, at (0,0,0) and (2/3, 1/3, 1/2). There are 14 crystal space lattices, called *Bravais lattices,* distinguished according to their symmetry properties or space group, that is, the group of translation, reflection, and rotational operations that takes the lattice into itself (e.g., Kittel [17]). The face-centered cubic, body-centered cubic, and hexagonal are the most common Bravais lattices for engineering materials (Figure 2.4).

One may consider all lattice points as lying on a series of planes in the crystal. A two-dimensional everyday analog is the rows of trees observed when passing the edge of a woodland planted in a regular array. The accepted description of the planes was originally proposed by W.H. Miller at Cambridge in 1839 to describe the faces of naturally occurring crystals. In this system, the planes are described by *Miller indices,* denoted *hkl.* The plane (*hkl*) intersects the axes of the unit cell at *a/h, b/k,* and *c/l,* respectively, as shown in Figure 2.5, where *hkl* are the smallest integers giving the correct ratio of intercepts. The perpendicular distance between lattice planes of type (*hkl*), one of which passes through the origin, is given by d_{hkl}, and is also shown in Figure 2.5.

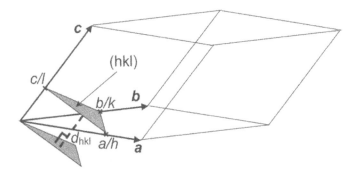

FIGURE 2.5
A schematic representation in real space of lattice planes described by the Miller indices (*hkl*)
for a general structure. The perpendicular spacing between planes d_{hkl} is shown.

2.3.2 Diffraction from Crystal Lattice Planes

Coherent elastic scattering, or diffraction, from atoms arranged in crystal
planes with spacing *d* is very strong only for certain specific, well-defined
scattering angles. This can be seen schematically by considering a plane wave
of radiation of wavelength λ incident at an angle θ^s to the atomic planes as
shown in Figure 2.6, and scattered at an equal angle. We have already seen
that individual atoms scatter both x-rays and neutrons in all directions. Consider the scattering from two atoms lying in successive atomic lattice planes
vertically above one another; each will scatter radiation coherently with a
phase relationship maintained between the incident and scattered waves. If
we consider the sum of their scattered amplitudes at a diffraction angle θ,
that is, a scattering angle of $\phi^s = 2\theta^s$, the two scattered waves will have
traveled distances different by $2d\sin\theta^s$, and will ordinarily be out of phase
with each other. However, if this extra path length for scattering from successive lattice planes is equal to a whole integral number, *n*, of wavelengths
$n\lambda$, then the scattered waves will be in phase and the total scattered amplitude
will simply be the sum from each scattering event. This occurs when

$$n\lambda = 2d \sin\theta^B \qquad (2.14)$$

which is a form of the familiar condition for diffraction known as *Bragg's Law*.
    In describing diffraction from planes with Miller indices *hkl*, it is conven-
tional to include the order *n* into the lattice spacing. For example, consider
the case that the lattice spacing *d* gives rise to a path length 2λ between
successive lattice planes for a scattering angle $\phi^s = 2\theta^s$, corresponding to
$n = 2$. We can consider this to arise from a first-order ($n = 1$) diffraction plane
of spacing $d/2$. For example, in a case where the cubic (111) reflection gives
a path length of 2λ, the (222) reflection gives a path length of λ. Therefore,
in general,

$$\lambda = 2d_{hkl} \sin \theta_{hkl}^B \qquad (2.15)$$

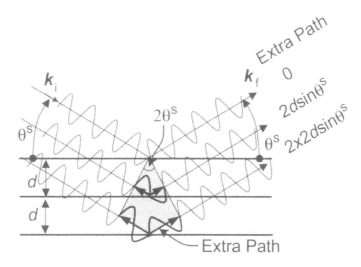

FIGURE 2.6
Schematic showing how the waves scatter and combine coherently from adjacent atomic lattice planes to give Bragg diffraction. Coherent reinforcement occurs when there is an integral number of wavelengths, $n\lambda$, in the path difference for planes separated by d. The situation shown here is for $n=1$.

Equation (2.15) is Bragg's Law, and is the basis of all diffraction measurements of lattice spacing, and hence of strain in polycrystalline materials. A more rigorous derivation of Bragg's Law is given in Section 2.3.3.

That the diffracted intensity is much larger at an angle $\theta^S = \theta^B$ satisfying Bragg's Law than at all other angles arises because the diffracted beam comprises many scattered wavelets, each reinforcing one another perfectly. Indeed, for a reasonably sized crystallite there will be many thousands of lattice planes contributing. We can estimate the angular width of the diffraction peak by considering the difference in path length between the *first* and the $(N + 1)^{th}$ lattice plane for a crystal $2N$ planes thick, for a scattering angle ϕ^s slightly larger, by $2\delta\theta$, than that giving a diffraction peak at $2\theta^B$. This will be:

$$2\,Nd_{hkl}\,\sin(\theta^B + \delta\theta) = 2\,Nd_{hkl}\,\sin\theta^B\cos\delta\theta + 2\,Nd_{hkl}\,\sin\delta\theta\cos\theta^B$$

$$\approx 2\,Nd_{hkl}\,\sin\theta^B + 2\,Nd_{hkl}\,\delta\theta\cos\theta^B$$

$$\approx N\lambda + 2\,Nd_{hkl}\,\delta\theta\cos\theta^B \tag{2.16}$$

The first term gives the coherence condition, but if the second term is equal to $\lambda/2$, the wavelets from the first and $N + 1^{th}$ planes will be exactly out of phase and will sum to zero. Likewise, the second and $N + 2^{th}$ will cancel, as will the Nth and $2N^{th}$, and so on, giving no diffracted signal at all. This occurs when

$$2\delta\theta = \lambda 2(Nd_{hkl}\cos\theta^B) = \lambda/(\ell_c\cos\theta^B) \tag{2.17}$$

As $2\delta\theta$ is inversely proportional to the thickness of the crystal, t_c, the range of angles over which the wavelets diffract constructively is very narrow for all but the smallest of crystallites. Diffraction is a scattering process in which all atoms within a crystallite cooperate to give a very strong cumulative effect.

The diffraction peaks thus occur only under very special conditions, that is, when the Bragg's Law equation is satisfied *and* the planes are oriented correctly at an angle θ^B to the incident beam, or what is equivalent, the normal to the plane *hkl* is parallel to *Q*, as shown in Figure 2.3b. Although there are some very important differences between diffraction and reflection, the diffraction peak corresponding to the lattice spacing *hkl* is often referred to as the *hkl reflection* because of the orientational condition.

2.3.2.1 Observation of Diffraction Peaks

In order to observe diffraction of a given wavelength of radiation from a single crystal, the crystal must be oriented such that a plane *hkl* satisfies the "reflection" condition. However, in the case of a polycrystalline sample, with a large number of small crystallites in random orientations, there will usually be a substantial fraction, or subset, of crystallites which are in this correct orientation. In brief, they will lie with their *hkl* planes at an angle of θ_{hkl}^B to the incident beam, but at all azimuthal angles around it. The diffracted intensity therefore occurs in a series of cones, known as *Debye–Scherrer cones* with semiangle $2\theta_{hkl}^B$ at the sample position about the direction of the incident beam as axis (Figure 2.7). Using a monochromatic incident beam (e.g., from a steady reactor source), the scattering angle corresponding to the peak intensity, $\phi^s = 2\theta_{hkl}^B$, is determined by scanning a detector angle $\phi^s = 2\theta^s$ through the peak, or by a fixed position sensitive detector as discussed in Sections 3.2.1 and 3.3.1.

At a spallation source, a single time pulse contains a continuous spectrum of wavelengths, and the wavelength of each detected neutron is established by measuring the time of flight, t, from the instant at which the pulse of thermal neutrons is generated in the moderator to the instant at which the neutron is captured in the detector. If the total flight path distance between the moderator and detector via the sample is L, then the wavelength is given from Equation (2.1) by

$$\lambda = ht/m_n L \tag{2.18}$$

The pulse of incident neutrons spreads out in time as it travels toward the sample, and the neutrons of highest velocity and hence shortest wavelength arrive first. As a result, each Debye–Scherrer cone moves from low scattering angles at short times (high energies/short wavelengths) to high angles at long times (low energies/long wavelengths) during each pulse. A detector at a spallation source records the time of arrival of all the scattered neutrons at a fixed scattering angle during the pulse, which arise as the Debye–Scherrer cones scan through its aperture. The advantage of this mode of operation

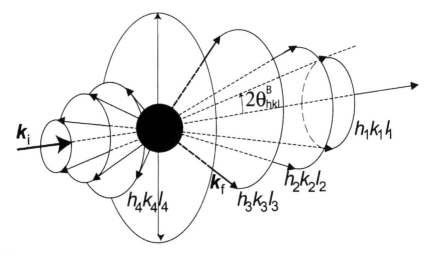

FIGURE 2.7
Neutron powder diffraction from a polycrystalline sample with random orientation of grains. The monochromatic neutron beam is Bragg diffracted by crystallites with the correct lattice spacing and orientation to give rise to Debye–Scherrer cones of scattered intensity, each corresponding to a lattice spacing *hkl* (adapted from Allen et al. [21]).

is that all lattice spacings d_{hkl}, and therefore strains, corresponding to different crystallographic planes *hkl* of the grains in a sample with plane normals in a direction Q defined relative to the sample, can be measured during the counting period. This may include several orders of reflection of the peaks, that is, *2h2k2l, 3h3k3l*, and so on, as well as *hkl*. These methods of measuring diffraction intensities are discussed further in Section 3.2.2.

2.3.3   Reciprocal Lattice Approach to Diffraction Theory

2.3.3.1   *Reciprocal Lattice*

An alternative approach to the theory of diffraction from a crystal lattice is to use the concept of reciprocal space rather than the more easily visualized real space. This section may be omitted on first reading, but is included since it leads more easily to several of the concepts introduced in the preceding sections. The concept of reciprocal space is essential for the detailed theory of radiation scattering events, and is also used in the theory of the behavior of electrons, and of excitations such as lattice vibrations, in crystalline materials with translational symmetry.

In reciprocal space, the variables wave vector and frequency replace real-space variables distance and time. Thus, the variables in the two spaces are related by reciprocals. Functions of these variables in the two spaces are related by a Fourier transform. Vectors such as k_i and k_f that describe the direction of travel of the neutrons before and after scattering, and the scattering vector Q, are directions in reciprocal space as well as real space. As a single nucleus is extremely small compared with the neutron wavelength, it can be represented effectively by a delta function at the origin of real space, the Fourier transform of which gives the variation of (static) nuclear scattering with Q, and is a constant function independent of Q in reciprocal space. On the other hand, magnetic scattering of neutrons by atoms, which arises from unpaired electron spins, has intensity enveloped by a magnetic form factor reflecting the Fourier transform of their spatial extent, in the same way as scattering of x-rays by the atoms' electrons. For our purposes, it is only necessary to consider the time-averaged spatial position of the atoms, that is $\hbar\omega = E_i - E_f = 0$ in Equation (2.11), corresponding to elastic scattering, and the relation between real spatial vectors r and scattering vectors Q.

For any lattice of points in real space, given by l (Equation (2.13)), a reciprocal lattice of points is defined by

$$G = hA + kB + lC \tag{2.19}$$

where h, k, and l are integers (with notation chosen anticipating their relationship to the Miller indices, as in Equation (2.27)). The vectors A, B, and C define the reciprocal lattice axes and are related to the axes defining a unit cell in the real lattice, a, b, and c, by

$$A = 2\pi \, b{\wedge}c/v_0, \ B = 2\pi \, c{\wedge}a/v_0, \ C = 2\pi \, a{\wedge}b/v_0 \tag{2.20}$$

where $v_0 = (a.b{\wedge}c)$ is the scalar triple product giving the volume of the unit cell of the real lattice. The set of vectors A, B, and C are orthogonal only for cubic, tetragonal, and orthorhombic lattices. These relations are valid for all unit cells in the real lattice, but they take on special significance if they are used with reference to the primitive real lattice unit cell. It is easily seen from these definitions that the real and reciprocal lattice vectors satisfy the relation

$$\exp[iG{\cdot}l] = \exp[i2\pi(hn_1+kn_2+ln_3)] = 1 \tag{2.21}$$

The scattering vector Q may be written as a vector in the reciprocal space of the crystal that is scattering radiation, as follows:

$$Q = Q_A \, A + Q_B \, B + Q_C \, C \tag{2.22}$$

where Q_A and so on are the components along A and so on, and are only integers when $Q = G$.

2.3.3.2 Neutron Coherent Elastic Cross-Section

In general, the neutron elastic scattering cross-section, now for the crystal as a whole rather than per atom, may be written as [3,4]

$$\frac{d\sigma}{d\Omega} = \sum_{i,j} \bar{b}_i \bar{b}_j \exp\left(i.Q{\cdot}\left(R_i - R_j\right)\right) \tag{2.23}$$

where R_i is the position vector of atom i, with average scattering length $\bar{b}_i$, in real space, and the sum is over all the atoms i, j in the crystal. We may write $R_i = l_i + r_u$. Here r_u is the position of the atom, with average scattering length $\bar{b}_u$, in the basis in the unit cell associated with each lattice vector l_i, and can be written

$$r_u = r_{ua}a + r_{ub}b + r_{uc}c \qquad (2.24)$$

where r_{ua}, r_{ub}, and r_{uc} are real but not necessarily integer numbers. Here we ignore small displacements of the atoms due to thermal vibration, but in Section 2.5 we discuss the effect of these vibrations and how they give rise to an extra factor, the Debye–Waller factor, in the nuclear structure factor defined in Equation (2.30) below. Substituting into Equation (2.23),

$$\frac{d\sigma}{d\Omega} = \sum_{i,u} \bar{b}_u \exp\left(i\mathbf{Q}\cdot(l_i + r_u)\right) \cdot \sum_{j,u} \bar{b}_u \exp\left(-i\mathbf{Q}\cdot(l_j + r_u)\right) \qquad (2.25)$$

where i,j are now summed over the lattice points and u over the basis. However, the summation over the basis atoms in each unit cell is the same for each lattice point, so that

$$\frac{d\sigma}{d\Omega} = \left[\sum_{i,j} \exp\left(i\mathbf{Q}\cdot(l_i - l_j)\right)\right]\left[\sum_u \bar{b}_u \exp\left(i\mathbf{Q}\cdot r_u\right)\right]\left[\sum_u \bar{b}_u \exp\left(-i\mathbf{Q}\cdot r_u\right)\right] \qquad (2.26)$$

In general, the phases of the first term will vary randomly, and the summation will be zero. However, if

$$Q = G \qquad (2.27)$$

it follows from Equation (2.21) that the exponential becomes unity, and the cross-section becomes nonzero. This is just the condition for coherent diffraction, and indeed expresses the Bragg condition. It gives rise to the term $\delta(\mathbf{Q} - G)$ in the total differential cross-section, which is summed over all G. In fact,

$$\sum_{i,j} \exp\left(i\mathbf{Q}\cdot(l_i - l_j)\right) = N_c \frac{(2\pi)^3}{v_0} \delta(\mathbf{Q} - G) \qquad (2.28)$$

where $\delta(x)$ is the Dirac delta function, which is unity if the argument $x = 0$ and zero otherwise, and N_c is the number of unit cells in the crystal of volume $V = N_c v_0$. From Equation (2.21), the condition $Q = G$ states that for coherent diffraction the components of Q along each of the reciprocal lattice vectors are the integers, *h, k*, and *l*. These integers are the Miller indices of the Bragg peak, or reflection, associated with this reciprocal lattice point. Indeed, there will be diffraction whenever *Q* ends on a reciprocal lattice point, *hkl*, although the intensity may yet be zero. The differential cross-section for scattering from the whole crystal into each peak *hkl* is then given by

$$\frac{d\sigma}{d\Omega}[hkl] = \frac{(2\pi)^3}{v_0} N_c.F(hkl).F^*(hkl) \tag{2.29}$$

where

$$F(hkl) = \sum_u \bar{b}_u \exp(i\mathbf{G}\cdot\mathbf{r}_u). \tag{2.30}$$

is called the *nuclear structure factor* for the unit cell, and involves the summation over all the atoms *u* in the basis, that is, over the atoms in the unit cell. The asterisk in Equation (2.29) denotes the complex conjugate. As noted above, the lattice vibration of each atom gives rise to an additional multiplicative factor of [exp($-W_u(Q)$] in the summation on the right side of Equation (2.30), which is discussed further in Section 2.5.

The nuclear structure factor has to be evaluated for $Q = G_{hkl}$, in which case (from Equation (2.30)),

$$F(hkl) = \sum_u \bar{b}_u \exp\left(i2\pi\left(h.r_{ua} + k.r_{ub} + l.r_{uc}\right)\right). \tag{2.31}$$

The corresponding expression, including an average Debye–Waller factor is given by Equation (2.49).

A few observations are worthwhile at this point. For a monatomic crystal with a basis of one atom at each real primitive lattice point, that is, for a Bravais lattice, there will be a finite Bragg intensity for all $Q = G_{hkl}$. The simplest example is a primitive, or simple, cubic structure with lattice constant a_0, in which case the reciprocal lattice is again a simple cubic array of points, with spacing $2\pi/a_0$. A primitive cell for the face-centered cubic (*fcc*) structure may be chosen, in which *a*, *b*, and *c* are the nonorthogonal vectors to each face center (Figure 2.4c). The reciprocal lattice of the primitive real lattice is, in fact, body-centered cubic (*bcc*). Conversely, the body-centered structure has a face-centered reciprocal lattice.

In practice, a nonprimitive "conventional" cubic lattice is often chosen to describe the face-centered structure, with side a_0 but with a basis of four atoms, one at the origin point and one at each face center. From Equation (2.31), one then finds that for the corresponding cubic reciprocal lattice of points separated by $2\pi/a_0$, not every *hkl* has intensity, that is, some reflections are systematically absent. Indeed, the structure factor for *fcc* is

TABLE 2.4

Summary of Structure Factors Evaluated for Unit Cell and Basis Calculated Using Equation (2.31) for Selected Simple Monatomic Crystal Structures

Structure	Unit Cell	Number of Basis Atoms	Selection Rule for Bragg Intensity	Example	$\lvert F(hkl) \rvert^2$
Primitive cubic	Primitive cubic	1	All hkl	100	$\bar{b}^2$
Face centered cubic (*fcc*)	Conventional cubic	4	All hkl even or all odd	111	$16\,\bar{b}^2$
			Otherwise	100	0
Body centered cubic (*bcc*)	Conventional cubic	2	$h+k+l$ even	200	$4\,\bar{b}^2$
			$h+k+l$ odd	111	0
Hexagonal	Primitive hexagonal	2	$h+2k=3n$, l odd	$11\bar{2}1$	0
			$h+2k=3n\pm1$, l even	$20\bar{2}0$	$\bar{b}^2$
			$h+2k=3n\pm1$, l odd	$20\bar{2}1$	$3\,\bar{b}^2$
			$h+2k=3n$, l even	$11\bar{2}0$	$4\,\bar{b}^2$

zero for all points, unless h, k, and l are either all odd or all even. These are the so-called *selection rules*. In other words, the reciprocal lattice points giving intensity do indeed lie on a body-centered cubic reciprocal lattice, but with sides of twice the cubic reciprocal lattice point spacing $(2.2\pi/a_0)$. The corresponding selection rules for a *bcc* structure indexed on a conventional cubic reciprocal lattice of side $2\pi/a_0$ are that $(h + k + l)$ must be an even integer. The derivation of the reciprocal lattice directly from the primitive real lattice for *fcc*, *bcc*, and hexagonal materials is given by Kittel [17]. It involves defining an appropriate orthogonal coordinate system to which the primitive lattice vectors can be referred, and calculating the reciprocal lattice axes using the expressions in Equation (2.20). Factors for some common structures are summarized in Table 2.4.

If we consider the coherent diffraction condition in reciprocal space (Equation (2.27)) a little further, we have

$$\lvert Q \rvert = 2\,k\,\sin\theta^B_{hkl} = \lvert G_{hkl} \rvert \tag{2.32}$$

Using vector algebra, it can readily be shown [17] that the vector G_{hkl} is normal to the plane hkl in the real lattice, as shown in Figure 2.3b, and that it has a magnitude given by the reciprocal relation

$$\lvert G_{hkl} \rvert = (G \cdot G)^{1/2} = 2\pi/d_{hkl} \tag{2.33}$$

where d_{hkl} is the lattice spacing of the hkl planes in the real space lattice.

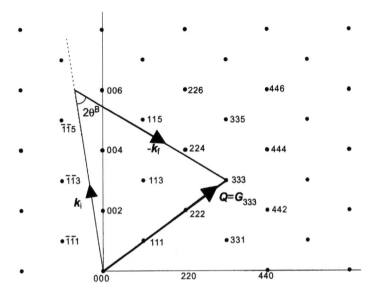

FIGURE 2.8

A section of the reciprocal lattice for a face-centered cubic crystal perpendicular to the [1$\bar{1}$0] direction. The incident k_i and scattered k_f neutron wavevectors are shown. The condition for diffraction from 333 planes is that the scattering vector Q should exactly equal the reciprocal lattice vector G_{333}.

Combining Equations (2.32) and (2.33), one finds that

$$2\,k\,\sin\theta^B_{hkl} = 2\pi/d_{hkl} \qquad (2.34)$$

and by substituting $k = 2\pi/\lambda$, one sees that the reciprocal space condition (Equations (2.27) and (2.32)), is indeed just the Bragg's Law expression in real space (Equation (2.15)):

$$2\,d_{hkl}\,\sin\theta^B_{hkl} = \lambda$$

Figure 2.8 shows a section of the reciprocal lattice representing a single crystal of a face-centered cubic material normal to the [1$\bar{1}$0] direction. Also shown are the incident and scattered neutron wave vectors and the condition that the scattering vector Q should exactly match the reciprocal lattice vector G_{333}. This illustrates in reciprocal space the diffraction condition for a single crystal, with the magnitude of Q and orientation of the crystal satisfying Equation (2.27). Q is perpendicular to the 333 lattice planes.

One important feature of Bragg's Law is that as the wavelength is increased above $\lambda_e^{hkl} = 2d_{hkl}$, Equation (2.15) can no longer be satisfied, and thus diffraction can no longer occur for the hkl planes with the result that the diffracted intensity drops to zero. This is accompanied by an increase in the transmitted intensity as fewer neutrons are diffracted out of the straight-through beam.

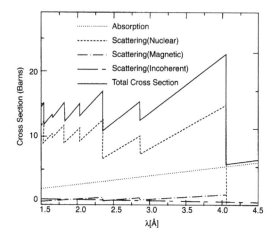

FIGURE 2.9

Total neutron cross-section of polycrystalline iron as a function of wavelength. The rapid variations in cross-section occur at the Bragg cut-off wavelengths at which a given plane can no longer diffract.

This effect is called the *Bragg cut-off* or *edge*, and successive Bragg cut-offs result in the pattern in the total cross-section for polycrystalline iron shown in Figure 2.9. The Bragg cut-off can be used for strain measurement with the transmission technique discussed in Section 3.7.4. However, it can also have an adverse effect on normal strain measurement if the incident wavelength is not chosen carefully, as discussed in Section 4.2.2.

In the reciprocal lattice representation of the case of diffraction from a random polycrystalline sample, all possible orientations of the grains are equally likely, so each reciprocal lattice point becomes an *hkl* sphere in reciprocal space, concentric about the origin. Depending on the crystal symmetry, some of these spheres may coincide, such as 333 and 115 for the face-centered cubic structure, since these have the same lattice plane spacings. The diffraction condition for some of the grains to diffract in a polycrystal then becomes simply

$$|Q| = |G_{hkl}| \tag{2.35}$$

since only the magnitude of the scattering vector has to match the radius of the sphere. However, only the subset of grains that are oriented so that their plane normals are, within the instrumental resolution, parallel to the scattering vector will diffract.

2.3.4   Lattice Plane Spacings

It is useful to list the lattice plane spacing d_{hkl} for a few common crystal structures. These may be calculated using Equations (2.19) and (2.33), and are summarized in Table 2.5.

TABLE 2.5

Expressions to Calculate Lattice Plane Spacings for *hkl* Lattice Planes in Some Important Crystal Structures

Cubic	$\dfrac{1}{d_{hkl}^2} = \dfrac{h^2 + k^2 + l^2}{a^2}$
Tetragonal	$\dfrac{1}{d_{hkl}^2} = \dfrac{h^2 + k^2}{a^2} + \dfrac{l^2}{c^2}$
Orthorhombic	$\dfrac{1}{d_{hkl}^2} = \dfrac{h^2}{a^2} + \dfrac{k^2}{b^2} + \dfrac{l^2}{c^2}$
Hexagonal	$\dfrac{1}{d_{hkl}^2} = \dfrac{4}{3}\left(\dfrac{h^2 + hk + k^2}{a^2}\right) + \dfrac{l^2}{c^2}$
Rhombohedral	$\dfrac{1}{d_{hkl}^2} = \dfrac{\left(h^2 + k^2 + l^2\right)\sin^2\alpha + 2\left(hk + kl + hl\right)\left(\cos^2\alpha - \cos\alpha\right)}{a^2\left(1 - 3\cos^2\alpha + 2\cos^3\alpha\right)}$
Monoclinic	$\dfrac{1}{d_{hkl}^2} = \dfrac{1}{\sin^2\beta}\left(\dfrac{h^2}{a^2} + \dfrac{k^2\sin^2\beta}{b^2} + \dfrac{l^2}{c^2} - \dfrac{2hl\cos\beta}{ac}\right)$

2.3.4.1 Orthogonal Crystal Structures

Since the axes of the unit cells are at right angles ($\alpha = \beta = \gamma = 90°$), the lattice spacings for cubic ($a = b = c$), tetragonal ($a = b \neq c$) and orthorhombic ($a \neq b \neq c$) crystal structures are particularly simple to calculate. The spacing, d_{hkl}, for the *hkl* lattice plane is given by

$$\frac{1}{d_{hkl}^2} = \frac{h^2}{a^2} + \frac{k^2}{b^2} + \frac{l^2}{c^2} \tag{2.36}$$

For cubic materials, this simplifies to

$$d_{hkl} = a/(h^2 + k^2 + l^2)^{1/2} \tag{2.37}$$

It is clear from this formulation that many cubic symmetry reflections have the same lattice spacing, as in $d_{121} = d_{211} = d_{12\bar{1}} = d_{2\bar{1}1}$, and so on. These 24 lattice planes are said to belong to a symmetry-related form, indicated by {121}. In a randomly oriented powder, all 24 such reflections will contribute to the diffracted intensity corresponding to these 24 different crystallite orientations. In general, the number of such equivalent peaks having the

TABLE 2.6

Multiplicities, j, of Reflections hkl for Random Polycrystalline Samples of Selected
Important Crystal Structures

Cubic	$hkl: j = 48$	$hhl: j = 24$	$hk0: j = 24$	$hh0: j = 12$	$hhh: j = 8$	$h00: j = 6$	
Tetragonal	$hkl: j = 16$	$hhl: j = 8$	$hk0: j = 8$	$hh0: j = 4$	$0kl: j = 8$	$h00: j = 4$	$00l: j = 2$
Orthorhombic	$hkl: j = 8$	$0kl: j = 4$	$hk0: j = 4$	$h0l: j = 4$	$h00: j = 2$	$0k0: j = 2$	$00l: j = 2$
Hexagonal and rhombohedral	$hkl: j = 24$	$hhl: j = 12$	$hk0: j = 12$	$hh0: j = 6$	$h0l: j = 12$	$h00: j = 6$	$00l: j = 2$
Monoclinic	$hkl: j = 4$	$h0l: j = 2$	$0k0: j = 2$				
Triclinic	$hkl: j = 2$						

same lattice spacing is termed the *multiplicity*, j, and j occurs as a factor in the expression for the diffracted intensity from a powder. The multiplicities of reflections hkl for various structures are given in Table 2.6.

Brackets are commonly used to denote planes and directions in the crystal lattice, which we illustrate here with reference to cubic systems. (hkl) denotes the plane hkl, and $\{hkl\}$ denotes the family of planes with the same symmetry-related indices that contribute to the hkl reflection of a polycrystalline assembly, or random powder, of grains. For example, in cubic systems, $\{002\}$ denotes the family (002), $(00\bar{2})$, (020), $(0\bar{2}0)$, (200), and $(\bar{2}00)$. Frequently in the literature, in unambiguous cases of powder diffraction, the simple notation hkl or (hkl) is used for the more correct $\{hkl\}$ to denote Bragg reflections. A direction in a crystal is denoted as $[uvw]$, where uvw is the set of smallest integers having the same ratios as the components of a vector in that direction resolved along the crystal lattice vectors. The full set of equivalent directions is denoted as $<uvw>$. Specifically, in the case of cubic crystals the $[uvw]$ direction has direction ratios u, v, and w, referred to the cubic axes, and the $[hkl]$ direction is perpendicular to the (hkl) plane, but this is not generally true for other crystal symmetries.

2.3.4.2   Hexagonal Structures

Hexagonal structures are defined as $a = b \neq c$ and $(\alpha = \beta = 90°; \gamma = 120°)$. For such nonorthogonal systems, the lattice spacings and multiplicities are less self-evident. While planes in all symmetries may be described by three indices, hkl, it is common to find planes in hexagonal structures described using a four-index notation $hkil$, where $i = -(h + k)$. This Miller–Bravais notation serves to identify planes of equivalent symmetry. For example, it is readily seen that $(10\bar{1}0)$, $(01\bar{1}0)$, and $(\bar{1}100)$ are related by symmetry, and have the same lattice spacing, whereas this would not be so evident from the corresponding three Miller indices (100), (010), and $(\bar{1}10)$.

2.3.4.3   Lower Symmetry Structures

For rhombohedral or trigonal structures $((a = b = c); (\alpha = \beta = \gamma < 120°, \neq 90°))$, monoclinic structures $((a \neq b \neq c), (\alpha = \gamma = 90°; \beta \neq 90°))$, and triclinic structures

$((a \neq b \neq c)$, $(\alpha \neq \beta \neq \gamma \neq 90°))$, the multiplicities decrease as the symmetry decreases, as seen in Table 2.6. As a result, the diffracted intensity of individual triclinic reflections tends to be low, as the total diffracted intensity is spread over more peaks.

2.3.5 Diffracted Intensity

From the conditions for diffraction and the form of the differential cross-section for coherent Bragg diffraction from a polycrystalline sample given in Section 2.3.3, the intensity of the diffracted beam from a particular sample may be calculated as the number of neutrons detected per unit time. With reference to a basic two-axis instrument on a continuous source, two sample–instrument configurations relevant to strain measurement are considered here, together with the special case of a plate sample. These are shown schematically in Figure 2.10. The first (Figure 2.10a) corresponds to the case where the strain profile inside the sample is being mapped, in which the whole of the instrumental gauge volume, V_g, defined by the beam apertures as described in Section 3.5.2, is filled by a part of the sample and contributes to the intensity. The second (Figure 2.10b) is when the complete sample, typically in cylindrical form, lies centered within the instrumental gauge volume and gives rise to the scattering. In each case, n_{hkl}, the number of neutrons from the *hkl* lattice planes per unit time falling on the detector, is given by

$$n_{hkl} = \Phi_0 V_v \frac{d\Sigma_{hkl}}{d\Omega} \Delta\Omega \tag{2.38}$$

where Φ_0 is the incident flux, V_v is the volume of the sample that is scattering the incident beam and is contributing to the detected intensity, and $\Delta\Omega$ is the element of solid angle subtended by the detector at the sample. We here assume the detector to be 100% efficient, but an efficiency factor can easily be included on the right side of the equation. The derivation of the expression for n_{hkl} is given by Squires [4], and involves determining the macroscopic cross-section for scattering by the polycrystalline sample into the whole of each Debye–Scherrer cone corresponding to $|G|$, or a scattering angle $\phi^s = 2\theta^B_{hkl}$. This is given by

$$\Sigma_G^{cone} = \frac{\lambda^3}{4v_o^2} \frac{1}{\sin\theta^B_{hkl}} \sum_{hkl} |F_{hkl}|^2 \tag{2.39}$$

where the summation is over all *hkl* with the same $|G|$, that is all planes contributing to the same cone at ϕ^s, such as the cubic 333 and 115 reflections mentioned in the previous section, with their respective multiplicities *j*. As defined previously, F_{hkl} is the structure factor, and v_0 is the volume of the unit cell.

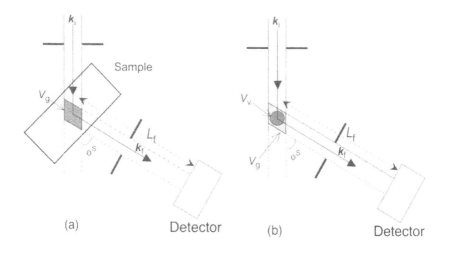

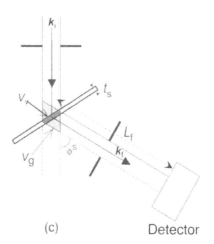

FIGURE 2.10
Schematic illustrating three commonly encountered sample configurations on a simple diffractometer shown in Figure 3.8, in plan view. The direction of the beams is defined by collimators (not shown). (a) Sample completely fills the instrumental gauge volume V_g ($=V_v$) (shaded), as might occur during strain profile mapping. (b) Cylindrical sample (shaded) centred in the instrumental gauge volume completely bathed by the incident beam, V_v is the sampled gauge volume. (c) A thin plate sample is placed symmetrically in the instrumental gauge volume in transmission geometry. The shaded sampled volume V_v contributes to the diffracted intensity.

A fraction of each cone given by $[h_d/2\pi L_f \sin 2\theta_{hkl}^B]$ is detected, where L_f is the distance from sample to detector, and h_d is the height of the detector. It is then found that for each set of contributing planes $\{hkl\}$,

$$n_{hkl} = \Phi_0 \frac{\lambda^3 h_d V_v}{8\pi L_f v_0^2} \frac{j|F_{hkl}|^2}{\sin\theta_{hkl}^B \sin 2\theta_{hkl}^B} A_{hkl}^{abs} \qquad (2.40)$$

We assume that the density of the sample is the theoretical density, but a simple ratio correction may be made if this is not so (see Bacon [1]). A_{hkl}^{abs} is an absorption factor that depends on the configuration of the sample. It should be noted that absorption in a sample can cause a shift of the effective scattering center of the sample and a spurious shift in the diffraction angle, as discussed in Section 3.6.

In the first case shown in Figure 2.10a, V_v is equal to the volume of the instrumental gauge V_g (see Section 3.5.2). A_{hkl}^{abs} has to be evaluated numerically by summing all path lengths, with the appropriate linear attenuation coefficient μ_a discussed in Section 2.4, in the sample for each wavelength and scattering angle. In the simple case shown schematically in Figure 2.10a, the sample is symmetrically in the transmission configuration, and all path lengths will be the same, but in most cases the sample will be irregular in shape.

In the second case of a cylindrical sample fully bathed by the beam (Figure 2.10b), V_v will be the volume of the sample in the beam, and the absorption factor A_{hkl}^{abs} has been tabulated for a range of sample radii and linear absorption coefficients [1,22].

One case that is sometimes encountered in microstrain measurement is a plate sample of thickness t_s measured in transmission and set symmetrically in the instrumental gauge volume, which is only filled in the vertical direction (Figure 2.10c). In this case, the absorption factor is constant for beam paths from each element of the sample, and $V_v = t_s A_i \sec\theta_{hkl}^B$, and $A_{hkl}^{abs} = \exp(-\mu_a t_s \sec\theta_{hkl}^B)$, where A_i is the area of the incident beam [1].

The flux and factors involving the instrument dimensions are constant for a given instrument configuration, and an intensity calibration constant can be obtained, if necessary, by carrying out a powder diffraction experiment from a standard sample of known structure factor, multiplicity, and unit cell volume, such as silicon, Si, alumina, Al_2O_3, or ceria, CeO. A comparison of the measured intensities under the same conditions allows the intensities from the sample to be placed on an absolute scale, since the ratio of the measured intensities is equal to the ratio of the theoretical intensities (Equation (2.40)), in which the constant instrumental parameters cancel out. Equation (2.40) shows that the diffracted intensity from a given set of planes increases with the volume of the sample in the gauge and as the cube of the wavelength, however the linear attenuation coefficient increases linearly with wavelength. The corresponding relation for an instrument on a pulsed source is given by David [23].

2.3.6 Scattering from Engineering Materials

It is instructive to consider the nature of the scattering from a few important engineering materials.

2.3.6.1 Scattering from Elements

Naturally occurring iron (Fe) is made up of the isotopes ^{54}Fe, ^{56}Fe, ^{57}Fe, and ^{58}Fe, with abundances of 5.8%, 91.7%, 2.2%, and 0.3%, respectively, which are distributed randomly on the atomic lattice each with its characteristic scattering length [13,24]. Of these isotopes, only ^{57}Fe possesses nuclear spin $I = 1/2$. Naturally occurring iron may then be described by an average value $\bar{b} = +0.945 \times 10^{-12}$ cm on each atomic site that gives rise to the Bragg diffraction intensity. The coherent scattering cross-section is 11.22 barns/atom. The incoherent scattering from iron atoms, with cross-section $\sigma^{incoh} = 0.40$ barns, only contributes to the isotropic background at all scattering angles.

Naturally occurring nickel (Ni) has six isotopes with various scattering lengths, and one isotope ^{61}Ni with nuclear spin $I = 3/2$. The average scattering length $\bar{b} = +1.03 \times 10^{-12}$ cm, and the corresponding coherent cross-section is 13.3 barns/atom. The deviations from this average, together with the nuclear spin incoherent scattering, give rise to a larger incoherent cross-section per atom than for iron of $\sigma^{incoh} = 5.2$ barns. Thus, nickel and its alloys give much larger background levels than iron.

2.3.6.2 Scattering from Alloys

In designing materials for engineering applications, elements are usually alloyed to impart strength, and there is often more than one type of atomic site in the unit cell. For example, in the simple cubic Ni_3Al phase (Figure 2.4a) that strengthens face-centered cubic (Figure 2.4c) Ni superalloys used in high-temperature aerospace applications, there is a basis of four atoms per primitive simple cubic cell, with Al atoms at the corner, A, sites and three Ni atoms at the face center, B, sites. The actual alloy used in practice has a number of other chemical constituents, some of which reside on the B sites and some on the A sites. In order to calculate the diffraction intensity, the average coherent scattering length must be calculated for each site: the average $\bar{b}_A$ over the elements occupying the A sites and the average $\bar{b}_B$ over the elements occupying the B sites. These mean values are then substituted in the expression for the scattering amplitude F_{hkl}, Equation (2.31) or (2.49), which in turn is used in Equation (2.40) to determine the Bragg diffraction intensity.

Any random occupation of the sites in an alloy also gives rise to diffuse elastic scattering from the random distribution of coherent scattering lengths Q. Considering an example of a binary alloy with atomic fractions a_1 and a_2 of elements, with mean scattering lengths $\bar{b}_1$ and $\bar{b}_2$, and incoherent scattering cross-sections σ_1^{incoh} and σ_2^{incoh}, respectively, situated on a

Bravais lattice, this coherent diffuse scattering, sometimes referred to as *Laue monotonic diffuse scattering*, is given by [3]

$$\sigma^{coh\text{-}diffuse} = 4\pi\, a_1 a_2 \left(\bar{b}_1 - \bar{b}_2 \right)^2 \tag{2.41}$$

The total coherent scattering therefore comprises the Bragg scattering, in which the average scattering length $\bar{b}_{Av} = a_1\bar{b}_1 + a_2\bar{b}_2$ replaces the monatomic scattering length, plus this diffuse term. The Laue diffuse scattering gives rise to an isotropic background which is in addition to the nuclear incoherent scattering described by

$$\sigma^{incoh} = a_1\sigma_1^{incoh} + a_2\sigma_2^{incoh} \tag{2.42}$$

An example occurs when iron atoms are replaced by silicon atoms with their different scattering lengths in an iron-silicon alloy, giving rise to additional isotropic background. Any small distortions of the lattice that might arise around the various atoms in an alloy, as well as thermal vibrations, will reduce the Bragg scattering and give additional factors or diffuse terms that can be angle dependent.

2.3.6.3 Application to Titanium Alloys

A case of practical importance is the study of components made from Ti-6Al-4V. Here, the addition of Al and V, with scattering lengths of 0.345 and -0.038×10^{-12} cm, respectively, to Ti with $\bar{b} = -0.344 \times 10^{-12}$cm, reduces the average scattering length in the alloy but augments the large intrinsic incoherent cross-section for Ti of 2.87 barns. In this case, the average coherent scattering length is only -0.263×10^{-12} cm, but the average incoherent and diffuse cross-section per atom is ~3.1 barns. As a result, even the strong reflections in randomly oriented polycrystalline titanium alloys are only two to three times higher than the background, and this makes strain studies very difficult in these materials as discussed in Sections 4.4.1 and 4.4.2.

2.3.7 Macroscopic Cross-Section

2.3.7.1 Elements

Total macroscopic cross-sections Σ^{coh}, Σ^{incoh}, and Σ^{abs} are particularly useful for calculating the attenuation of the neutron beam through a sample due to coherent scattering, incoherent scattering, and absorption, respectively [24]. They are listed in Appendix 4 for all the elements. For a single elemental material, the corresponding microscopic cross-section of each element, σ_j, is multiplied by the number of atoms per unit volume. For an element j, with density ρ_j and atomic weight W_j, the number of atoms per unit volume is given by $N_A \rho_j / W_j$, where N_A is Avogadro's number (6.022×10^{23} mole^{-1}). Then the corresponding macroscopic cross-section is given by

$$\Sigma_j = \sigma_j \, N_A \, \rho_j / W_j \qquad (2.43)$$

2.3.7.2 Alloys

For a material of more than one element (compound or alloy), the macroscopic incoherent scattering cross-section and the macroscopic absorption cross-section may be calculated directly from a weighted summation of the contribution from the individual constituents. The corresponding macroscopic cross-sections are given by

$$\overline{\Sigma} = N_A \overline{\rho} \sum_j \frac{w_j \sigma_j}{W_j} \qquad (2.44)$$

where $\overline{\rho}$ is the mean density, σ_j the microscopic incoherent or absorption cross-section, w_j the weight fraction, and W_j the atomic weight of atom j. Alternatively, they can be expressed in terms of atomic fraction, a_j, of each constituent:

$$\overline{\Sigma} = N_A \overline{\rho} \frac{\sum_j a_j \sigma_j}{\sum_j a_j W_j} \qquad (2.45)$$

The total macroscopic coherent scattering cross-section of a compound or random alloy may also, in the short wavelength limit, be given by the above expressions in terms of the total microscopic coherent cross-sections. In the case of random alloys, this total coherent scattering will include both the Bragg and Laue diffuse terms discussed in the previous section. However, the total coherent cross-section will in general be wavelength dependent as discussed in Section 2.3.3, in particular due to the variation in the coherent cross-section that arises from the Bragg cut-off for those atomic planes hkl for which $\lambda > 2d_{hkl}$ as shown in Figure 2.9. When the short wavelength limit cannot be applied, the total coherent diffraction cross-section, $\overline{\Sigma}^{coh}$, must be found by integration of the expression given in Equation (2.39), with the appropriate structure factor containing averaged scattering lengths, over all the Debye–Scherrer cones corresponding to different Bragg peaks up to the Bragg cut-off. For random alloys, the Laue diffuse scattering, discussed in the previous section, must also be included.

Together, the three macroscopic cross-sections contribute to the total linear attenuation as described in the next section.

2.4 Penetration of Neutron Beams

As already remarked with reference to Table 2.2, one of the major attributes of neutrons compared to other radiation probes is their high penetration in most engineering materials, making them a useful tool for making measurements deep within samples. When embarking on a series of strain measurements, it is important therefore to be able to predict the extent to which the incident and diffracted beam will be attenuated over the total path length in the material. The variation of the neutron flux Φ_l compared to the incident flux Φ_0 with path length, l, through a sample is given in terms of the linear attenuation factor, μ_a, by

$$\Phi_l = \Phi_0 \cdot e^{-\mu_a l} \tag{2.46}$$

The linear attenuation factor is equal to the sum of all the macroscopic cross-sections that remove neutrons from the beam, that is, coherent and incoherent scattering and absorption, as discussed in the previous section:

$$\mu_a = \overline{\Sigma}^{coh} + \overline{\Sigma}^{incoh} + \overline{\Sigma}^{abs} \tag{2.47}$$

An attenuation length, l_μ, can be defined as the distance over which the incident neutron flux declines by a factor $1/e$. This is the reciprocal of the linear attenuation factor and is quoted in centimeters:

$$l_\mu = 1 / (\overline{\Sigma}^{coh} + \overline{\Sigma}^{incoh} + \overline{\Sigma}^{abs}) \tag{2.48}$$

This is shown schematically in Figure 2.11. The macroscopic cross-sections are calculated as described in the previous section, and are given with values of l_μ in Appendix 4. As they depend on nuclear properties, the attenuation lengths of different nuclei vary erratically from element to element in the periodic table, as shown in Figure 2.2. In perfect crystals, extinction effects cause the coherent diffraction cross-section to dominate the attenuation as discussed in Section 2.6.1.

Figure 2.9 shows the total cross-section per atom for iron, including coherent diffraction, nuclear as well as magnetic, incoherent scattering and absorption, as a function of wavelength [25]. For iron, the dominating factor depleting the beam is the coherent diffraction term. Note that at a wavelength slightly greater than 4.1 Å, all reflections have exceeded the Bragg cut-off (Section 2.3.3) so that Bragg diffraction is no longer possible and the transmitted intensity increases sharply. Upon further increasing the wavelength (lowering the energy), the transmission falls, due mainly to the increase in true absorption.

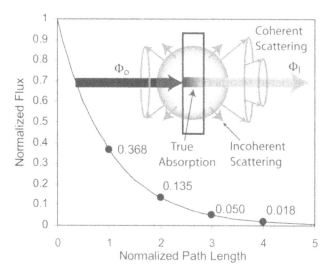

FIGURE 2.11

The transmitted neutron flux normalized by the incident flux (Φ_I/Φ_0) as a function of the normalised path length (l/l_μ). As indicated by the circles, path lengths of one, two, three, and four times the attenuation length l_μ correspond to reductions in the transmitted flux to 36.8, 13.5, 5.0, and 1.8%, respectively.

The penetration in millimeters for which a neutron beam is attenuated to 5% of its initial value, calculated from Equation (2.48), is given in Table 2.3. It is usually still possible to carry out strain measurements at this level of attenuation. More sophisticated estimates of the maximum path length at which it is feasible to make strain measurements are given in Section 4.4.2.

2.5   Effects of Lattice Vibrations

As mentioned in Section 2.3.3, atoms are not positioned rigidly within the crystal structure, but instead vibrate about their mean lattice site r_u with an amplitude that increases with temperature. Thus, the time-averaged atomic nuclei positions are smeared and this gives rise to a term in intensity analogous to the atomic form factor for x-rays or neutrons, but as the spatial extent is much smaller than that of the atomic electrons, the Q dependence of this term is usually very much less. The full theoretical treatment of the thermal vibration of the atoms [26] gives rise to a temperature factor that multiplies the exponential term for each atom in the summation in the structure factor. However, an average term for harmonic vibrations may be taken out of the summation so that the structure factor is given by

$$F(hkl) = \exp(-W(Q)).\sum_u \bar{b}_u \exp\left(i2\pi\left(h.r_{ua} + k.r_{ub} + l.r_{uc}\right)\right). \tag{2.49}$$

The extra term therefore appears as a factor, $\exp(-2W(Q))$, called the Debye–Waller factor, in the expression for the cross-section and intensity of diffraction in Equation (2.40).

The quantity $W(Q)$ may be written in terms of the mean square amplitude of vibration of the atoms, $<u^2>$, and for cubic or spherical site symmetry is

$$W(Q) = Q^2<u^2>/6 = \frac{8\pi^2}{3}\left(u^2\right)\frac{\sin^2\theta}{\lambda^2} \tag{2.50}$$

The Debye model may be used to calculate $<u^2>$ for a simple lattice of atoms of mass M (see, for example, Ghatak and Khotari [26]), resulting in

$$W(Q) = \frac{3\ ^2Q^2}{2Mk_B\theta_D}\left\{\frac{1}{4} + \phi\left(\frac{\theta_D}{T}\right)\right\} \tag{2.51}$$

Here, θ_D is called the *Debye temperature*, and $\phi(\theta_D/T)$ is the *Debye function* [27]. The first term shows that the factor causes the Bragg peak intensity to decrease as the scattering vector Q increases. The Debye function gives the temperature dependence; it increases as T increases causing the intensity to decrease with increasing temperature. However, the Bragg peak width does not broaden and is independent of temperature. The Debye-Waller factor $\exp(-2W(Q)$ must also be included in the incoherent scattering cross-section, and so gives rise to a small Q dependence to this scattering.

The atomic lattice vibrations in a crystal may be described in terms of quantized excitations called *phonons*. In an inelastic scattering process, in which both the neutron energy and wave vector are changed, the neutron may excite or deexcite lattice vibrations, that is, cause the creation or annihilation of phonons. Inelastic scattering is weaker than Bragg diffraction, and from a polycrystalline sample contributes mainly under and near the Bragg peaks, and is referred to as *thermal diffuse scattering* (TDS). As the temperature increases, the TDS increases and, through the Debye–Waller factor, the Bragg peak intensity decreases.

These effects of lattice vibrations must be included in the analysis of Bragg peak intensity data aimed at determining crystal structures, and indeed give important additional information on the structure. However, they are not usually of great importance in neutron diffraction strain measurements, although fitting a Debye–Waller factor may be used to establish the temperature of the measurement [27,28].

2.6   Extinction, Texture, and Multiple Scattering

We now mention three other important effects that may alter a sample diffraction intensity from that expected in the ideal case. These are extinction, texture, and multiple scattering.

2.6.1   Extinction

If a monochromatic neutron beam falls on a highly perfect single crystal oriented in the reflection condition for Bragg diffraction, the incident beam will be diffracted by successive lattice planes and consequently will diminish in intensity as it penetrates into the crystal, as shown schematically in Figure 2.12a. The whole of the crystal will therefore not experience the same incident intensity, which may even be exhausted in the near-surface region. This attenuation due to diffraction is termed *primary extinction*. As an example, an estimate of the extinction distance over which the attenuation is appreciable for 2 Å neutrons diffracted by the 111 reflection from Ni at a scattering angle of ϕ^s ~52° is ~2.2 μm [4]. Since this is only a very small volume of crystal, there is a consequent size broadening of the diffraction peak of ~3 s of arc (see Equation (2.17)), and the peak profile becomes flatter. However, most crystals contain dislocations that break the perfect regions up into a mosaic of small volumes at small angles to each other, making primary extinction small. Indeed, a mosaic structure is important for crystal monochromators, as discussed in Section 3.2.1. As a result, only a relatively few successive lattice planes diffract before the incident beam meets a new mosaic at a slight angle. It is not until the beam meets another mosaic volume at exactly the correct angle that extinction attenuation will continue as illustrated in Figure 2.12b. Such extinction is termed *secondary extinction*. The intensity of the diffraction peak from the mosaic crystal may be affected, but the width of the peak is determined by the angular mosaic spread.

Although extinction is of prime importance in single crystal diffraction, it is usually very much less important for powder diffraction and can often be neglected. However, if the grains in a polycrystalline sample are sufficiently large and perfect, then extinction could affect the diffraction peak intensities. For example, austenitic steel weldments often have a large grain size in the weld itself.

Extinction may also affect the wavelength composition of the beam penetrating a large sample, since one component in a distribution of λ may be strongly diffracted out of the beam by a large grain. Secondary extinction may also be present over a long path length in the polycrystalline sample, since although the grains may be randomly oriented in contrast to lying small angles in a mosaic crystal, there is still a chance of the beam encountering further grains in the same orientation.

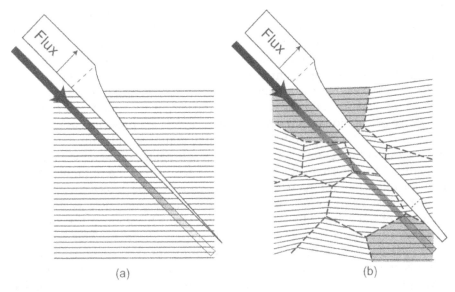

FIGURE 2.12

Schematic showing the loss of transmitted neutron flux due to (a) primary and (b) secondary extinction on passing through a single crystal. In (b), the grain has polygonized into slightly misoriented subgrains so that it is broken up into a mosaic crystal, only the shaded subgrains are oriented to satisfy the diffraction condition.

In accurate structure determination, the effects of extinction should be included in procedures that fit the complete powder diffraction pattern such as the Rietveld method, described further in Sections 3.2.2 and 4.5 [29]. In strain measurements performed at a reactor, it would be hard to detect extinction if only one order of reflection is measured. In time-of-flight measurements, however, where several orders of a given reflection may be collected simultaneously, the relative intensity of different orders of a reflection will reveal the presence of extinction. Further discussion of extinction is provided by Bacon [1] and Sears [7].

2.6.2 Texture

The crystallographic texture is a measure of the degree to which the grains in a polycrystalline sample are not randomly oriented. Engineering components are often textured, since manufacturing processes generally involve plastic deformation, and this has the effect of modifying the crystallographic orientation of the grains, causing preferred orientation of the grains along certain macroscopic sample directions, such as the case of rolled plate in the directions of rolling and normal to the rolling plane. Other examples include extruded wires and tubes, and welds where there has been recrystallisation from the melt. Texture can also arise from other causes, such as oriented growth in the casting of metals, and controlled vapor deposition of ceramics.

As a result of texture, the relative intensities of diffraction peaks are changed from those anticipated on the basis of the multiplicity of the planes, j, given in Table 2.6, which were based on random grain orientation.

In order to measure the degree of texture in a sample using a continuous monochromatic neutron beam from a reactor, the detector on the diffracto-meter is positioned at the scattering angle corresponding to the center of a diffraction peak *hkl*, and the diffracted intensity is measured as a function of sample orientation over a hemisphere. The data are presented as contours of intensity of the Bragg reflection as a function of sample direction along the corresponding Q on what is termed a *pole figure*; an example is shown in Figure 2.13. Alternatively, particularly when using time-of-flight diffrac-tion, the intensities for many different reflections can be collected at the same time for just one sample direction. The data are presented as points or contours of intensity of the different Bragg peaks lying along the direction of Q on a triangular plot of the crystallographic directions. This is termed an *inverse pole figure*, and is particularly useful for fiber textures in which there is one main direction of crystallite alignment in the sample. Different orders of reflection, such as 111 and 222, will display the same angular dependence of intensity since they correspond to the same grain orientation. A quantitative description of the distribution of crystalite orientations, the degree of texture, in a polycrystalline aggregate is given by the orientation distribution function (ODF) which is determined from the measured pole figures. The study of crystallographic texture is a well-developed field with many applications in materials science that have been dealt with recently in the textbook by Kocks et al. [30]. The avoidance of, or correction for, extinc-tion effects can be important when determining texture from the variation in peak intensities with sample orientation.

When determining crystal structures from a powder diffraction pattern, using for example, the Rietveld method, a texture coefficient may be included as a parameter to correct for weak texture effects. When measuring strain, and in converting strain data to stress, texture in a sample can have important consequences. Strong texture may prevent the observation of one Bragg reflection in all the directions in the sample for which strain data are required. Consequently, more than one reflection may have to be used. The conversion of strain to stress, discussed in Chapter 5, requires a knowledge of the texture if it is not random as it can affect the elastic constants to be used. Texture may arise from, and be an indication of, plastic deformation of a sample. As plastic deformation may cause intergranular stresses, texture indicates when careful data analysis for intergranular effects may be required, as discussed in Section 4.2.2 and Chapter 5. Finally, texture gradients in a sample can affect data interpretation, as discussed in Section 3.6.4.

2.6.3 Multiple Scattering

The most intense scattering process from most polycrystalline samples is the coherent diffraction into Debye–Scherrer cones. In general, strain

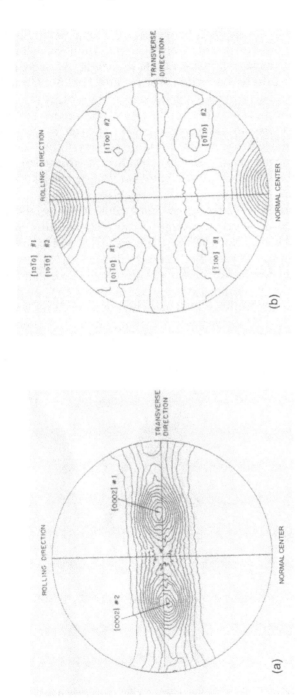

FIGURE 2.13
The (a) (0002) and (b) (10$\bar{1}$0) pole figures for a highly textured rolled Zircaloy-2 plate. The two regions of intensity in the (0002) pole figure between the normal and transverse directions correspond to the two "families" of grains in the material. There is correspondence between sample direction and crystallographic plane in a strongly textured sample such as this plate. The contour levels are 0.1, 0.6, 1.1, etc., multiples of random distribution.

measurements are made on large samples with long paths through the material, and as a consequence neutrons scattered from the incident beam into Debye–Scherrer cones will have a finite probability of being diffracted a second time into a different direction. The net effect of double and higher-order scattering in a polycrystalline solid is to add to the incoherent background between the peaks. For a single crystal, however, double Bragg scattering can give rise to extra weak peaks, as well as changing the intensity of the primary Bragg peaks. In a large polycrystalline sample with a large coherent cross-section and a low incoherent cross-section such as iron, the major source of background between Bragg peaks is likely to be multiple diffraction.

The extent of multiple scattering has been considered for conventional powder diffraction for which the samples are usually ~10 mm thick. It has been calculated by Vineyard [31] for plate geometry, and for fully bathed cylinders (see Figure 2.10b) by Blech and Averbach [32], as described by Bacon [1]. The latter results are shown in Figure 2.14. Note that for a 2 cm high copper cylinder, multiple scattering accounts for 85% of the diffuse scattering, while for vanadium, which has a higher incoherent contribution, it is only around 20%. However, when making strain measurements, the sampled volume is typically smaller than the entire sample volume, which is the configuration shown in Figure 2.10a. In this case, it is more difficult to calculate the multiple scattering contribution.

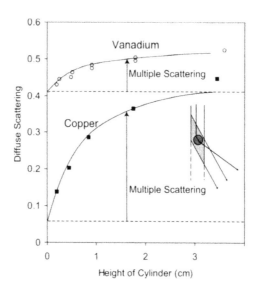

FIGURE 2.14
Relative multiple scattering calculated for the sample configuration shown in Figure 2.10b from 1 cm radius cylinders of copper and vanadium of various heights. The broken horizontal lines show the calculated sum of the incoherent and thermal diffuse scattering, while the contribution from predicted multiple scattering has been added to give the full-line curves, which are in good agreement with the experiment (adapted from Bacon [1]).

References

1. G.E. Bacon, *Neutron Diffraction*, 3rd ed., Clarendon Press, Oxford, 1975.
2. P.A. Egelstaff, *Thermal Neutron Scattering*, Academic Press, New York, 1965.
3. S.W. Lovesey, *Theory of Neutron Scattering from Condensed Matter*, International Series of Monographs on Physics No. 72, Vols. 1/2, Clarendon Press, Oxford, 1984.
4. G. Squires, *Theory of Neutron Scattering*, Cambridge University Press, Cambridge, 1982.
5. C.G. Windsor, *Pulsed Neutron Scattering*, Taylor & Francis, London, 1981.
6. K. Sköld and D.L. Price, Eds., *Methods of Experimental Physics*, vols. 23A-235, Neutron Scattering, Academic Press, New York, 1986.
7. V.F. Sears, *Neutron Optics*, Oxford University Press, Oxford, 1989.
8. R.B. Von Dreele, Neutron diffraction, in *Materials Science and Technology*, E.J. Kramer, Ed., VCH , Weinheim, 1994, 562–610.
9. B.D. Cullity, *Elements of X-ray Diffraction*, 2nd ed., Addison-Wesley, Reading, MA, 1978.
10. I.C. Noyan and J.B. Cohen, *Residual Stress: Measurement by Diffraction and Interpretation*, in *Materials Research and Engineering*, B.I. Ischner and N.J. Grant, Eds., Springer-Verlag, New York, 1987.
11. V.F. Sears, Neutron scattering lengths and cross sections, in *Methods of Experimental Physics* vol. 23A *Neutron Scattering*, K. Sköld and D.L. Price, Eds., Academic Press, New York, 1986, 521–550.
12. V.F. Sears, Neutron scattering lengths and cross sections, *Neutron News*, 3, 26–36, 1992.
13. S.F. Mughabghab, *Neutron Cross Sections*, vol. 1B, Academic Press, New York, 1984.
14. P.J. Withers, Use of synchrotron x-ray radiation for stress measurement, in *Analysis of Residual Stress by Diffraction using Neutron and Synchrotron Radiation*, M.E. Fitzpatrick and A. Lodini, Eds., Taylor & Francis, London, 2003, 170–189.
15. P.J. Withers, Monitoring strain and damage by neutron and synchrotron beams, *Mat. Sci. Technol.*, 17, 759–765, 2001, and *Adv. Eng. Mat.*, 3, 453–460, 2001.
16. B.C. Larson, W. Yang, G.E. Ice, J.D. Budai, and J.Z. Tischler, Three-dimensional X-ray structural microscopy with submicrometre resolution, *Nature*, 415, 887–890, 2002.
17. C. Kittel, *Introduction to Solid State Physics*, 6th ed., John Wiley & Sons, New York, 1986.
18. N.W. Ashcroft and N.D. Mermin, *Solid State Physics*, Saunders, Philadelphia, 1976.
19. M.J. Buerger, *X-ray Crystallography*, John Wiley & Sons, New York, 1953.
20. A.D. Krawitz, *Introduction to Diffraction in Materials Science and Engineering*, John Wiley, New York, 2001.
21. A.J. Allen, C. Andreani, M.T. Hutchings, and C.G. Windsor, Measurement of internal stress within bulk materials using neutron diffraction, *NDT Int.*, 15, 249–254, 1981.
22. K.D. Rouse, M.J. Cooper, E.J. York, and A. Chakera, Absorption corrections for neutron diffraction, *Acta Crystl.*, A26, 682–691, 1970.

23. W.I.F. David, Powder diffractometry, in *Neutron Scattering at a Pulsed Source*, R.J. Newport, B.D. Rainford, and R. Cywinski, Eds., Adam Hilger, IOP Publishing, Bristol, UK, 1988, 189–232.

24. M.T. Hutchings and C.G. Windsor, Industrial applications, in *Methods of Experimental Physics*, vol. 23C *Neutron Scattering*, K. Sköld and D.L. Price, Eds., Academic Press, New York, 1987, 405–482.

25. T.C. Hsu, F. Marsiglio, J.H. Root, and T.M. Holden, Effects of multiple scattering and wavelength-dependent attenuation on strain measurements by neutron scattering, *J. Neutron Res.*, 3, 27–39, 1995.

26. A.K. Ghatak and L.S. Khotari, *Introduction to Lattice Dynamics*, Addison Wesley, London, 1972.

27. T.M. Holden, J.H. Root, D.C. Tennant, D.E. Kroeze, and D. Leggett, Non-invasive measurement of strain and temperature by neutron diffraction, in Proceedings of Gas Turbine and Aeroengine Congress, Brussels, Belgium, 1990 (ASME 90-GT-324).

28. T.M. Holden, J.H. Root, D.C. Tennant, R.R. Hosbons, R.A. Holt, K.W. Mahin, and D. Leggett, Industrial research and development and the role of neutron diffraction, *Physica B*, 180–181, 1014–1020, 1992.

29. H.M. Rietveld, Line profiles of neutron powder-diffraction peaks for structure refinement, *Acta Cryst.*, 22, 151, 1967.

30. U.F. Kocks, C.N. Tomé, and H.-R. Wenk, *Texture and Anisotropy*, Cambridge University Press, Cambridge, 1998.

31. G.H. Vineyard, 2nd Order scattering correction in neutron and x-ray diffraction, *Phys. Rev.*, 91, 239–245, 1953.

32. I.A. Blech and B.L. Averbach, Multiple scattering of neutrons in vanadium and copper, *Phys. Rev.*, 137A, 1113–1116, 1965.

3

Diffraction Techniques and Instrument Design

In this chapter the practical instrumentation for neutron diffraction strain measurement is described, based on the general principles of neutron scattering developed in the previous chapter. Details of the types of sources available are outlined, followed by a description of the basic elements of the types of instruments and a description of the various detector arrangements used in practice. Methods are divided into those appropriate for either a continuous or time-pulsed flux source, although many of the elements are common to both. The aim is to provide the reader with a basis for the selection of the optimal experimental arrangement for a given application.

3.1 Neutron Sources

At the present time, the two principal types of neutron sources that have sufficient neutron flux to be suitable for the measurement of strain, and hence stress, are nuclear reactors incorporating neutron beam tubes, and spallation neutron sources. These are major facilities and are located at national or international laboratory sites. However, access to these neutron facilities is now relatively straightforward for anyone with an innovative experiment to carry out, and expert assistance is usually available. Most laboratories are open to applications for beam-time usage from a range of participating countries, and simply require a proposal to be written that goes before a peer review committee held at regular intervals. Providing that the proposal is generic and sound, available beam time is granted, sometimes with a contribution toward the expenses of visiting the site. Industrial users can usually perform measurements with full commercial confidentiality by paying for the beam time. Although perhaps a little daunting at first sight to those used to working in a small laboratory, the environment of the neutron beam facility soon becomes familiar and once the safety formalities are overwith, one is concerned only with the instrument for strain measurement, which is often no more complex or larger than a piece of conventional laboratory equipment. It is indicative of the rise in the use of neutron diffraction for engineering and materials problems associated with the measurement of residual stress that at almost every neutron source facility a state-of-the-art dedicated strain measurement instrument is either in use or is being planned.

An inevitable question from all interested in stress measurement is whether portable neutron sources are ever likely to become available that can make measurements *in situ*. Unfortunately, at present and for the foreseeable future, these are too low in neutron intensity, and are suitable only for radiography and other unsophisticated uses. Nevertheless, the flux of portable sources has increased markedly in recent years, and while falling far short of that needed for detailed strain and stress measurement, may one day approach the required levels for stress measurement in specialized cases.

3.1.1   Steady-State Reactor Neutron Source

The most intense continuous beams of neutrons, in terms of the total integrated neutron flux, are those emanating from the beam tubes of a nuclear reactor producing a continuous high flux of neutrons from fission of U^{235}, which has been specially designed for neutron research [1–3]. Most early research reactors, or "materials testing reactors" (MTRs), were designed to provide a range of facilities with many vertical tubes in and around the reactor core in which irradiation experiments could be performed and isotopes produced. Often, horizontal beam tubes were provided as an additional facility for irradiations, as well as to provide external beams of thermally moderated neutrons. The beam tubes view a source inside the reactor situated in the moderator/reflector in a region of high thermal flux. However, they were rarely optimized for maximum external neutron flux, and in some cases viewed the reactor core, and thus their beams had a high fast neutron and γ-ray content.

As the importance of neutron scattering techniques has gained recognition, research reactors designed primarily to provide high-flux neutron beams have been constructed. These have horizontal, and sometimes a few angled, beam tubes that are tangential to the core, to reduce the fast neutron and γ-ray component of the beam. Their source originates in an area of the moderator/reflector specially designed to optimize the thermal neutron flux. A typical layout of such a reactor is shown in Figure 3.1. Sometimes the beam tubes pass right through the reactor, and such "through" beam tubes house source blocks positioned in the region of maximum thermal flux to scatter out the neutrons. The beam tubes transport the neutrons from the source to the region beyond the outer shielding of the reactor where neutron scattering instruments are situated. Early beam tubes were circular in cross-section, but as most instruments use relaxed vertical resolution, elliptical or rectangular cross-sections are often used to enhance the flux on the instrument. The beam tube usually contains an absorbing shutter to switch off most of the beam, or can be flooded with water to reduce the beam intensity

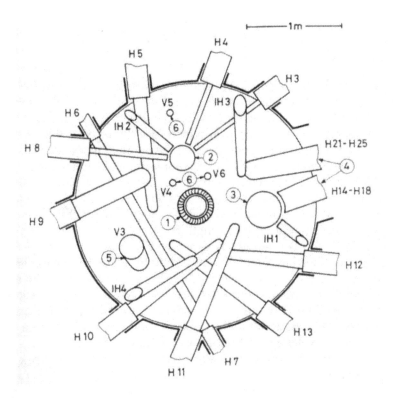

FIGURE 3.1
The arrangement of beam tubes and moderator sources in high-flux reactor at the Institut Laue Langevin, Grenoble. (1) Core. (2) Hot source. (3) Cold source. (4) Neutron guide tubes. (5) Vertical beam tubes. (6) Pneumatic post for irradiations [2]. A second cold source has now been added.

to very low levels. The latter allows work to be carried out safely in the instrument's beam exit region, where monochromating crystals or choppers may be located.

3.1.1.1 Moderation

The energy of the neutrons in the core of a reactor (~2 to 3 MeV) is much too high to be useful for diffraction experiments. The neutron energies are therefore thermalized by a moderator. When in equilibrium with the moderator at a thermal temperature T, the distribution of neutron velocities in the beam is given by the Maxwell–Boltzmann distribution (Appendix 1). The flux distribution, $\Phi(E)$, can then be determined as a function of neutron energy so that $\Phi(E)dE$ is the flux of neutrons with energy between E and $E + dE$. There is also a contribution from the unmoderated reactor spectrum that gives rise to a high-energy $1/E$ epithermal "tail" to the spectrum. Thus, the flux distribution is given by:

$$\Phi(E) = \Phi_F \left[\frac{E}{(k_B T)^2} e^{\frac{-E}{k_B T}} + \eta_f \frac{1}{E} \right] \qquad (3.1)$$

Here k_B is Boltzmann's constant, Φ_F is the total thermalized flux, and η_f (typically <1/100) the fraction of epithermal flux. The maximum thermal neutron flux $\Phi(E)$ occurs at an energy $E \sim k_B T$. However, it should be noted that "undermoderation" may occur, in which case the peak in flux corresponds to a slightly higher temperature than the nominal moderator value.

When the moderator and source block are at reactor ambient temperature, that is, when the moderator temperature T lies between 300 K and 430 K, depending on the type of moderator, the distribution is called "thermal." Typically, for $T \sim 330$ K, the thermalized neutron energy is of order $E \sim 28$ meV, and the corresponding neutron wavelength is $\lambda_T \sim 1.7$ Å.

The flux distribution is often given as a function of neutron velocity or wavelength, as described in Appendix 1. It should be noted that the function $\Phi(\lambda)$, which gives the flux per wavelength interval $d\lambda$ rather than energy interval dE, peaks at a wavelength $\lambda_f(maxflux) = \lambda_T \sqrt{(2/5)}$, where λ_T is the wavelength corresponding to the peak in $\Phi(E)$ at $E(maxflux) \sim k_B T$. Thus, the peak in a thermally moderated flux distribution, typically shown in Figure 3.2, is at a wavelength of $\lambda_f(maxflux) = 1.7 \times 0.632 \sim 1.1$ Å. Relations between the flux distributions expressed in terms of the different variables are given in Appendix 1.

Consideration of the flux distribution with respect to wavelength rather than energy is more appropriate when considering a crystal monochromator, since the latter selects a range of wavelengths $d\lambda = 2d_{hkl} \cos\theta^M d\theta^M$ out of the spectrum corresponding to the effective collimation angle $d\theta^M$, as discussed in Section 3.2.1.

Specialized source blocks may be provided in the reactor at the end of the beam tube, which are cooled, typically by liquid hydrogen at 20 K ("cold source"), or warmed, typically by nuclear heating to 2300 K ("hot source"),

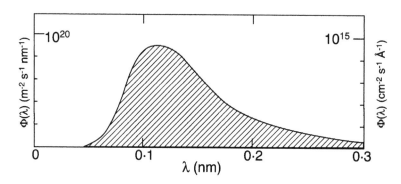

FIGURE 3.2
Maxwellian flux distribution for a moderator at 300K calculated according to Equation 3.1. Note that the maximum in the wavelength distribution is around 1.1 Å (see Appendix 1).

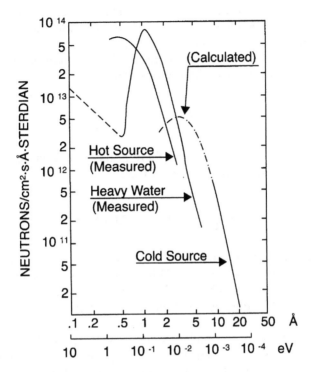

FIGURE 3.3
The neutron flux spectra, $\Phi(\lambda)$, from the thermal (heavy water ~330K), hot (~2300 K), and cold (~20 K) moderator sources at high-flux reactor, Institut Laue Langevin, Grenoble, France [2]. The bottom abscissa scale denotes the neutron energy corresponding to the neutron wavelength of the top scale (Table 2.1).

to give a spread of wavelengths biased toward longer or shorter values, respectively. Typical neutron flux distributions from these sources, given in the form $\Phi(\lambda)$ versus λ, are shown in Figure 3.3.

Many dedicated beam reactors are available or are under construction around the world; that at the Institut Laue-Langevin (ILL) in Grenoble, France, is a prime example of a 58-MW reactor with a very high neutron flux of 3×10^{15} ncm^{-2}s^{-1}. The High Flux Isotope Reactor at the Oak Ridge National Laboratory (Oak Ridge, Tennessee) has a flux close to that of the ILL, whereas medium-flux reactors — such as those at the Laboratoire Léon Brillouin in Saclay, France; the National Institute for Standards and Technology (NIST), Gaithersburg, Maryland; the University of Missouri, Columbia, Missouri; the NRU reactor at Chalk River, Ontario, Canada, and the FRM II reactor in Munich, Germany — have fluxes in the range of 2 to 3×10^{14} cm^{-2}s^{-1} (Figure 1.2).

3.1.1.2 Neutron Guide Tubes

Neutron guides provide an efficient means of transporting the useful moderated neutrons over short or long distances. They act as filters in that they alter the composition of the neutron's spectrum since their transmission is very wavelength dependent. The principle of the guide is to utilize the total external reflection of neutrons impinging at low angles on a smooth surface, which occurs because the refractive index of the surface material, $n(\lambda)$, for neutrons of wavelength λ is slightly less than unity:

$$n(\lambda) = \left[1 - \lambda^2 \rho_n \bar{b} / 2\pi\right] \tag{3.2}$$

For a given neutron wavelength, there is therefore a critical angle $\theta_c(\lambda)$ between the incident beam and the surface.

$$\theta_c(\lambda) = \left[2(1 - n(\lambda))\right]^{1/2} = \lambda \left[\rho_n \bar{b} / \pi\right]^{1/2} \tag{3.3}$$

Here $\theta_c(\lambda)$ is in radians, and the surface material has a number density of atoms ρ_n with scattering length $\bar{b}$. This angle is very small, of the order of one degree, for neutrons because $n(\lambda)$ is only slightly less than unity. There is an analogy with light rays that are totally reflected below the *Brewster angle* at an optically dense to rare interface.

Neutrons impinging at smaller angles, $\theta < \theta_c(\lambda)$, are totally externally reflected at the wall surface, whereas those at greater angles enter the surface. Guide tubes are usually of rectangular cross-sections and, as the walls must be optically flat, made of float glass usually coated with a metal such as nickel. The neutrons are transmitted along the guide by multiple reflections at the coated walls. Some typical neutron guides are shown in Figure 3.4. By using a slightly bent guide, of several kilometers radius, unwanted γ and fast neutron background can be removed from the useful beam, being transmitted through the walls into a biological shielding absorber surrounding the guide. The effective collimation angle of a natural nickel guide, α, can be calculated using Equation (3.3) with $\rho_n = 0.0914 \times 10^{24}$ cm^{-3}, $\bar{b} = 1.03 \times 10^{-12}$ cm. Thus, as a rule of thumb,

$$\alpha = \theta_c = 0.1\ \lambda, \text{ or } \alpha = 0.6/k \tag{3.4}$$

where α is in degrees, λ is in Å, and the wave vector, k, is in Å^{-1}. For example, $\alpha = 0.3°$ for 3 Å neutrons. A characteristic guide wavelength, λ_c, may be defined, below which there is low transmission, given by:

$$\lambda_c^2 = 2W\pi / (\rho_n \bar{b} R) \tag{3.5}$$

where W is the width of the guide and R the radius of curvature [4,5].

Using ^{58}Ni coatings gives a 17% increase in θ_c, because it has a higher scattering length. However, through the use of many deposited layers of graded thickness of various materials, notably Ni or Ni-C, and Ti, the value

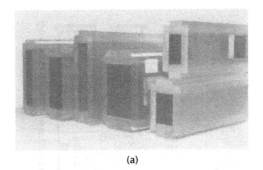

(a)

(b)

FIGURE 3.4

(a) Neutron guides with various cross-sections, and (b) utilization of guides at Orphée Reactor, Laboratoire Léon Brillouin, CEA Saclay, France. (Photos courtesy of Celas.)

of θ_c and λ_c may be increased by factors of 2 to 3, in what are sometimes referred to as "m = 2 or 3" guides [4]. Although most efficient for cold neutrons, these "supermirror" guide tubes can be fabricated to transport thermal neutrons efficiently.

The use of guides enables neutron beams to be transported to "guide halls," which are located outside the main reactor shell. Here, many instruments can be located on each guide in a region of low radiation levels and low instrument background, as shown in Figure 3.5 [6]. The various instruments on the same guide may take different vertical sections of the guided beam, or the part of the beam transmitted through the monochromator of an upstream instrument. In principle, less shielding is required in the guide hall region than in the reactor shell. Multilayer neutron guides may also be used on time-pulsed sources, discussed in Section 3.1.2, to transport a beam

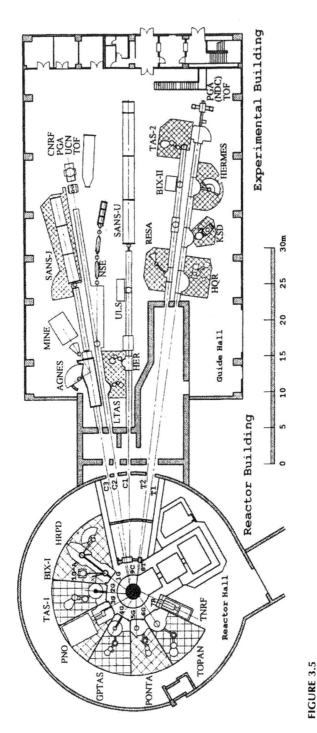

FIGURE 3.5
Instrument layout at the JRR-3M reactor at the Japanese Atomic Energy Research Institute, Tokai, Japan, illustrating the use of guides to transport beams to a guide hall [6].

away from the shielded moderator region to a distant instrument. Typically, useful fluxes of neutrons of wavelengths ~0.7 Å and above may be delivered to an instrument.

By using multilayers of ferromagnetic materials, short bent guides a few tens of centimeters long may be used to provide beams of polarized neutrons. The surface of such "polarizing supermirror" guides may be built up from several hundred layers of magnetized Fe/Si, Ni/Ti, or Co/Ti. However, the use of polarized neutrons in instruments for strain measurement has only recently begun to be developed for specialized cases of very high strain resolution using neutron Larmor precession techniques [7].

3.1.1.3 Neutron Lenses

Recently, much attention has been focused on the design of neutron "lenses," with the aim of focusing the neutron flux over a range of wavelengths and a large beam area into a smaller area of the size of the sample being studied. Two possibly useful methods are being developed. The first is to use many curved, hollow, glass capillary tubes that act as miniature neutron guides to focus the beam onto a spot. The second is to use curved multiple thin layer films acting as stacked bent microguides. Staggered [8] or stacked [9] lengths of nickel deposited on aluminum, or single crystal Si, have been used as the mirror to give a focused line of intensity. A variation of these is the "lobster's eye" system of stacked square capillary tubes, with deposits on two sides [10].

A simpler way of concentrating a guided beam is to have a curved super-mirror "nose" on the end of the guide before the instrument. This can be designed to concentrate the vertical extent of the beam onto the sample at the expense of increasing the vertical divergence, and is most suited to spallation source instruments. In the case of monochromated beams, bent or tilted monochromator crystals can be used to focus the beam onto the sample as discussed in Section 3.2.1.

3.1.2 Time-Pulsed Neutron Source

Beams of short bursts of neutron intensity (a few microseconds) in time separated by much longer periods (tens of milliseconds) may be produced in a number of ways, and may comprise pulses of a polychromatic spectrum of neutrons or of monochromated neutrons [2,11]. The beam from steady reactor sources may be time pulsed using velocity selectors or "choppers" (Section 3.2.2). The reactor itself may be time pulsed into criticality, such as the IBR-I reactor at Dubna near Moscow, where a critical part of the fuel was rotated through a subcritical core on rotating arms, producing pulsed polychromatic beams of neutrons every time criticality was reached. The newer IBR-II reactor has rotating neutron reflectors that cause the reactor to achieve criticality each time they pass the core. In this way, extremely intense beams corresponding to 10^{16} neutrons/cm^2 s at the moderator surface are produced for a relatively low mean power rating (~2 MW). However, the 5-Hz or 25-Hz pulse width of up to 320 μs is long for high-resolution diffraction work.

Linear electron accelerators (LINACs) can be used to produce pulses of electrons that impinge onto a heavy metal target such as tungsten to produce bursts of γ-rays or *bremsstrahlung*, which in turn may be used to produce neutrons by a (γ, n) reaction in the metal target, or from a uranium "booster" target.

The steady-state reactor as a neutron beam source has nearly reached its limits at the ILL facility, due to the requirement of dissipating more than 60 MW of power by continuous cooling. A proposal for "the ultimate reactor" or "king" reactor at Oak Ridge National Laboratory, which would have operated at ~100 MW, was considered too expensive. The latter has been supplanted by a new spallation neutron source, known as the SNS. Such pulsed sources now seem likely to provide the most intense neutron beams in the future, since the time-pulsed nature of the beams can be used to advantage to rival those from steady-state reactors, and the mean power dissipation requirements are in the region of only hundreds of kilowatts (Figure 1.2).

Spallation neutron sources use the spallation of neutrons from a heavy metal target such as tantalum or tungsten, by a beam of time-pulsed high-energy particles, usually protons, from a synchrotron source. The process is more efficient if natural uranium is used as the target, although irradiation and thermal damage effects are more severe, which foreshorten target life. Future sources are likely to use circulating liquid mercury targets. In the spallation process, the neutrons are stripped off the metal nucleus. These neutrons have very high energies and require moderation to thermal energies before they can be used for diffraction. The ISIS spallation source at the Rutherford Appleton Laboratory (Oxfordshire, UK) is currently the world's most intense, or "bright," spallation neutron source [12]. At this facility, the protons start life as negative hydrogen ions that are accelerated in an injector column and linear accelerator to ~70 MeV. The electrons are then stripped off by passage through a 0.25-μm alumina foil, and the protons are accelerated to ~800 MeV in a 52-m diameter proton synchrotron to give a time-pulsed beam of protons containing 2.5×10^{13} protons/pulse at 50 Hz. Each pulse has a time extent of 0.4 μs. On impact with a tantalum target, each proton produces on the order of 25 neutrons. These neutrons have a broad spectrum of high energies, on average 3 to 4 MeV, and are moderated for use in beam experiments by moderators fabricated from hydrogenous materials that must be carefully designed to slow down the neutrons but to not significantly increase the pulse width in time. At the ISIS source, they are typically of size ~10 × 10 × 5 cm^3, and of water at 316 K, liquid methane at 100 K, and liquid hydrogen at 20 K, so that the neutron intensity in each pulse will peak near the corresponding energies. It should be noted that the useful lower-energy neutrons in beam experiments at pulsed sources are detected at significant time intervals after the production of each pulse. As a result, the initial gamma and fast neutron bursts can be discounted, and

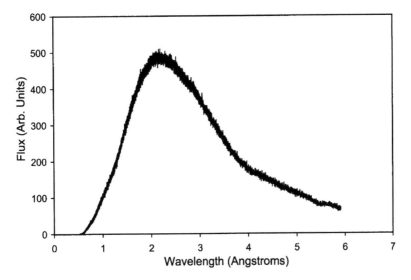

FIGURE 3.6
Incident flux spectrum measured on the ENGIN-X beamline at the ISIS pulsed neutron source.
(Courtesy of M.R. Daymond.)

therefore do not contribute to the measured background levels. A typical flux distribution in a beam arising from a liquid methane (CH_4) moderator at 100 K at the pulsed source, ISIS, is shown in Figure 3.6.

The general layout of the ISIS facility is shown in Figure 3.7 [13]. A second target station will be operational in 2006. This will take every fifth proton pulse to impinge on two types of cold moderator to provide high flux pulses of long-wavelength neutrons for an additional suite of new instruments [14]. However, these are unlikely to include an instrument for strain measurement. Other current spallation sources include the IPNS facility at the Argonne National Laboratory (Argonne, Illinois) [3,15], LANSCE at Los Alamos National Laboratory (Los Alamos, New Mexico) [16], and KEK (Tsukuba, Japan) [17]. Next-generation sources are now being built, including the SNS at Oak Ridge National Laboratory, a joint KEK/JAERI facility at Tokai, Japan, and the abovementioned second target station at ISIS (United Kingdom). The new Swiss Spallation Neutron Source (SINQ) facility at the Paul Scherrer Institut (PSI), Villigen, Switzerland, is a novel variation on the pulsed source concept. This spallation source provides long time pulses used to give an essentially continuous beam, and the instruments use the average flux as on a reactor source without time resolution [18]. The time-averaged flux of thermal neutrons in a spallation source ranges from about 5×10^{11} n $cm^{-2}s^{-1}$ at the KEK source in Japan and the IPNS source at Argonne National Laboratory, to 5×10^{12} n $cm^{-2}s^{-1}$ at ISIS and LANSCE at Los Alamos National Laboratory. The peak thermal flux of several pulsed sources is shown in Figure 1.2 along with the flux in continuous reactor sources.

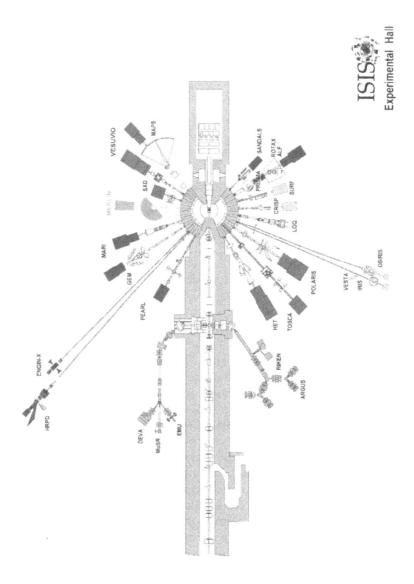

FIGURE 3.7
Layout of instruments on target station 1 of the ISIS spallation neutron source [13].

3.1.3 Safety

In using any large facility to carry out experiments, the safety of all staff and visiting scientists is of overriding importance. The hazards of working with radiation are very well known and documented, and the resident health physicists will ensure that rules and regulations are in place. Needless to say, these must be strictly adhered to. Usually regular users of neutron facilities must be "registered radiation workers," which means that they will have undergone a basic course on working with radiation and have had an annual general health examination. Less regular users may be given temporary status. All users are given an introduction to the notices and regulations to be observed at each site, and issued with local radiation dose monitors and/or film badges.

Usually the danger of contamination at a neutron facility is minimal, and that from radiation is of most concern. In carrying out experiments it is essential to be aware of radiation levels around the instrument, and radiation monitors are to be found at most instruments with which to undertake a survey. Often access to the instrument is impossible unless the beam is switched off through the use of interlocks. One overriding feature of instruments situated on a pulsed beam from a spallation neutron source is the very high level of neutron intensity in the pulse often accompanied by high γ-ray levels. This has led to safety features that do not allow any human access to the sample area, usually in an interlocked "hut", while the beam is open. Consequently, all movements of the sample and variation of the apertures must be made remotely. On the more accessible reactor based instruments, the temptation to return to adjust apertures and samples with the beam "on" must be resisted. In general it is always advisable to seek the advice of the health physicist and scientist responsible for the instrument before undertaking any nonroutine action. The background level of radiation at pulsed sources and experimental guide halls are often well below those natural levels tolerated in certain geographical regions. However, this and the fact that thermal neutron irradiation carries relatively low levels of risk should not be allowed to breed an atmosphere of complacency. All radiation should be regarded as potentially dangerous, and exposure kept to a minimum. In brief, the ALARA (as low as reasonably achievable) principle of exposure to radiation must be adhered to.

At the end of every set of measurements, the sample should be monitored for any activity induced by neutron absorption before removal from the instrument area. Certain elements, which may be present in an alloy, are particularly prone to induced activity with long lifetimes. Some typical examples are given in Table 3.1 [19]. At most sources, the likely activity profile of the sample after irradiation will be assessed by a health physicist, who will either allow its immediate removal from the site, or request that it remain for a period of activity decay. In the latter case, it will be placed in a radioactive material store until the activity, usually short-lived, has died away before it can be removed from the facility. Typically, the sample will not be allowed

TABLE 3.1

Selected Typical Activities Induced in 5 cm Cube of Parent Material After 1-Day Exposure in ISIS Neutron Beam 2 Minutes Following Shutter Closing

Parent Element	Activated Isotope	Natural Abundance (%)	Induced Isotope	Induced Activity (Bq)	Half-Life
Al	^{27}Al	100	^{28}Si	1×10^5	2.2 min
Mn	^{55}Mn	100	^{56}Fe	7×10^6	2.6 h
Fe	^{54}Fe	6.0	^{55}Mn	8×10^1	27 years
	^{58}Fe	0.3	^{59}Co	5×10^1	45 days
Co	^{59}Co	100	^{60}Co	4×10^6*	5.3 years
Ni	^{62}Ni	3.6	^{63}Cu	1×10^4	2.5 h
	^{64}Ni	0.9	^{65}Cu	~1.0	100 years
Cu	^{63}Cu	69	^{64}Zn/^{64}Ni	2×10^6	12.7 h
	^{65}Cu	31	^{66}Zn	6×10^5	5.1 min

Note: Where more than one isotope is activated, an additional row is given [81]. * = estimated.

to leave the facility until its activity has fallen to give a radiation level of 0.1 μSv at its surface. If it is really necessary to remove a sample that has a higher activity, then there are certain shielding, packaging, and certification requirements that must be met to comply with international regulations. Typically, a package must have less than 0.1 μSv at the package surface.

In addition to radiation levels, other safety aspects should be kept in mind, particularly those concerned with lifting and handling heavy engineering samples for strain measurement.

3.2   Diffractometers for Strain Measurement

The first measurements of strain were made using slightly modified conventional neutron powder diffractometers on steady-state sources, although some special designs were tried from the outset. Much of the early development of strain measurement was carried out in this way (see, e.g., Allen et al. [20,21]). However, it soon became apparent that the special needs of defining the beam size, observing only a restricted range of scattering angles, and positioning and moving sometimes heavy samples through the beam, would require dedicated instruments, and these are now installed at most neutron facilities. The demands of strain measurement have provided a new challenge to the instrument designer, and have required a new approach and a more comprehensive understanding of the details of the diffraction principles, particularly concerning beam profiles over the gauge volume. The result has been the practical implementation of a large number of new ideas in neutron scattering.

Although the principles used in the design of continuous beam and pulsed beam instruments differ, many of the instrument components are the same.

The reactor-based instrument will be discussed first in the next section, followed by the time-pulsed beam instrument in Section 3.2.2. Neutron detectors that are largely common to both types of instrument are described in Section 3.3.

3.2.1 Reactor-Based Continuous Beam Instrument for Strain Measurement

3.2.1.1 Basic Instrument

The simplest instrument for strain measurement on a continuous beam from a reactor is shown schematically in Figure 3.8. The principal feature of the thermal neutron beam from a reactor source is its wide spectrum of neutron flux as a function of wavelength, shown in Figure 3.2. From this continuous distribution of wavelengths, a small range of wavelengths λ is extracted by a large single-crystal monochromator using the principle of Bragg's Law described in Section 2.3:

$$2d_{hkl}^{M} \sin \theta_{hkl}^{M} = \lambda \qquad (3.6)$$

A monochromator crystal or array of monochromator crystals that is large enough to fill as much of the beam as possible is oriented so that a chosen

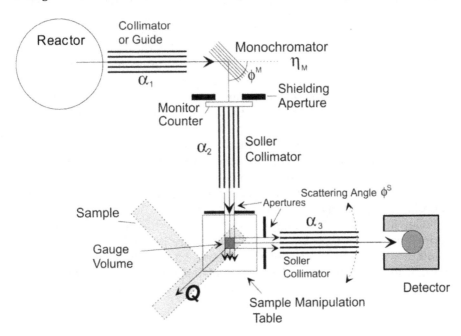

FIGURE 3.8
Schematic layout of a standard two-axis instrument for strain measurement on a reactor neutron source.

lattice plane (hkl), with lattice spacing d_{hkl}^M, contains the vertical direction, and is in the reflecting condition. In this way, it can diffract a beam of mean wavelength λ down a collimated beam path at a scattering angle $\phi_{hkl}^M = 2\theta_{hkl}^M$ to the incident reactor beam in the horizontal plane, as shown in Figure 3.8. The spread of wavelengths selected will depend on the collimation angles before and after the monochromator, α_1 and α_2, and the angular mosaic spread of the monochromator crystal, η_M, discussed in Section 3.4.1. Typically, the spread is about ± 0.02 Å and only includes about 0.5% of the total number of neutrons falling on the monochromator crystal. On most instruments the monochromator take-off angle ϕ_{hkl}^M is variable to allow a range of wavelengths to be selected. A low efficiency monitor counter is usually placed in the beam after the monochromator in order to monitor the neutron flux incident on the sample. An estimate of the typical monochromated beam intensity may be made as follows. The size of the beam from the core of the reactor to the monochromator is typically about 12×12 cm^2 by section, and the distance, r, from the source to the monochromator is about 6 m. The element of solid angle, $d\Omega/4\pi r^2$, subtended at the monochromator is then about 3×10^{-5}. If the neutron flux in the core is 3×10^{14} neutrons cm^{-2}s^{-1}, the number of monochromatic neutrons leaving the monochromator is about 3×10^7 neutrons s^{-1}.

The monochromated beam direction is defined by a collimator, and/or by apertures to pass over the sample axis about which the detector rotates. The detector counts neutrons scattered through an angle ϕ^s, with the scattered beam again defined in direction by a collimator between sample and detector. Both the beam incident on the sample and the beam entering the detector are usually defined in area by the horizontal and vertical extent of apertures in neutron absorbing masks, made for example of cadmium. The "gauge volume" or "volume sampled" is defined by the intersection of the incident and scattered beam, as shown in Figure 3.9 and discussed further in Section 3.5. As discussed in Section 1.4.2, the quantity measured depends on the ratio of the sampled volume V_v to the scale of the variation of the property being measured. If the sample is placed wholly within the gauge volume, then only properties averaged over the whole sample can be determined and other shorter-range variations will be averaged out. On the other hand, a large sample may be moved through the gauge volume in order to obtain a profile of a property such as the macrostrain.

Following Section 2.3, Bragg's Law may be rewritten in terms of the scattering vector $Q = (k_i - k_f)$, where k_i and k_f are the incident and scattered neutron wave vectors, as shown in Figure 2.3. For Bragg reflection, Q is normal to the diffracting planes, and of magnitude equal to that of the reciprocal lattice vector G of the planes, $|Q| = |G_{hkl}|$. Thus,

$$2k\sin\theta_{hkl}^B = Q = |G_{hkl}| . \tag{3.7}$$

Polycrystalline materials give rise to Debye–Scherrer cones of diffracted neutrons at angles $\phi_{hkl}^s = 2\theta_{hkl}^B$ about the incident beam on the sample as

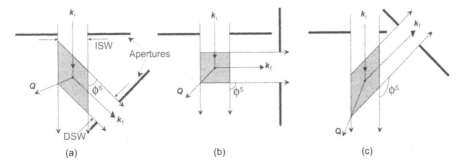

FIGURE 3.9
Sections of the nominal gauge volume defined in the scattering plane by incident and diffracted beam slit apertures of widths ISW and DSW, respectively, shown shaded. Three values of scattering angle ϕ^s are shown: (a) $\phi^s < 90°$, (b) $\phi^s = 90°$ and (c) $\phi^s > 90°$. The most symmetric shape is at $\phi^s = 90°$, but other shapes may be appropriate in certain cases.

illustrated in Figure 2.7, where d_{hkl} is the lattice spacing of planes with Miller indices *hkl*. The cone of scattering arises from a subset of the randomly oriented crystallites, or grains, in the sample that satisfy Bragg's Law, and which are oriented so that the planes are at an angle θ^B_{hkl} to the incident beam. It is usually assumed that the grains of the sample are sufficiently small that within the gauge volume there is a sufficient number of them to give a uniform intensity around the cone. Although this is normally the case in conventional powder diffraction, it may not be so for small gauge volumes within engineering samples having large grain size.

In the simplest case, a single detector is scanned in ϕ^s through the cone of *hkl* Bragg reflection scattering intensity in the "focused" configuration discussed in Section 3.4.1. Intensity, which is measured in detected counts per fixed number of monitor counts or per fixed unit of time, is recorded as a function of ϕ^s. Normalization to monitor counts is more usual in order to correct for any variation in reactor power and hence incident flux. Typical Bragg peaks give count rates from 0.1 to 1 count per second as recorded by a single detector, and in general adequate statistics are normally acquired on a time scale of minutes. In the case of a large sample, the observed peak corresponds to the average lattice spacing, d_{hkl}, of the grains in the sampled gauge volume (SGV). Usually a fitting routine incorporating least-squares minimization is used to determine the peak center angle and angular width from the profile of detected count intensity variation with angle ϕ^s, which is often represented by a Gaussian function, as discussed in Section 4.3. The average elastic lattice macrostrain in the volume sampled is then given in terms of the diffraction peak shift ($\phi^s_{hkl} - \phi^{s0}_{hkl}) = 2(\theta^B_{hkl} - \theta^{B0}_{hkl})$ by

$$\varepsilon^{hkl} = \left(d_{hkl} - d^0_{hkl}\right)/d^0_{hkl} = -\left[\cot(\theta^{B0}_{hkl})\right]\left(\theta^B_{hkl} - \theta^{B0}_{hkl}\right) \tag{3.8}$$

where d^0_{hkl} is the lattice spacing of a stress-free sample of the same material composition, and $2\theta^{B0}_{hkl}$ the corresponding Bragg scattering angle in radians. Note that as most lattice strains are small, $\sim 10^{-4}$ to $\sim 10^{-2}$, we adopt the "engineering" strain concept where the change from zero strain d^0_{hkl} is measured, rather than an incremental strain. Typical angular widths of Bragg peaks are $\sim 1.0°$ ϕ_s under focusing conditions discussed in Section 3.4.1. Peak shifts can be measured to about 1% of the angular width in θ^B, that is to $\pm 0.005°$ for a well-constructed spectrometer, corresponding to an accuracy in strain measurement of ~ 100 $\mu\varepsilon$, where the unit of $\mu\varepsilon$ is 10^{-6}. The direction in which the lattice strain ε^{hkl} is measured is parallel to the scattering vector Q. The measured strains in at least three directions are combined with the elastic constants of the sample material to determine the stress as discussed in Chapter 5. If the sample is stressed in the elastic regime, then the strain-stress relationship is straightforward. However, if the elastic limit is, or has been, exceeded, complications can arise due to intergranular stress effects, which are described in Chapter 5.

Since Type II and III microstrains (see Section 1.4.1) vary over distances much smaller than the sampled volume, V_v, the mean lattice microstrain in the sampled volume is related to the angular peak width. This also has contributions from the instrument angular resolution width, possibly from small grain size, and from steep strain gradients causing the angle ϕ^s_{hkl} to vary over the gauge volume. The observed peak width will be a convolution of all these contributions as discussed in Section 5.7. If an accurate value of $2\theta^{B0}_{hkl}$ can be measured, the Type I macrostrain determined is absolute. However, obtaining a true $2\theta^{B0}_{hkl}$ can prove difficult in practice, as discussed in Section 4.6.

In order to determine the strain in different directions in the sample, the sample must be rotated accurately about the center of the gauge volume so that each required direction lies along Q. This requires very careful alignment and centering of the sample, and is a major challenge to the instrument designer and to the experimentalist. This is discussed further in Section 3.6.3.

At $\phi^s = 2\theta^B = 90°$ the diffracting Debye–Scherrer cone becomes a plane disk (see Figure 2.7). This is also the scattering angle for definition of the most symmetric (i.e., cuboidal) and simplest gauge volume, as demonstrated in Figure 3.9b. Clearly, the gauge volume should be as closely as possible the same for all orientations of the sample if error from combining strains in different directions to give stress is to be minimized. The resolution in strain increases with ϕ^s, and is highest toward $\phi^s \sim 180°$, but the gauge then becomes an elongated diamond. As discussed in the Section 3.4.1, using $\phi^s \sim 90°$ scattering at both monochromator and sample is often a good compromise between resolution and gauge shape. However, an angle of ϕ^s greater or less than 90° may be used to advantage in cases where it can provide more ready access to a point in the sample at which the strain is to be measured.

These conventional angle-scanning diffractometers may be calibrated by measurement of several diffraction peaks from a standard powder sample, such as Si, CeO, Al_2O_3, or CaF_2, which have accurately known lattice spacings. With the aid of Bragg's Law, a precise measurement of the selected wavelength can be obtained, together with the zero-angle reading of the ϕ^s arm encoder. As strain measurements are always made relative to a stress-free sample, the wavelength and zero-angle reading need not be known precisely, as long as they remain constant. However, such accurate standard sample measurements are very useful for checking consistency, and for normalization of data taken under different conditions or at different times. If different instruments are used, any such normalization using a standard sample will require accurate wavelength and zero angle determination.

An alternative mode of operation of a two-axis diffractometer is to scan the incident wavelength through the Bragg reflection keeping the scattering angle fixed. In this way, the measurement becomes similar to that made using a time-pulsed polychromatic neutron beam discussed below, albeit using one Bragg peak only. Although less often used in practice, this mode of operation may be advantageous if the scattering angle must be fixed for special reasons, such as viewing the gauge volume through a small aperture cut in the sample or sample containment vessel in order to reduce attenuation of the incident or diffracted beam.

3.2.1.2 Instrument Construction

The construction of instruments has varied considerably over the past 40 years. In particular, the use of beam guide tubes to guide halls has reduced the necessity of massive monochromator shielding, needed when the instrument is adjacent to the reactor. The latter requires complex lifting segments to allow the monochromator take-off angle to vary. In early instrument design, the sample table rotated around the monochromator, and the detector rotated about the sample table, on heavy counterbalanced arms or on wheels running on tracks. Even gun mountings were used! However, the use of the hovercraft principle pioneered by the United Kingdom Atomic Energy Authority (UKAEA) (Harwell, UK), in which heavy components are supported on air cushions running on smooth steel or marble floors, has made the controlled movement of the instrument components much easier. This "tanzboden" principle has been developed extensively on many instruments at the Institut Laue-Langevin (Grenoble, France), and is now found at most facilities.

A further important logistic development has been the use of "optical benches," or accurately machined metal beams, on which components such as collimators, aperture masks, and shielding apertures can be positioned accurately and reproducibly. These are placed on the rotating arms between monochromator and sample, and between sample and detector.

Computer control of the instrument has developed in sophistication along with the availability of the increasing power and decreasing size of computers. Originally a central mainframe computer was often used to control several instruments, with obvious disadvantages. Now dedicated computers control each instrument, and not only the angles of the instrument components and the position of the sample, but also the size and positions of

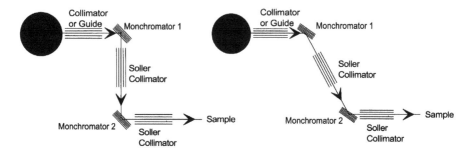

FIGURE 3.10

The double-crystal monochromator configuration at two take-off angles. Note that the sample axis and instrument reference point can be fixed.

apertures, can often be set automatically for each measurement. This means that complete measurement schedules can be defined to acquire data automatically 24 h/day. The data may be analyzed using peak profile fitting, "online," and the results used to define subsequent measurements. Thus, automatic alignment of the instrument is possible. Usually the data are backed up on a central computer.

3.2.1.3   Monochromator and Choice of Wavelength

Single-crystal monochromators are most commonly used, but the double-crystal monochromator configuration, shown in Figure 3.10, has advantages in certain cases. In this configuration, the beam onto the sample is deflected sideways, so that heavy shielding can be placed after the first monochromator to cut incident fast neutron and γ-ray background. The instrument is more compact and the sample table does not need to move. However, beam intensity is lost at the second monochromator reflection, and the range of available wavelengths is limited.

The choice of monochromator crystal is very important in strain measurement. As a rule of thumb, one requires the lattice spacing of the monochromator to be close to that of the material of which the sample under measurement is made. This is the "focusing" condition for the best resolution in powder diffraction, described in Section 3.4.1. This means that the take-off angle from the monochromator should be approximately equal to the sample scattering angle, $\phi^M = \phi^s$. However, a larger ϕ^M may optimize strain measurements at $\phi^s = 90°$, as discussed in Section 3.4.1. Often a monochromator crystal is cut in a manner that allows a range of useful diffracting planes, and therefore wavelengths, to be selected. This is accomplished by rotation about an axis of a chosen lattice direction (the zone axis $[uvw]$) — usually vertical — allowing access in the horizontal plane to all the reflections hkl whose lattice planes contain this axis. As mentioned in Section 2.6.1, single crystals are rarely perfect. Instead, they tend to be broken up into little blocks, each having slightly different orientations; the spread of these orientations is called *mosaic spread*. Besides considering the range of lattice spacings available, there are at least four important characteristics to consider when selecting the most appropriate material for a monochromator:

- Easy-to-grow large crystals
- High scattering length density
- Low absorption and incoherent scattering
- Uniform mosaic angular spread that matches the collimation angles.

Some examples of commonly used monochromator crystal planes are given in the Table 3.2, and their selection to provide a wavelength appropriate for a measurement is discussed in Section 4.2. One material that satisfies most of the above criteria is pyrolytic graphite (PG), which has an unusual layer structure in which uncorrelated two-dimensional arrays of carbon atoms are stacked regularly along the c-axis. This gives rise only to sharp 00l reflections, where l is even, and powder-like "rings" of reciprocal lattice points in the planes perpendicular to the c-axis. The mosaic can be accurately controlled during growth, and monochromators with areas up to 50×75 mm^2, and 1 to 2 mm thick, are grown commercially. The high scattering length and low absorption make PG an ideal monochromator for relatively long wavelengths.

Arrays of carefully aligned crystals may be used to extend the dimensions of a monochromator. For example PG crystals are often used in such arrays, and if the relative angles of each component crystal can be adjusted, often under computer control, "focusing" monochromators can be constructed. As this increases the divergence of the incident beam on the sample, a compromise between an increase in intensity and a decrease in resolution must be made. By curving such an array about a horizontal axis to act like a concave mirror, focusing in the vertical direction can be attained, and as the vertical resolution can often be relaxed, such a monochromatic beam focused onto the gauge volume can increase the intensity several fold. Focusing in the horizontal plane is also used in specialized instruments, and examples are given in Section 3.7.2. Such a focused array is shown in Figure 3.11 [22].

The mosaic angular spread of the monochromator crystal together with the angular spread of the incident beam determines the range of wavelength selected. Unlike pyrolytic graphite, many crystals grow with a small mosaic spread. Germanium, for example, which has a range of lattice spacings very suitable for producing useful wavelengths, is normally grown with a mosaic spread of only a few minutes. This gives excellent wavelength definition but very little intensity, as only a very narrow region of the incident flux is diffracted into useful beam. A great deal of effort has therefore been made to uniformly increase the angular mosaic spread of monochromator crystals by controlled cooling during crystallization, alloying, and treatment of pressure and temperature. Until recently, germanium crystals with appropriate mosaic spread could only be produced by applying pressure at high temperature, with varying results that were only rarely successful although Chalk River Laboratories have had considerable success. However, methods

TABLE 3.2

Lattice Spacings for Selected Typical Monochromator Planes and Resulting Wavelengths at $2\theta^M = \phi^M = 90°$[a]

Monochromator Plane (hkl)	d_{hkl} Spacing (Å)	Wavelength λ (Å) at $\phi^M = 90°$	ϕ^s Fe bcc (110) $a_0 = 2.877$ Å $d_{110} = 2.0343$ Å	ϕ^s Fe bcc (112) $a_0 = 2.877$ Å $d_{112} = 1.1745$ Å	ϕ^s Fe fcc (111) $a_0 = 3.592$ Å $d_{111} = 2.0738$ Å	ϕ^s Fe fcc (113) $a_0 = 3.592$ Å $d_{113} = 1.0830$ Å
PG 002	3.3528	4.7416	—	—	—	—
PG 004	1.6764	2.3708	71.28	—	69.72	—
Germanium 111[b]	3.2663	4.6192	—	—	—	—
Germanium 113[b]	1.7058	2.4124	72.73	—	71.13	—
Germanium 115[b]	1.0888	1.5398	44.48	81.92	43.58	90.62
Silicon 111[b]	3.1356	4.4344	—	—	—	—
Silicon 113[b]	1.6375	2.3158	69.39	160.71	67.883	—
Silicon 115[b]	1.0452	1.4781	42.61	77.99	41.76	86.06
Copper 002	1.8074	2.5560	77.84	—	76.09	—
Copper 220	1.2780	1.8074	52.75	100.60	51.67	113.11
Beryllium 0002	1.7916	2.5337	77.03	—	75.31	—
Beryllium 11-20	1.1428	1.6162	46.81	86.94	45.87	96.51
Aluminum 002	2.0248	2.8635	89.46	—	87.32	—
Aluminum 111	2.3380	3.3064	108.71	—	105.72	—

[a] The Bragg scattering angles ϕ^s at these wavelengths for two planes of ferritic and austenitic steel are also given.
[b] No second-order ($\lambda/2$) contamination.

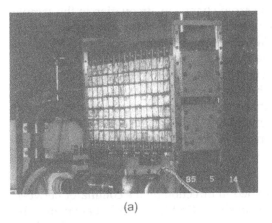
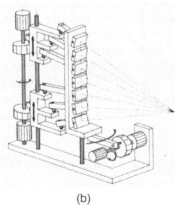

(a)   (b)

FIGURE 3.11
(a) An example of an array, typically ~140 × 300 mm², of 7 × 15 of 2 cm² monochromator crystals of pyrolytic graphite. In this case, both the vertical and horizontal curvatures are motor controlled. (b) The mechanism for one of 15 columns of crystals is shown on the left [22].

of increasing the mosaic spread of germanium crystals have been developed at Brookhaven National Laboratory [23,24] and reproduced with success elsewhere [25]. Thin slices of perfect crystals of area ~10 × 70 mm and a few hundred microns thick are cut in the required reflection plane, 511 or 311, from a perfect crystal. Each slice is heated to ~870°C, close to the melting temperature of 938°C, and subjected to plastic bending in two opposite senses with flattening after each bend before cooling. This plastic bending treatment increases the mosaic spread to ~10' to 15'. Around five of these slices are then stacked on top of each other and bound together by tin foil to form a small monochromator with the desired mosaic spread. Each such multiple wafer is aligned and then stuck onto formers to stack them in arrays. The curvature of these arrays about horizontal or vertical axes can be computer controlled, thus forming an extended monochromator that can provide a focused beam. Such arrays are quite expensive to produce, but a factor of 3 or 4 in intensity is obtained at a price much cheaper than that of increasing the flux of a reactor by the same factor!

In the orientation required to give a neutron wavelength of λ from the *hkl* lattice planes, the monochromator crystal may also diffract neutrons of wavelength $\lambda/2$, $\lambda/3$, and so on, if there exist lattice planes (2*h*, 2*k*, 2*l*) and (3*h*, 3*k*, 3*l*) and so on in the structure. These higher orders will have different intensities from that of wavelength λ depending on their position in the flux distribution in the beam. Although of major concern in powder diffraction measurements concerning structure determination, which demands that these higher orders are reduced to a minimum using filters or crystals of a particular structure, they are usually of no consequence to macrostrain measurement, and indeed may add to the useful intensity.

In choosing the wavelength for a set of measurements, it is important to check that it does not lie close to an absorption Bragg edge for a lattice plane in the sample. This is because the rapid change in attenuation in the sample near the edge can give anomalous effective shifts in the Bragg peak diffraction angle used to measure the lattice strain at different depths. As a general rule, the wavelength must not lie within 2% of such a wavelength [26]. This effect is discussed further in Section 4.2.2.

One further consideration if the beam is shared with another instrument nearer the source is to ensure that the monochromator on that instrument does not vary in a manner that might affect the wavelength composition of the beam on the instrument being used for strain measurement.

3.2.1.4   Collimation

The monochromated beam is defined in direction by the collimator between the monochromator and sample position (M-S). The collimation may be made by an arrangement of closely spaced parallel sheets of absorbing material, termed a "soller" collimator. Early collimators were made from many cadmium sheets or gadolinium coated steel sheets, and separated by a distance d_c and of length l_c, to give a collimation angle of $\alpha = d_c/l_c$. The transmission of such collimators, particularly for small or "tight" collimation angles, can be low on account of the thickness of the sheets relative to the apertures between them. Thin stretched Mylar films coated with gadolinium oxide are therefore now used extensively. Coated very thin steel sheets are now also being used. An alternative possibility is to collimate by using two apertures of greater width d_c placed a larger distance l_c apart. Usually horizontal collimation is achieved by soller collimators, whereas vertical collimation occurs via apertures or indeed a fixed dimension such as the monochromator and/or sample height. The angle of the scattered beam, ϕ^s, is defined in a similar manner by a collimator between the sample and detector (S-D) if a single detector is used. A typical collimation arrangement is shown in Figure 3.8. Each soller collimator must have some angular adjustment about a vertical axis so that it may be accurately aligned along the direction between axes of rotation of the monochromator and sample table, and of the sample table axis and detector. A convenient way of mounting them, along with other beam components such as apertures, is by clamping them to an accurately machined "optical bench."

In conjunction with position-sensitive detectors (PSDs), discussed later in Section 3.3, a single aperture slit may be used, although it can give rise to severe problems in defining the gauge volumes precisely as discussed in Section 3.5.2. It is usually advantageous to use "radial," sometimes referred to as "focusing," collimators in which the vertical absorbing thin sheets are angled so that they converge on a point at, or close to, the center of the gauge volume [27,28] (see Figure 3.18). These may be oscillated over a small range of angle ϕ^s with time so that the PSD is uniformly illuminated. Typical instrument setups are shown in Figure 3.12. Radial collimators must be used if PSDs are used in conjunction with a time-pulsed polychromatic beam source [29]. Such collimators can also be used between monochromator and sample when using a monochromator focusing in the horizontal plane.

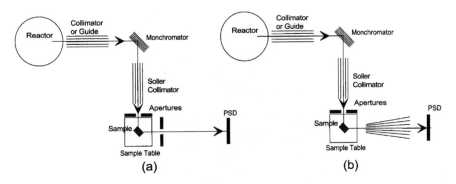

FIGURE 3.12
Schematic showing the use of a position-sensitive detector with a standard two-axis instrument for strain measurement, with (a) single slit, and (b) radial collimator.

3.2.1.5 Apertures and Beam Area Definition

The beam may be defined in area by carefully machined apertures in masks made of neutron-absorbing material. These are usually placed in mounts on optical benches, in the incident beam after the soller collimator before the sample, and in the diffracted beam after the sample. These apertures define the gauge volume as discussed later in the section. The traditional aperture design is an aperture cut in a standard sized cadmium plate that slots accurately into a holder, or four separate thick cadmium plates that are moveable in pairs in slots using screw threads. However, while these are useful for cutting background and are often used close to the monochromator, strain measurement requires, in addition, finer adjustment and accurate reproducibility of a sharp-edged aperture to define the beam size. For very fine apertures sharp edges are required and cleaved BN crystals have been used. It is usually advantageous to place these apertures reproducibly as close to the sample as possible, allowing for the necessary movement of the sample. If the adjustments are manual, a vernier slide on the holder is used to enable the apertures to be positioned along the beam directions, and the aperture can be positioned transversely and vertically by use of lockable calibrated screw threads. In this way, the beam can be adjusted to pass centrally over the reference point, as discussed in the next Section 3.2.1.6 [30]. Ideally, all adjustments of aperture size and position should be computer controlled for repeatability and to allow automated acquisition schemes.

For measurement of strain very close to a vertical surface of a sample, the aperture must be narrow in width and placed very close to the sample in order to minimize penumbra effects. As the angle of approach is often ~45°, a pointed design in which the absorbers tend to a narrow opening of horizontal width as small as 0.25 mm may be necessary. Clearly, the vertical edges of the sample and apertures must be very accurately aligned. Some typical aperture systems are shown in Figure 3.13.

(a)

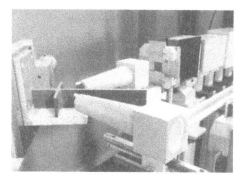

(b)

(c)

FIGURE 3.13

Some typical slit systems used to reduce background and to define the instrument gauge volume. (a) A motorized adjustable aperture for course beam definition. (Grenoble Modular Instruments). (b) An early fine slit system, illustrating the use of an optical bench for alignment of instrument components (Risø). (c) An adjustable fine slit system enabling close approach to the sample (Studsvik).

3.2.1.6 Shielding

There is rarely a measurement where good shielding to reduce background counts to an absolute minimum is not beneficial, and it is often essential. Indeed, it is usually the background level that determines the maximum depth to which one can probe as discussed in Section 4.4.1. Shielding is best accomplished as near to the neutron source as possible, so most monochromator housings have heavy and often complex shielding enclosing them, particularly if close to the reactor. Materials are usually borated concrete or borated polymer resin with lead shot, so that fast neutrons are first slowed by moderation by the hydrogen and then absorbed by the boron, and the γ-rays are absorbed by the lead. It is essential that all the moderated fast neutrons are absorbed; otherwise, the moderating shielding can actually increase the background in a thermal neutron detector. The presence of a detected neutron background originating from fast neutrons in the incident beam on a sample can be tested by placing a cadmium sheet in the incident beam. As cadmium is transparent to fast neutrons, but absorbs thermal neutrons, any background still present must originate from fast neutrons.

Shielding along the beam path in the instrument is usually by borated polymer resin, either built into the form of the components such as aperture holders, flexible sheets, or solid blocks. Borated material without hydrogenous binders is available in the form of sintered BC castings (called "crispy"), and should be used in regions where any slight amount of scattering from hydrogen would be harmful to background. Cadmium is also used extensively for apertures and some shielding, however, the absorption of neutrons by cadmium gives rise to γ-rays, and cadmium should never be placed directly in the primary beam from the neutron source.

The use of guide tubes to locate the instrument in a remote guide hall reduces the source of background considerably, but one should not be led into a false sense of immunity. Some background can arise from that part of the beam not transmitted by the guide. Air scattering from the beam incident on the sample can be appreciable and needs to be shielded against by use of borated resin tunnels or sheets. Nearby instruments, and even nearby samples if radioactive, can also give rise to background and stray beams. A simple test with their beam shut off will reveal whether a problem exists. The detector is a very important component requiring careful shielding, and its mounting and surrounding shielding material should be carefully designed. As discussed in Section 3.3, the detector electronics must be set to detect neutrons only, and to discriminate against electronic noise and γ-rays.

In general, every set of measurements should be preceded by a few checks to ensure that the background is minimized. The beam may be blocked off at different points along its path in order to see how the background level is made up. At all instruments, care must be taken to ensure that the unwanted beam leaving the monochromator is absorbed as close as possible to the monochromator, and the only beam on the sample is that defining the desired gauge volume. It is therefore important to check that when this beam on the sample is blocked, for example, by a piece of cadmium over the incident beam aperture, no counts above background are recorded in the detector. Another useful check is to ensure that with the beam open into the gauge volume when no sample is present, only the expected background level is measured. Both these checks may require measuring counts while rotating the detector through the sample diffraction peak angle ϕ.

3.2.2   Time-Pulsed Beam Instrument for Strain Measurement

Time-pulsed beams use the neutron's velocity, which is measured by timing its passage over a known distance, to determine its energy and wavelength. The beams are usually polychromatic in wavelength — often termed "white" — for diffraction experiments (Figure 3.6), and may originate from pulsing a continuous flux reactor beam or from a pulsed neutron source. Each pulse therefore contains a wide range of wavelengths. A detector placed to collect neutrons diffracted from a sample at a fixed angle to the incident pulsed beam will build up, over many pulses, a complete diffraction pattern of Bragg peak intensity as a function of time of arrival, or wavelength, as described in more detail in the following sections. In general, a single Bragg peak can be measured faster using a constant flux monochromatic beam from a reactor, but if more than one peak is required pulsed beam instruments outperform constant flux instruments.

3.2.2.1   Time-Pulsed Neutron Beams on Reactors

An alternative to the continuous flux crystal-monochromator instrument, used on most reactor facilities, is to vary the intensity of the polychromatic neutron beam from the reactor with time. This is accomplished by using a rotating "chopper" velocity selector, which absorbs the neutrons except for a short duration when its window is open. Such a velocity selector may take the form of a thin disk, made of absorbing material, which rotates about a horizontal axis parallel to the neutron beams such that the beam passes through slots in its outer periphery. Alternatively, it may be a thick cylinder with helical slots cut in its outer surface. A different configuration is to use a thicker disk rotating about a vertical axis through which is cut a curved slot to allow neutron pulse through once each cycle, called a "Fermi" chopper. Depending on the nature of the diffraction instrument, one or two choppers, separated by a short flight path, are usually used. Two phased choppers can be used to monochromate the neutrons with good wavelength resolution, the rotating discs acting like phased traffic lights so that neutrons of only a desired velocity are transmitted. Alternatively, the slots in a single disk may be used to vary the intensity profile with time, as discussed further in Section 3.7.3. Some typical chopper devices are shown in Figure 3.14a, b, and c.

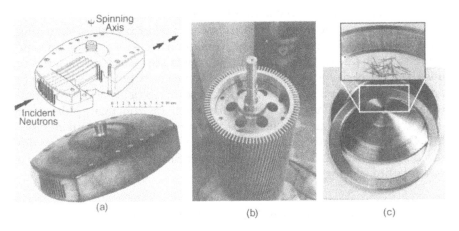

FIGURE 3.14
Some types of neutron chopper design: (a) Fermi chopper [Copyright UKAEA, 31], (b) helical velocity selector [32], and (c) Fourier disk chopper [33].

In the following discussion, we take the case where a single chopper is used to provide a pulsed "polychromatic" beam of a wide range of wavelengths. As discussed in Section 2.3.2, the wavelength of the neutron corresponding to its flight time t over a distance L is

$$\lambda = h\, t/(m_n\, L) \tag{3.9}$$

As thermal neutrons have velocities of the order of hundreds of meters per second (Table 1.1) their flight time can be readily measured, however, neutron detectors with as near instantaneous response and as little dead time as possible are required, and usually scintillator counters are used. Typical precision of measurement of peak positions in time-of-flight (tof) diffraction is 5 μs in 15,000 μs.

Neutron diffractometers that use time-pulsed neutrons are called tof instruments, and by using sufficiently long flight paths, L, in neutron guides, very high resolution can be attained through the accurate definition of wavelength. The flight path $L = L_i + L_f$, where L_i is the length of the incident path from chopper, or moderator in a pulsed source, to sample, and L_f the path from sample to detector.

3.2.2.2   Time-Pulsed Neutron Beams from Pulsed Sources

Although a time-pulsed beam can be engineered on reactors, such a polychromatic beam is provided directly by pulsed neutron sources. The brightest sources of this type, the spallation neutron source was described in Section 3.1.2. The very high intensity and high time resolution available from such sources can enable transient phenomena, such as stress relaxation processes, to be observed.

The diffraction pattern from a sample obtained using a time-pulsed poly-chromatic beam is recorded as the diffracted intensity at fixed scattering angle, the so-called "fixed geometry" configuration, as a function of time after a start signal. The detected neutron counts are recorded in a series of time channels, each corresponding to a certain velocity or wavelength. Only rarely does a single pulse provide sufficient scattered neutrons for quantitative analysis, instead the integrated profile is built up from many pulses over a period of time. A single detector at a scattering angle ϕ^s, will record intensity ($I(t)$) versus time, corresponding to intensity versus wavelength giving diffraction peaks hkl in $I(t)$ at time t_{hkl} whenever Bragg's Law is satisfied by a certain lattice spacing:

$$d_{hkl} = \lambda_{hkl}/[2 \sin\theta^B_{hkl}] = h.t_{hkl}/[2\, m_n\, L\, \sin\theta^B_{hkl}]. \tag{3.10}$$

Here L is the total flight path distance over which the pulse is timed, and t_{hkl} is the flight time of the hkl peak center. The start of timing is usually triggered by the source beam pulse on a pulsed source, or a pulse recorded in a low-efficiency monitor counter placed near the chopper in a reactor instrument. Typical diffraction patterns are shown in Figures 3.15a and b. As the diffraction pattern is collected at just one angular setting and is built up over a period of time, the counting period can be adjusted to give the desired measurement accuracy. The more intense the original polychromatic beam of neutrons, the more quickly the pattern is obtained. For conventional powder diffraction, the detection rate can be accelerated by using a large number of detectors spread over many angles. Usually these are placed around the back reflection angles, of $\phi^s \sim 180°$, since these give the greatest resolution in the measured d_{hkl}. The resolution of tof instruments is discussed in Section 3.4.2. At the fixed scattering angle ϕ^s, the strain is given by

$$\varepsilon^{hkl} = \left(d_{hkl} - d^0_{hkl}\right)/d^0_{hkl} = \left(t_{hkl} - t^0_{hkl}\right)/t^0_{hkl} \tag{3.11}$$

where t_{hkl} and t^0_{hkl} are the measured tof for the hkl peak of the sample and stress-free reference, respectively. Note that

$$\Delta d/d = \Delta\lambda/\lambda = \Delta t/t = v.\Delta t/L = h\, \Delta t/(L.m_n.\lambda) \tag{3.12}$$

In order to obtain good resolution, a short time pulse, a long flight path L, and large scattering angle ϕ^s are necessary. The flight path, L_i, from the source to the instrument is usually along a curved neutron guide to remove fast γ-rays. Figure 3.16 shows schematic layouts of some typical time-of flight instruments.

The use of many detectors poses a problem of collating the data obtained from each detector, or detector element, because due to their specific locations having different scattering angles and path lengths, a different pattern

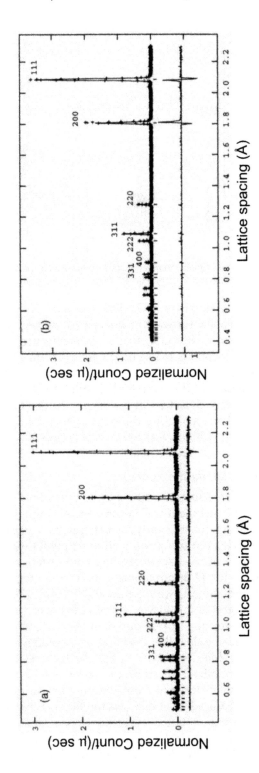

FIGURE 3.15

A typical time-of-flight diffraction pattern from stainless steel under uniaxial loading of (a) 5 MPa tensile stress, and (b) 340 MPa tensile stress. In each case, an average lattice parameter is fitted using the Rietveld technique. The fitted background has been subtracted, and the upper curves are the experimental (crosses) and fitted (line) profiles. Underneath each is the difference curve on the same scale. It is seen from (b) that the fitted curve using an average lattice parameter does not account for the elastic anisotropy (see Section 5.5) of the 111 and 200 peaks, which show a variance in the opposite sense as expected [41].

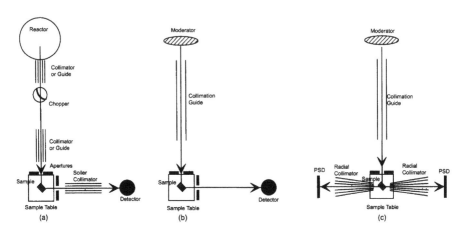

FIGURE 3.16
Schematic layout of three time-of flight instruments: (a) on a reactor using a single detector; (b) on a pulsed source with single detector; and (c) with PSD with focusing radial collimators.

of intensity versus time is obtained for each. Two methods of data combination are used. The first is to use software to convert the pattern from each detector to the required form of binning, that is, intensity versus λ or d, before adding together. The second is to use hardware to arrange the spatial layout of the detectors to give "time-focusing," described further in Section 3.4.2. In practice, a combination of these methods is often used. When combining information from many detectors in a detector bank, it should be remembered that each will have a slightly different Q vector, and thus the strain measurement direction will be averaged over a region of solid angle.

3.2.2.3   Instruments on a Spallation Source

As the most intense neutron sources in the future will likely be of the spallation type, this type of instrument will become increasingly important. Similar to steady-state instruments, the early measurements of elastic lattice strain in a sample were carried out using conventional powder diffractometers that often had detectors in the back-scattering configuration ϕ^s ~180°. These gave a scattering vector Q approximately along the incident beam direction. Furthermore, the sample was contained within an evacuated chamber suspended from, and accessed via, a manhole cover from above. This has been described as working upside down, in the dark, in a vacuum, down a manhole! It soon became clear that the ideal arrangement for making engineering measurements required more space, better access, and an arrangement of detectors optimized for strain measurement. Today's dedicated instruments are radically different from the powder diffractometers from which they evolved. ENGIN [34] at ISIS was the first (~1994) of several dedicated pulsed source instruments that have now been constructed, the most recent of which include ENGIN-X at ISIS [35–37] and SMARTS at LANSCE [38], with VULCAN at the design stage at the SNS. The detectors are usually clustered in banks set at the most favorable angles, usually ϕ^s ~90°, for the required strain directions. The use of many detectors for strain measurement is limited by the acceptable range of Q, and hence strain, direction relative to the sample, and by the definition of the gauge volume. This has led to use of radial collimators, rather than apertures, in the scattered beam to define the gauge volume [29]. Detection of scattering both sides of the incident beam is possible as there is no resolution focusing effect to consider, in contrast to instruments using a monochromator crystal (see Section 3.4.1). As a consequence, simultaneous measurement of strain with Q in two directions in the sample at 90° to each other can be made. For large samples one detector bank can be removed. Such an arrangement, as adapted on the ENGIN-X instrument at ISIS, is shown in Figure 3.17a and b.

The ENGIN-X beam line is on a 50m-curved supermirror guide. This optimum length of guide was determined using the Figure of Merit concept discussed in Section 4.4.2, maximizing strain resolution through a combination of high intensity and good time (wavelength) resolution [39,40]. With only thermal neutrons reaching the sample, the shielding requirements in the instrument blockhouse are closer to those found on a reactor guide hall instrument than typical pulsed source instruments. Large arrays of scintillator detectors are centered on $\phi^s = \pm 90°$. An array of interchangeable radial collimators can be easily and accurately swung into place in front of the detectors, which provides gauge dimensions of 0.5 to 5 mm. Moveable sintered boron carbide shielding is placed between the collimators and the detectors to maintain a low background. The positioning table is capable of moving 500 kg to within 10 microns over 100 mm, with translation ranges of ±250 mm in X and Y, 600 mm in Z and 370° rotation. The incident beam is defined in area using fully motorized boron carbide slits, which can define the beam from 0.3 to 20 mm vertical and 0.3 to 10 mm horizontal. A remotely steerable web camera allows monitoring of experimental progress.

In principle, strains in three perpendicular directions can be measured simultaneously at one sample setting if three banks of counters are used. This can be seen by considering a cube in which the incident beam is in the [111] direction and the banks of detectors are positioned at angles of ~109.5° to this incident beam, corresponding to $\phi^s = 70.5°$, in each of the <110> planes containing this [111]. In this arrangement the strain is measured in each of the three <001> directions. However, shape considerations and path lengths would make it unlikely that many suitable engineering component samples would be found to benefit from measurement using such an arrangement.

The profile of intensity versus time for a Bragg peak on a spallation source instrument is more complex than the Gaussian function usually adequate to describe the peak profile on a continuous beam instrument [2,11]. This is because the moderating process introduces a time broadening, which is in addition to the broadening introduced by the instrument components (discussed in Section 4.3.1). Any additional intrinsic broadening such as that from microstrain is usually represented by a Gaussian function convolved with this profile. If only one peak *hkl* is of concern, the strain measurement

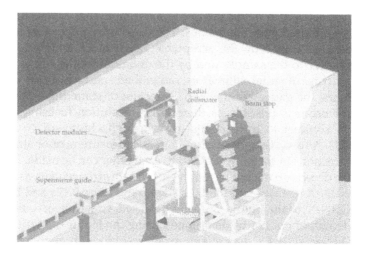

(a)

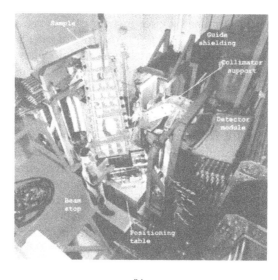

(b)

FIGURE 3.17
(a) A CAD image of the ENGIN-X instrument at ISIS, showing the two detector banks at 90° scattering angle, the radial collimators and the x-y-z-Ω sample positioner. (b) Photograph of the ENGIN-X instrument at ISIS showing a large aircraft component sample undergoing strain measurement. (Courtesy of M.R. Daymond.)

is obtained from Δt as in Equation (3.11), and this difference can be determined more accurately than the absolute time or d-spacing by fitting the same profile function to the both sample and reference peak. The error in the latter only affects the overall error in strain.

The detected counts are normalized to the intensity of the incident spectrum of neutrons, and to account for detector efficiency and put on an
absolute scale if required, by comparing the measured intensity to that from
an incoherent scatterer, such as vanadium, which has a known cross-section
of $\sigma^{incoh}/4\pi = 0.403$ barns/steradian atom. Subtraction of containment cans
and background must be made for both the sample and the vanadium placed
at the same sample position. Likewise, by measurements of the complete
diffraction pattern from a standard powder sample such as Si or CeO at
exactly the same position, the true angles of the many counters on tof instruments and the true time of origin of the neutron pulse may be calibrated.

Although the whole diffraction pattern of many peaks, and therefore many
elastic lattice plane strains, can be measured at one setting of the sample,
time considerations often mean that the desired accuracy can only be
attained for the most intense of these peaks. The asymmetric nature of the
peak profiles means that longer counting times are required than for symmetric peaks. Nevertheless, the fact that many peaks are collected, gives a
decided advantage in this respect over a steady-state instrument, since as
discussed in Chapter 5, the response to stress of different lattice planes can
give important information on texture, elastic anisotropy and plasticity. An
alternative is to fit an average lattice spacing to all the measured diffraction
peaks, and this has been found to give a value close to the macroscopic strain
in both elastic and plastic regimes, and hence related by the macroscopic
bulk elastic constants to the stress in the sample component [41]. Such
analysis is accomplished using the Rietveld [42] technique in which the
theoretical intensity profile of the whole diffraction pattern $I(t)$ versus t —
including peak profiles, unit cell parameters, structure and multiplicity factors, and absorption and background — is fitted to the observed data. A
texture factor is also included in the theory to allow for preferred orientation
of the grains in the sample. The Rietveld method is used extensively in
structure determination from powder diffraction patterns. The mean lattice
parameters are obtained from such fitting, shown in Figure 3.15, and can be
compared with those from a "stress-free" sample to enable strains to be
deduced. Further details on the Rietveld analysis technique are given in
Section 4.5.

An alternative fitting procedure is not to constrain the calculated intensities
of the peaks to the theoretical functions but to allow each to be fitted independently, with only the positions in time (wavelength) constrained to give
the unit cell parameters required for strain measurement. This procedure
was first used by Pawley, and is known as a "Pawley–Rietveld" analysis; it
is very useful for strain measurement when the intensity of the peaks is not
of prime importance [43].

3.2.2.4 Collimation

To utilize fully the time-pulsed polychromatic neutron beam on a spallation
source as many detectors as can give useful information are used. Since a

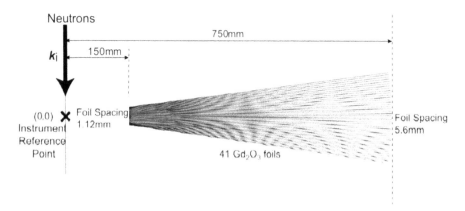

FIGURE 3.18
Plan view diagram of the first radial collimator built for the ENGIN instrument at ISIS based on Withers et al. [29].

sample placed in a polychromatic pulsed beam will diffract at all angles as a function of time, the scattering angles (and the gauge volume) are defined using a radial soller when using arrays of detectors [29]. A typical "focusing" or "radial" collimator is shown in Figure 3.18. The vanes of GD- or Gd_2O_3-coated stretched Mylar or shim steel are accurately set at angles to converge on the instrument reference point. The only drawback is that a different radial collimator is required for each size of gauge volume. In practice the radial collimators can be slightly defocused so as not to converge on the reference point to increase the gauge volume size.

3.3   Neutron Detectors

Neutrons cannot be detected directly and must be detected by their nuclear interaction with an absorbing nucleus to produce charged particles with energies in the MeV range. These can then be detected by gas counters acting in the proportional mode, or by production of a photon detected by a scintillation detector. The principal reactions used follow:

$$n + {}^{10}B \rightarrow \alpha + {}^{7}Li + \gamma + 2.3 \text{ MeV} \qquad \text{I}$$

$$n + {}^{3}He \rightarrow p + {}^{3}H + + 0.77 \text{ MeV} \qquad \text{II}$$

$$n + {}^{6}Li \rightarrow \alpha + {}^{3}H + 4.79 \text{ MeV} \qquad \text{III}$$

Reactions I and II are used with ${}^{10}B$ in BF_3 gas detectors or ${}^{3}He$ gas detectors [3,15]. In these detectors, the anode is a wire held at high voltage relative to the surrounding sheet cathode in cylindrical or rectangular shape. Although BF_3 has the advantage of relative insensitivity to γ-rays, owing to its toxicity it is now almost fully superseded by ^{3}He, which is somewhat more γ-ray sensitive. The γ-rays may be isolated by use of a voltage discriminator since they give smaller voltage pulses than neutrons. The detector must absorb most of the neutrons in order to be efficient, and therefore must contain gas at high pressure and have dimensions of a few centimeters along the neutron path. Early detectors had ceramic end windows and were used in an axial configuration surrounded by shielding material of borated paraffin and steel. However, the more that the diffracted beam cone is intercepted by the detector, the higher the measured intensity; thus, it is common now to use detectors containing the higher absorbing ^{3}He at pressure of ~8 bar with their axis vertical. In the simplest diffractometer design, the neutrons are detected by a single detector tube (as shown in Figure 3.8) placed after the collimator in the diffracted beam.

Scintillator detectors [44] are used extensively in instruments on pulsed neutron beams, where tof techniques demand good time resolution. The principle of these detectors is the absorption of the neutron to give rise to a pulse of electromagnetic radiation in the visible or ultraviolet region. There are two types of scintillation neutron detectors commonly used. The first comprises ^{6}Li-loaded glass containing a Ce activator. The second comprises bound powders of ^{6}LiF mixed with ZnS doped with Ag. In both the energetic products of reaction III excites an atom of the surrounding material into an excited state, which decays by emission of a visible or ultra violet photon. This photon pulse is fed by fiber optics to the photocathode of a photomultiplier that converts it to a pulse of electrons and amplifies it, resulting in a voltage pulse. The speed of conversion is important to give good time resolution. The glass scintillator has better pulse pair resolution but higher γ-ray sensitivity and background than the LiF/ZnS powder scintillator. The γ-ray and neutron pulses must be separated by height and shape discrimination.

In using any type of detector, it is important to ensure that the dead time after each detected neutron is sufficiently short not to affect the accuracy of the measured count. The extreme situation is saturation, when the count rate recorded can actually reduce because of the high rate of incoming neutron counts. In such cases, usually when making initial instrument alignments with intense beams, absorbers with calibrated attenuation must be placed in the beam. Especially high count rates are encountered when making transmission measurements, and special detectors are required as discussed in Section 3.7.4 [45].

3.3.1 Position-Sensitive Detectors and Detector Arrays

In recent years, position-sensitive detectors (PSDs) have been developed to accept a wide angular range from the sample so that the angular scanning of such a detector on a monochromated beam instrument is necessary over only a small range or is not necessary at all [46,47]. The earliest PSDs took the form of a horizontal gas-filled detector with a resistive anode wire, with the charge resulting from the detection of a neutron at each end, which is measured by low-impedance charge-sensitive amplifiers. From the size of the voltage pulses at each end, the position along the detector at which the neutron was detected can be deduced — the so-called charge division technique. A different technique is to use the rate at which charge is accumulated at each end, which is measured by timing circuits incorporating delay line techniques.

Stability is the key problem with such detectors, but with modern electronics this problem is now largely overcome, and indeed two-dimensional arrays of resistive wires forming anodes and cathodes are used on most small-angle neutron scattering (SANS) instruments. The spatial resolution can be better than the spacing of the wires. Two-dimensional PSDs can be used for diffraction instruments by summing the neutron count over the vertical cells. Ideally, on a monochromated beam the horizontal angular width of the cone of scattering is intercepted by the PSD with an angular aperture in ϕ^s large enough to cover it and some adjacent background. This has the obvious advantage of speed of measurement. However, the wider aperture can lead to a higher background count.

An alternative form of PSD is to use an array of small-diameter vertically oriented gas counters, or an equivalent array of vertical anode wires in a single encasing cathode housing — the so-called "banana" configuration [46]. Arrays of horizontal linear detectors are used at the Missouri University Research Reactor [48]. Most recently microstrip gas detectors have been developed using photolithographic techniques, in which thin strips of electrodes are deposited on glass or polymer bases. Originally developed for x-ray work, these detectors are being increasingly used for neutron diffraction [46,49]. Figure 3.19 illustrates some typical detectors.

PSDs can be used on time-pulsed polychromatic beam instruments using an array of scintillator counters, and these are generally favored for strain measurement. The ^{6}LiF/ZnS scintillator elements of the PSDs on ENGIN-X at ISIS are oriented like venetian blind shutters. Each of the two detector banks is made up of five modules of 240 scintillator elements. Each scintillator element is 3 mm wide, and approximately 150 mm high [36]. The efficiency of such detectors can be as high as 60% for 1 Å neutrons. The electromagnetic radiation pulses are viewed by bundles of optic fibers, connected to 32 photomultiplier tubes per module, and coded in such a way that the signal from any pair of tubes uniquely identifies the activated detector element. The total detector coverage is roughly ±15° horizontally and ±20° vertically around $\phi^s = \pm 90°$. Software corrects for the variation of the instrumental contribution to the tof, $L\sin\theta^s$, to each cell in the detector. These values of $L\sin\theta^s$ are determined empirically by diffraction from a standard sample, usually Si, Al_2O_3, or CeO.

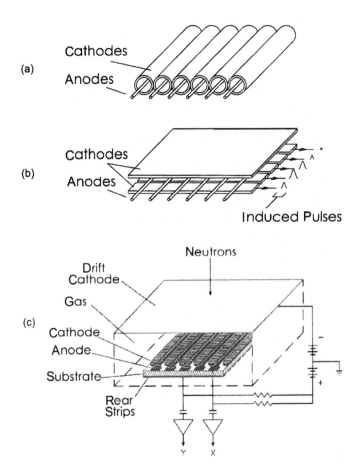

FIGURE 3.19
Schematic drawings of selected neutron gas detector systems: (a) linear PSD consisting of an array of individual detectors; (b) two-dimensional PSD consisting of a number of anode wires and a segmented second cathode; and (c) a microstrip detector. Charges on the cathode or anode strips encode the X spatial coordinate. Charges induced on the rear strips, orthogonal to the microstrip anodes and cathodes, allow the Y spatial coordinate to be encoded [46].

3.4 The Instrumental Resolution

The uncertainty in the determination of the angle of the Bragg reflection, $\phi^s = 2\theta$, or the uncertainty in timing of a detected neutron, determines the error in the measured strain. This needs to be determined to an accuracy of $\Delta d/d \sim 10^{-4}$, or 100 $\mu\varepsilon$, in order to give an accuracy of $\sim$20 MPa in stress in steel, for example. The contributions to the error from fitting a function to the peak shape and background are discussed in Chapter 4. One of the most important factors is the inherent instrument contribution to the angular, or time, width of the Bragg peak. This will be briefly discussed in this section as it gives rise to a number of important points to be taken into consideration when designing instrumentation and carrying out measurements.

Measurement of Type I macrostrain, defined in Section 1.4.1, is concerned with accurate measurement of the shift, or difference in angle or time of the Bragg peak center from that from the reference sample, rather than the absolute peak center. Good instrument resolution is important in the region of scattering angle ϕ^s used, because the uncertainty of peak position determination is proportional to the full width at half maximum (FWHM) of the peak, but improves only as the square root of the integrated intensity (Section 4.4). It is also particularly important in order to isolate the reflection being measured from others close by in angle, or time, and it may be vital if there are several closely related structures in which a material may exist, for example the different polytypes of SiC in metal matrix composites. Depending on the type of sample-induced background, good resolution can increase the "signal height to background" ratio of a peak. The measurement of Type II or III microstrain through the measurement of Bragg peak widths requires excellent instrument resolution, so that the inherent sample broadening can be more easily isolated.

The aim of an instrument designed for strain measurement is to maximize the useful intensity while giving the required resolution. This will involve relaxing resolution elements where they are not critical, such as in the vertical, out-of-scattering plane direction. Only the resolution in the horizontal scattering plane will be considered here.

3.4.1　Continuous Beam Instrument

A simple approach that indicates the main factors governing resolution is given by differentiating Bragg's Law equation for the sample (Equations (1.1) and (2.15)), and dividing by λ:

$$\Delta\lambda/\lambda = \Delta d/d + \cot\theta^s\, \Delta\theta^s,\ \text{or}$$

$$\Delta d/d = \Delta\lambda/\lambda - \cot\theta^s\, \Delta\theta^s = \cot\theta^M\, \Delta\theta^M - \cot\theta^s\, \Delta\theta^s \qquad (3.13)$$

Thus, the variation of d depends on the angular spread of the beams from the monochromator and entering the detector. For a random variation of each term, they may be combined by convolution, in which case the widths add as the sum of squares:

$$(\Delta d/d)^2 = (\Delta\lambda/\lambda)^2 + (\cot\theta^s\ \Delta\theta)^2 = (\cot\theta^M\ \Delta\theta^M)^2 + (\cot\theta^s\ \Delta\theta)^2 \qquad (3.14)$$

This simple equation shows that the resolution improves as θ^s gets larger at the monochromator and at the sample, with the best resolution in a "back scattering," or a $\phi^s = 2\theta^s = 180°$ configuration. Some of the first strain measurements were made in this way [20,21].

A detailed treatment must consider the correlation between angular spread and wavelength, since the above terms are not strictly independent. The instrument resolution in the scattering plane is of major concern. There are contributions from the horizontal angular spread, FWHM, of the neutron beam onto the monochromator, α_1, between the monochromator to sample, α_2, and sample to detector, α_3, together with the mosaic spread of the mono-chromator, η_M (Figure 3.8). There are corresponding but uncorrelated angular spreads in the vertical plane, which are not very important in strain measurement, especially if $\phi^s \sim 90°$. The combination of these angular collimations gives rise to a spread in the angle and wavelength of the rays that make up the neutron beam incident on the sample. There is a spatial correlation among angle, wavelength, and position of the rays throughout the instrumental gauge volume (IGV), and, from Bragg's Law, with those rays diffracted into the detector.

There has been much discussion of the instrument peak widths in conventional powder diffraction instruments where the spatial correlation between angle and wavelength across the beams are less important, as the usually large IGV is filled by the sample. The same theoretical approach may be adapted to the case of strain measurement, but there is the important additional consideration that the IGV plays a more vital role in that it may be only partially filled, or that the strain may vary over the SGV. A partly filled instrument gauge volume can give rise to a systematic shift in peak angle and can have a critical effect on the error in strain determined, as discussed in Section 3.6.4.

For a fully filled IGV centered at the instrument reference point, the conventional treatment in real space was first provided by Caglioti [50,51]. One may consider each component in the neutron beam path as giving rise to a probability function that a neutron will be transmitted. The transmission of each component i, with collimation angle α_i, is written in the Gaussian approximation in the form

$$G(\phi') = \exp[-(\phi'/\alpha_1')^2], \qquad (3.15)$$

where $\alpha_1' = \alpha_i/[2(\ln 2)^{1/2}]$. Here ϕ' is the deviation in angle from the principal straight through ray. These functions are multiplied together and the angles related by Bragg's Law. An integration is then made over all in-pile ray wavelengths and directions. The result is a function for the FWHM of the powder diffraction peak in ϕ^s. The expression given by Caglioti [50,51] is

$$(FWHM)^2 = U\,(\tan\theta^s/\tan\theta^M)^2 + V\,(\tan\theta^s/\tan\theta^M) + W \qquad (3.16)$$

where

$$U = 4 \cdot \left(\alpha_1^2 + \alpha_2^2 + \alpha_1^2 \cdot \eta_M^2 + \alpha_2^2 \eta_M^2 \right) / \left(\alpha_1^2 + \alpha_2^2 + 4\eta_M^2 \right)$$

$$V = -4 \cdot \alpha_2^2 \cdot \left(\alpha_1^2 + 2\eta_M^2 \right) / \left(\alpha_1^2 + \alpha_2^2 + 4\eta_M^2 \right)$$

and

$$W = \left[\alpha_1^2 \cdot \alpha_2^2 + \alpha_1^2 \cdot \alpha_3^2 + \alpha_2^2 \cdot \alpha_3^2 + 4 \cdot \eta_M^2 \left(\alpha_2^2 + \alpha_3^2 \right) \right] / \left(\alpha_1^2 + \alpha_2^2 + 4\eta_M^2 \right)$$

Furthermore, the intensity of the peak is proportional to the instrument factor *L*, where

$$L = \alpha_1 \alpha_2 \alpha_3 \eta_M / \left(\alpha_1^2 + \alpha_2^2 + 4\eta_M^2 \right)^{1/2} \tag{3.17}$$

Several points emerge from the above equations.

1. The peak width is narrower for the "focused" condition where diffraction at the sample is such that the diffracted beam from the sample is in a direction close to parallel to that incident on the monochromator, so that the second term, $V(tan\theta^s/tan\theta^M)$, is then negative.

2. The best choice of collimation angles with a view to optimizing the intensity/resolution compromise has been discussed by many authors. Often the mosaic width η_M is difficult to adjust, as discussed in Section 3.2.1. The FWHM has a minimum in the region of $\theta^M = \theta^s$, the so-called "focusing" condition. The value of α_2 may often be relaxed without a great effect on the resolution. Margaça [52] and others have considered the setup giving optimum resolution and intensity for strain measurement when $2\theta^s = 90°$. For their assumptions made, it is found that on a guide with $\alpha_1 \sim 10'$, $2\theta^M$ should be $\sim 127°$, whereas on a beam tube with large α_1 it should be $\sim 150°$.

3. In the special case of $\alpha_1 = \alpha_2 = \alpha_3 = \eta_M = \alpha$, FWHM $\propto \alpha$, whereas $L \propto \alpha^3$. This shows that the intensity varies as the cube of the resolution, and illustrates the fact that there is always a compromise to be made between resolution and intensity. Each component must be "matched" to the others to give the best compromise solution. As discussed in Section 4.4.2, it has been suggested that a good "quality factor" or Figure of Merit (FoM) for a diffractometer in cases where the background is negligible compared to the height of the diffraction peak is given, for a required uncertainty in angle, by the ratio

$$FoM \propto \frac{1}{t_M} \frac{I_C}{\left(FWHM\right)^2} , \qquad (3.18)$$

where I_C is the integrated count, and t_M is the measurement time (Equation (4.17)). The FoM should be maximized and is inversely proportional to the time to make a measurement to the specified uncertainty. In the above (very) special case, FoM $\propto \sigma/t_M$.

An alternative approach to considering the resolution has been given by Cooper and Nathans [53,54], who treated the problem in reciprocal space. They define a resolution function of the instrument which gives the probability that a scattering vector $Q_0 + \Delta Q$ will be detected when the instrument is set to measure a scattering vector Q_0. Using the same Gaussian transmission functions of each component, they show that this resolution function takes the form of a three-dimensional ellipsoid, or a two-dimensional ellipse in the scattering plane, centered on the tip of Q. During a conventional powder scan, this ellipsoid tracks through the thin spherical shells of reciprocal lattice points, and the intensity at any setting is the convolution of the resolution function and the scattering cross-section for each shell. This resolution function approach is particularly useful when considering the optimization of required resolution in certain directions, and in which directions it may be relaxed to increase intensity.

3.4.2 Time-Pulsed Beam Instrument

The resolution of tof instruments is discussed in some detail by Windsor [11]. In a tof instrument, the lattice spacing d is given by Equation (3.10), and on differentiating,

$$\Delta d/d = \Delta t/t - [\Delta(L.\sin\theta^s)/(L.\sin\theta^s)] = \Delta t/t - [\Delta L/L + \cot\theta^s \Delta\theta^s] = -\Delta Q/Q \quad (3.19)$$

The resolution is given by

$$(\Delta d/d)^2 = (\Delta t/t)^2 + [\Delta(L.\sin\theta^s)/(L.\sin\theta^s)]^2$$

$$= (\Delta t/t)^2 + (\Delta L/L)^2 + (\cot\theta^s \Delta\theta^s)^2 = (\Delta Q/Q)^2 \qquad (3.20)$$

However, this is an approximation since the terms, as in the case of the steady-source instrument, are not independent. One feature of a tof instrument is to use "time focusing" in which the distance to, and angle of, each detector is arranged so that distance term cancels the angle term in the expression for $\Delta d/d$. In other words, the terms in square brackets in Equation (3.19) are zero. For all such time-focused detectors, each time channel corresponds to the same $|Q|$. However, in practice it often is easier to use software to calculate the Q corresponding to each time channel for each detector, and to bin the counts in increments of Q, to give $I(Q)$. The data can be similarly binned to give $I(d)$ or $I(\lambda)$.

In the above expression, the uncertainty in time t arises from the neutron energy moderation process and from the time channel bin width. The uncertainty in length L has contributions from the sample size and position, and from the detector thickness. The angular uncertainty arises from the angular collimation. The resolution can be increased by increasing the overall length of the instrument L. However, because the flight time increases, frame overlap can occur. This happens when the fastest neutrons of a new pulse arrive at the detector before the slowest neutrons from the previous pulse. Should this occur, it is necessary to reduce the utilized pulse frequency from 50 to 25 Hz (one in two pulses) or 16.67 Hz (one in three pulses), by the use of two "frame-definition" phased disk choppers a few meters apart early in the flight path from the moderator. On the high-resolution powder diffractometer (HRPD) at ISIS, the flight path is very long and only one pulse in five is used. The new ENGIN-X strain measurement instrument at the 50-m flight path has a measurable d spacing range of 1.1 Å at 50 Hz or 2.2 Å at 25 Hz. These can be shifted in time to cover various intervals, such as 2.3 to 1.2 Å, which correspond to (111) to (311) reflections for aluminum.

Frame overlap can be used to advantage by increasing the duty cycle of an instrument, but necessitates the correct isolation and identification of each peak. An instrument that uses frame overlap in this way is the POLDI diffractometer on the SINQ neutron source at PSI described in Section 3.7.3.

3.5   Instrument Gauge Volumes

In this section, some important concepts and procedures are discussed with regard to the positioning of the sample on the instrument, as well as the definition of that part of it that gives rise to diffracted intensity. Consequences for the measured diffraction angle are discussed in Section 3.6. These aspects of strain measurement require a full understanding of how the diffracted beam intensity arises from grains in different parts of the sample that are located in the incident beam, particularly when the sample only partly fills the instrument gauge volume. This has necessitated new applications of diffraction theory and experimental work. Such detailed considerations are not usually important in conventional powder diffraction, where the sample is usually placed symmetrically and fully bathed by the beam (e.g., Figure 2.10b). Although neutron diffraction is unique because of its penetration power for accessing information on the strain and stress state within the bulk of a crystalline sample, the technique has also been developed to include investigations in near-surface regions by using small gauge volumes. Measurement of these near-surface strains gives important information on surface stress modification processes like shot-peening, or by various kinds of coatings, made in order to enhance structural properties.

3.5.1   Instrument Reference Point

For every instrument it is necessary to define a *reference axis* and a *reference point* on this axis which is at the center of the instrumental gauge volume (IGV), and to which the position of the sample may be referred. The usual convention is to adopt as reference axis the vertical axis about which the sample table and the sample to detector arm rotates. The reference point on this axis is the center of the incident beam in the vertical direction. The gauge volume is then arranged to have this reference point as its center by carefully positioning the apertures in the horizontal and vertical planes. There are a number of ways of accomplishing this [30] using an x-y translator, or "slide" on the sample table (described in Section 3.6.3).

One method is to first center an accurately machined vertical pin on the sample rotation table so that its tip defines the reference point. This centering can be performed by viewing the pin through a theodolite, or using a micrometer contact gauge, and adjusting the position of the x-y translator until there is no movement of the tip while rotating the sample table Ω-axis. The x and y encoder values then define the reference axis position. In general, the reference point should be defined in position to ~10% of the minimum gauge volume dimension [55]. The pin and aperture are then viewed by a theodolite aligned along each of the incident and scattered beams in turn, and the corresponding aperture adjusted for coincidence of its midpoint with the axis. The use of a fine laser beam aligned along the beam and mounted on the M-S and S-D optical benches can assist this procedure. A theodolite mounted in a similar fashion on the bench can be used to view the pin through the aperture.

3.5.2   Gauge Volume

The extent of the gauge volume is defined by the size of the apertures before the sample in the incident beam, and after the sample in the diffracted beam, together with the angular divergence of the incident and diffracted beam and the wavelength spread of the incident beam.

The neutron beam itself may be used to set or verify the center of the gauge volume at the reference point by observing the diffraction from a cylindrical sample of diameter close to the aperture dimension, with its axis on the reference axis. With the detector at the correct scattering angle, the sample is tracked in position in a direction perpendicular to the incident and diffracted beam in turn, and the corresponding horizontal aperture position adjusted to ensure that the center of the intensity profile versus the translator position is at the encoder values corresponding to the reference axis position. If the aperture position is capable of being scanned, an alternative method is to center the sample at the reference axis position and then scan each aperture position. The positioning in the vertical direction can be carried out in a similar fashion.

The alignment of radial collimators with the reference point can be made using a nylon thread which scatters neutrons incoherently, first positioned on the reference axis, and then scanned in position along the incident beam. The collimators are adjusted to give maximum summed detector intensity when the thread is at the central axis position. Positioning of radial collimators, and of the IGV, is particularly important on a time-pulsed source instrument, since any change in flight path may be interpreted as a strain. In fact, the path length $L = L_i + L_f$ will vary over the gauge volume, and calibration using a centered standard Si, or other reference sample, will give the average value.

Three distinct gauge volumes defined in accordance with VAMAS, TWA20, [55] (see Section 4.1) are discussed in the following subsections:

- The nominal gauge volume (NGV)
- The instrument gauge volume (IGV)
- The sampled gauge volume (SGV).

3.5.2.1   Nominal Gauge Volume

Perfectly parallel beams, that is, with perfect collimation and of uniform wavelength, would give rise to a volume defined in horizontal area by the incident slit width (ISW) and the diffracted slit width (DSW), as shown in Figure 3.20, and in height by the height of the incident aperture. This geometrically defined volume is called the nominal gauge volume (NGV), and may be calculated simply from the aperture dimensions. The centroid of the NGV may be made to coincide with the instrument reference point using the optical methods described above.

It should be noted that as the diffracted beam is by nature a cone of intensity emanating from every part of the sample within the gauge volume, it is not easy to define the height of a gauge volume by an aperture in the diffracted beam. In practice, the height of the outgoing aperture and the height of the detector will limit the overall extent of the vertical beam detected. In order to increase intensity, it is usual to accept as large a portion of the cone as possible consistent with the acceptable spread on the definition of the strain direction measured. An angular extent of $\pm\beta_v$ in the vertical plane will give rise to a component of $\Delta Q_\perp = \pm k.\sin\beta_v$ out of the plane in the scattering vector Q. In standard powder diffraction, there is a problem of defining the angle of scattering ϕ^s if the vertical spread of the detected beam is too high because of the curvature of the cone, which is particularly important at low angles. For strain measurement, values of ϕ^s are usually around 90°, the angle at which cone becomes close to a disk of intensity, and this problem is minimized. Also, in strain measurement one is always measuring small shifts in angle under the same conditions, so that the absolute value

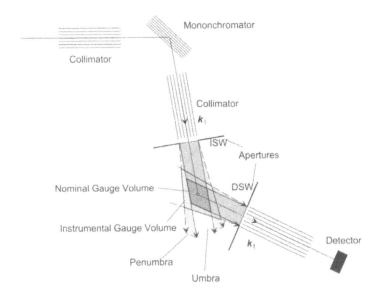

FIGURE 3.20
Schematic illustration of the section of the 'instrument gauge volume' in the scattering plane arising from the angular spread in the beam definition, and its relation to the 'nominal gauge volume.'

of the angle ϕ^s is not so critical as in structure determination, and the vertical extent of the detected beam may be relaxed.

3.5.2.2 Instrumental Gauge Volume

Since the angular divergence of the beams and the wavelength spread across them give rise to a penumbra, as shown diagrammatically in Figure 3.20, there is a gradual decline in the contribution to the diffracted peak intensity from grains in the sample situated at the edge of the volume. The effective gauge volume over which the average strain is measured in a sample filling the gauge volume is larger than the NGV, and is taken to be the definition of the instrument gauge volume (IGV). Conventionally, the IGV is defined in terms of a contour map, and the edge is taken to be where the contributing intensity has fallen to a fraction of that recorded at the center, such as 1/e, that is, by 63%. The vertical divergence of the incident beam is often greater than that in the scattering plane, giving a larger vertical penumbra. This is altered in shape if a vertically focusing monochromator is used, when it is essential to focus the beam onto the reference point. For some applications it may be necessary to reduce the vertical divergence or avoid focusing the incident beam. The IGV contours may be calculated by analytic consideration of the range of possible beam paths contributing to the diffracted beam profile, as shown in Figure 3.21 [56], or by Monte Carlo–type techniques. In the case of a monochromated continuous beam diffractometer the IGV will change very slightly as the detector arm rotates through a Bragg peak scattering angle, but this variation is usually neglected.

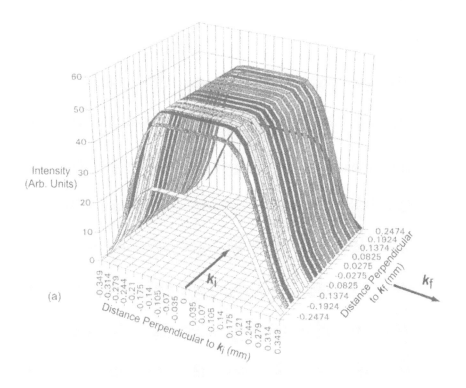

FIGURE 3.21
Illustration of the IGV profile calculated analytically: (a) for a two-axis diffractometer on a continuous beam, with beam defined by a pair of slit apertures with ISW=0.5 mm and DSW=0.5 mm [56]; (b) Monte Carlo calculation of the IGV profiles on a pulsed source instrument, ENGIN, with the radial collimator shown in Figure 3.18 for an open incident aperture; and (c) measured intensity contours on ENGIN with a 2 mm incident slit aperture [29]. In each case, the origin (0,0) is at the instrument reference point and the scattering angle is 90°. The contours give the contribution to the scattered intensity as a function of position.

The IGV contours can be measured experimentally in a number of ways. The most important horizontal section may be measured by tracking through it a scatterer of much smaller dimension, using an x-y translation stage. Ideally, the scatterer should be of the same material as the sample under measurement, for example a thin wire, but the grain size must be relatively very small in order to diffract uniformly. An incoherent scattering sample such as a nylon thread can be used in the case where there is a soller collimator in the outgoing beam, but in this case any wavelength dependence will not be reproduced. An anomalous large grain of say 50 μm in an otherwise small grain polycrystalline sample may also be used. If such a grain is carefully orientated to give maximum intensity when positioned at the center

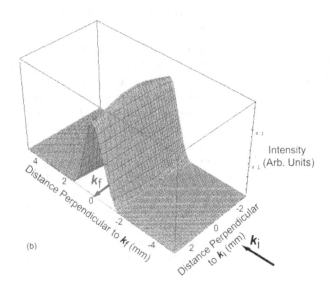

(b)

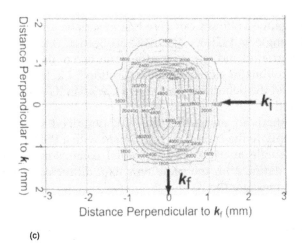

Distance Perpendicular to k_f (mm)

(c)

FIGURE 3.21 (CONTINUED)

of the gauge volume, the intensity is high and the scans can be made very quickly. However, only those wavelengths and angular direction of rays in the beam that satisfy Bragg's Law will be selected. Although not ideal, these methods give a good estimate of the horizontal dimension of the gauge volume. The vertical dimension may be measured by using similar horizontal wires or thread.

3.5.2.3 Definition of Instrumental Gauge Volume When Using a Position-Sensitive Detector

There is a potential difficulty over the definition of the gauge volume when using a PSD. If only one aperture in the diffracted beam is placed between the sample and detector, this and the diffracted beam angle define the peak shape, the gauge volume, and indeed its center as recorded on the PSD. This is because the aperture selects a part of the diffracted beam from the sample along the incident beam at angles defined by Bragg's Law and the incident collimation. This is shown schematically in Figures 3.22a and b. Figure 3.22b illustrates the loss in definition of the IGV with increasing distance of the aperture from the sample. If such an aperture is placed as close to the sample as possible, errors introduced in the definition of the gauge volume are usually small. However, the shape of the gauge volume is also dependent on the angular FWHM of the diffraction peak, which arises from incident collimation, wavelength spread, and the sample itself (e.g., from microstress), as well as the aperture size and aperture–detector distance. If the diffraction peak angular FWHM is large, then neutrons can find their way to the detector from a larger region of the sample. This is illustrated in Figure 3.22a and b. As a result, the IGV would vary when investigating a component having regions with different levels of work hardening, such as peened or forged components.

Partially filled gauge volumes can give rise to serious errors in the measured scattering angle ϕ^s [57], as discussed in Section 3.6.4. One way of reducing the background and improving the definition of the gauge volume is to use a radial collimator as mentioned in Section 3.2.2. In this case there is no umbra; instead the penumbra describes a Gaussian IGV profile, as seen in Figure 3.21b and c.

Not all the methods of measuring the IGV described in the previous subsection will apply to the case of an instrument with a PSD. For example, in the case of a single aperture and no radial collimator, an incoherent scatterer will not define ϕ^s in the same way as a diffracted beam, and so cannot be used to measure the IGV.

3.6 Sampled Gauge Volume and Effective Measurement Position

3.6.1 Sampled Gauge Volume

The SGV is the part of the sample in the IGV from which a measurement of the diffraction peak is obtained. It is the volume of sample over which the strain measurement is averaged, and is affected by [55]:

- Partial filling of the gauge volume
- The wavelength distribution of neutrons across the incident beam

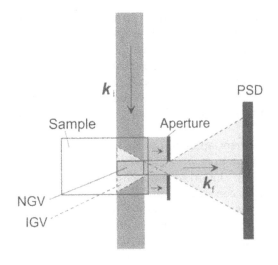

(a)

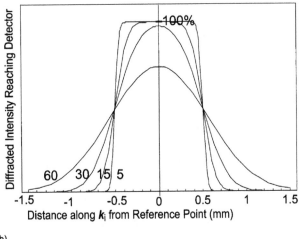

(b)

FIGURE 3.22

(a) Schematic showing the use of a PSD with a slit aperture, showing how the aperture size and the angular peak width together define the instrument gauge volume and peak profile on a continuous source instrument. The solid line is for no angular width, the dashed line for a large (exaggerated for clarity) angular width. (b) The percentage contribution to the total peak signal (ordinate) detected by a PSD arising from diffraction at positions within the sample along k_i as a function of that position relative to the centre of the NGV (abscissa). The angular divergence of the diffracted beam is taken to be 1° FWHM in ϕ^s, and the beam is defined by a 1 mm slit aperture positioned at 5, 15, 30, and 60 mm from the instrument reference point.

- Attenuation of neutrons within the sample
- A high gradient in texture in the sample

A sample placed on the instrument sample table may wholly or partly fill the IGV. For a sample wholly filling the IGV and exhibiting no texture variation or attenuation of the beams, the SGV will coincide with the IGV, and the point of measurement will be at their centroid, which, by adjustment, is usually the instrument reference point. However, if the sample only partly fills the IGV, the SGV will depend on the fraction filled, and the actual, or *effective*, measurement point in the sample will be the *geometric* centroid of the SGV.

In cases where attenuation across the gauge is significant, or the wavelength distribution is uneven across the gauge, these effects must also be included in the contribution to the measured intensity, to give a weighted average, or *effective*, centroid of the SGV. A high gradient of texture in the sample can give rise to a similar effect. The average strain measured will then be that corresponding to the position in the sample at the effective centroid. Even if the sample fills the IGV, these contributions will mean that the effective centroid of the SGV no longer lies at the centroid of the IGV.

The SGV is illustrated schematically in Figure 3.23 for the case of a sample translated into the IGV in reflection geometry with no beam divergence so that the IGV is the same as NGV. The effects of attenuation on the position of the SGV centroid are discussed in the next section. As well as affecting the *position* of the representative strain measurement, the fact that the effective centroid is not at the instrument reference point causes an anomalous shift in the measured diffraction *angle*, which if not corrected for can give rise to spurious values of strain. This correction to the measured scattering angle is discussed in Section 3.6.4 with particular reference to the partially filled IGV. The SGV and its effective centroid, the location in the sample at which the strain averaged over the SGV is measured, can only be determined by calculation, the anomalous angular shift can also be calculated although its accurate value is often difficult to determine. It can be minimized or compensated for by the experimental techniques discussed in Section 3.6.4.

3.6.2 Effects of Beam Attenuation in Sample

The intensity from that part of the sample filling the IGV may be modified by the attenuation of the beam through absorption and scattering, both coherent and incoherent. As a result, the intensity of the beam will decrease along each incident and diffracted path. Some typical attenuation lengths, l_μ, are given in Table 2.2 and Appendix 4. The attenuation imposes a limit on the size of gauge volume used. It has been recommended that the maximum dimensions of the NGV should be restricted so that the ratio of intensity from the front to the back of the gauge volume due to attenuation is less than $1/e$; in other words, the gauge dimensions in the scattering plane should be smaller than l_μ [55]. Thus, for example, 8x8 mm^2 in the horizontal plane is a practical upper limit for large steel samples and little is to be gained from anything larger. Clearly, in such a case a calculation of the effect

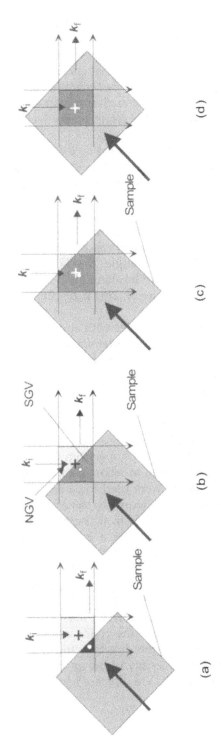

FIGURE 3.23
Illustration of sampled gauge volume (dark) due to partial filling of the nominal gauge volume (lightly shaded) by the sample as it is translated in the direction of the arrow in reflection geometry. The distance between the cross and the spot shows the difference between the locations of the geometric centroid of the nominal gauge volume (NGV) and the centroid of the sampled gauge volume as the NGV is increasingly filled. In reflection geometry, even when completely filled, the two centroids will not be coincident unless the attenuation over the gauge dimensions is negligible (see Figure 3.25).

of attenuation on the position of the centroid of the SGV is essential. In order to reduce the size of the correction and increase its accuracy, an NGV dimension of ~$l_\mu/3$ is more typical. Since the sample must be reoriented, usually about the instrument reference point, in at least three directions at right angles, in order to measure the strain components and calculate the stress, as discussed in Chapters 4 and 5, the attenuation related shift of the SGV centroid may mean that each component of strain is effectively measured at a slightly different position in the sample.

A subtle effect on the wavelength distribution over the gauge volume may occur if the polycrystalline sample contains a few large grains. If they are correctly oriented to give Bragg diffraction from a wavelength component of the beam, they may diffract out a large part of that component from the rays incident on them. This effect is called extinction and is discussed in Section 2.6.1. The resulting change in wavelength composition of the beam can affect the diffraction from the subsequent crystallites in the beam path, causing an anomalous shift in diffraction angle. Another way in which the wavelength composition of the beam can be altered is if it contains a wavelength close to a Bragg edge of the material. This is discussed further, with examples in Section 4.2.2.

It should also be mentioned that in samples with appreciable attenuation due to absorption or scattering, the change in attenuation due to the variation in the average diffracted beam path length in the sample over a scan of scattering angle ϕ^s, even over only a few degrees, can also give rise to small distortions of the peak profile causing an apparent shift in the center. A correction for this effect is easily calculated.

On tof instruments, the effect of attenuation can cause an error in the results of routines used for binning the data into intervals of wavelength or d-spacing. In normal powder diffraction, the spectrum from a nonattenuating Si, or similar reference sample, is used to calibrate and set up the binning. However, if such a reference sample were used to calibrate an instrument to be used to measure strain in a highly attenuating material, the attenuation-related shift of the centroid of the SGV from its nominal position would introduce errors in the binning. Typically, for a 2×2 mm^2 gauge volume area, errors in positioning are estimated to be of the order of 0.1 mm for steel (see Figure 3.25) and 0.2 mm for nickel [55]. Even if binned correctly, such a variation in the path length could be interpreted as a change in lattice spacing, but the effect on the strain is generally small.

In view of all these effects, the IGV shape and size must resulting be chosen with due consideration paid to the size of the grains in the sample and the attenuation of the sample, as well as the direction and magnitude of the strain and strain gradient, and the intensity of Bragg scattering and consequent measurement time. These points are discussed further in Chapter 4.

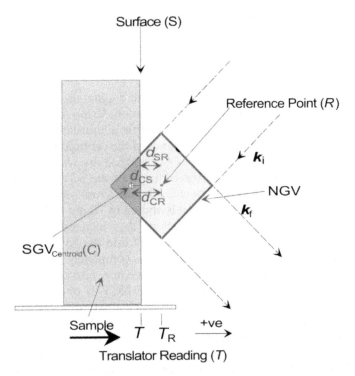

FIGURE 3.24

Schematic illustration showing the surface of a sample translated in reflection configuration into the nominal gauge volume (NGV) (lightly shaded) with the centroid of the NGV as instrument reference point. The relation between the distances among the sampled gauge volume centroid, reference point, sample surface, and translator reading are discussed here in terms of a convention where distances to the right of the reference point are positive.

3.6.2.1 Geometric and Attenuation Shifts of Effective Centroid of Sampled Gauge Volume

We have seen that the centroid of the SGV represents the effective location of a measurement of strain within the sample. The geometric and attenuation related shifts of the centroid are illustrated by considering the translation of the surface edge of a sample through the NGV in a "reflection" configuration, as in Figure 3.23. Figure 3.24 shows in detail the relative location of the sample surface edge, the NGV with the instrument reference point at its center, and the centroid of the SGV. The sample translator reading, T, records the position of the sample and this is usually made relative to the translator reading when the edge is at the reference point, $T = T_R$. Often this is reset to zero in the software for convenience. The way in which T_R can be determined is described in Section 6.5.2. In Figure 3.24, distances to the right of the reference point are taken as positive, so that $[T - T_R]$ as shown is negative, and d_{CS} the distance of the SGV centroid from the surface edge will always be negative in this convention. We then have d_{SR}, the distance of the surface edge to the reference point, and d_{CR}, the distance of the centroid of the SGV to the reference point, related by

$$d_{SR} = T - T_R = d_{CR} - d_{CS} \tag{3.21}$$

In the situation shown in Figure 3.24, d_{CR} and d_{SR} are also negative, but as the NGV fills, for a sample with no attenuation, C moves toward R, and once the NGV is filled, $d_{CS} = -d_{SR}$ as the sample is translated to the right.

The general relation between the distance of the centroid from the surface and the translator position is given in Appendix 2. By way of an example, the separation of the position of the centroid of the SGV and the surface edge $|d_{CS}|$ has been calculated for a 2×2 mm² cross-section NGV and a scattering angle $\phi^s = 90°$. This distance is shown in Figure 3.25 as a function of d_{SR}. With no attenuation, the separation $|d_{CS}|$ increases linearly with a slope of one-third as the sample edge moves into the NGV until it is half filled, it then increases more rapidly until it is completely filled after which it increases linearly with a slope of unity. If we now also consider the effect of attenuation, this will be small when the NGV is only partially filled because in reflection the scattered neutrons originate predominantly from near the surface. However, the attenuation related shift becomes more marked, reducing the separation, when the NGV is more nearly filled as shown in Figure 3.25, since increasing attenuation with path length causes the weighted centroid to be biased toward the free surface. For a plate sample in the transmission configuration, only geometric shifting occurs, as given by the $\mu = 0$ line, because the attenuation path lengths in the sample are the same for all diffracting positions in the SGV.

In practice, the correction for the geometrical shift of the SGV centroid, the zero attenuation line in the example of Figure 3.25, is relatively easy to calculate and is usually made. However, the correction for attenuation is seldom made. In fact, it must be considered for measurements made on highly attenuating materials, or in cases where high positional accuracy is required. From Figure 3.25, it is clear that with the surface centered on the IGV centroid, that is, the translator positioned at zero, neglect of attenuation results in error of ~100 μm in the position of the calculated SGV centroid, compared with an error of ~400 μm for deeper measurements in highly attenuating materials when the IGV is filled by the sample.

3.6.3   Sample Positioning

3.6.3.1   *Sample Positioning, Orientation, and Movement*

An accurate knowledge of the effective position in the sample at which the strain is measured is equally as important as the accurate measurement of strain itself. This involves (a) positioning a nominal point of measurement in the sample, which is in the correct orientation, accurately with respect to the instrument reference point at the center of the IGV, and (b) calculating

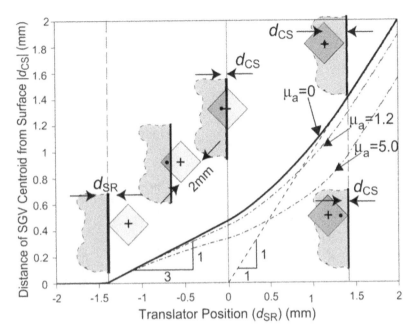

FIGURE 3.25

The relationship between the distance of effective centroid of a diffracting samples SGV and the position of the surface edge of the sample $|d_{CS}|$ and the translator position, d_{SR}, for various sample materials with different linear attenuation coefficients, μ_a. The distances d_{CS} and d_{SR} are defined in Figure 3.24. The calculation is for a 2×2mm square NGV ($\phi^s = 90°$), with the sample in the reflection configuration. With $\mu_a = 0.087$ cm^{-1}, corresponding to Al, the relation is essentially coincident with the $\mu_a = 0$ cm^{-1} line. The $\mu_a = 1.2$ cm^{-1} curve is representative of Fe [29].

the effective point of measurement at the centroid of the SGV relative to the nominal point, as in the example above. Thus, the mounting of the sample and its movement through the IGV is one of the most important aspects of strain measurement, and calls for a quite different approach from that for structural powder diffraction when the sample is usually symmetric and fully bathed in the neutron beam.

Positioning requires appropriate coordinate axes and an origin to be chosen in the sample, such as the cylindrical coordinates shown in Figure 3.26. Also, a reference point in the sample — the *sample reference point* — must be defined, which may or may not be the origin of sample coordinates to which all positions of measurement can be referred. This must be easily accessible, and marked to the accuracy of the desired positioning. A well-defined feature of the sample, or at the intersection of two lines scribed onto the surface, is usual for samples with simple geometries.

As well as the normal rotation Ω of the sample table about a vertical axis, it is necessary to be able to translate the sample in three mutually perpendicular directions relative to the incident beam. Ideally, such an "x-y-z" translator, and the sample rotational axis on which it is mounted, should be capable of moving a weight of ~100 to 200 kg and to position the sample to

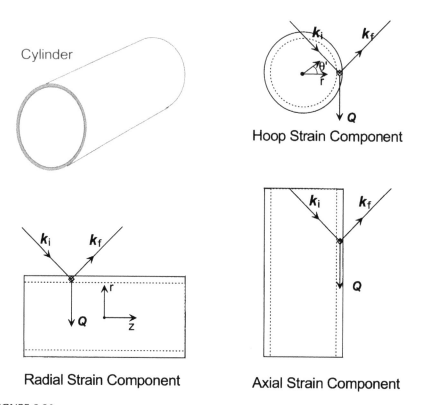

FIGURE 3.26
Definition of axes, and the three orientations of a cylindrical sample relative to the *Q* vector necessary to measure the principal strains assumed to lie along the symmetry axes.

better than 10% of the minimum gauge volume dimension [55]. This means to 0.1 mm for many cases, but even 0.01 mm if very-fine-gauge volumes are used, such as for measurement near the surface of a sample, or if there are high strain gradients in the sample. A positional error of 0.05 mm would give a strain error of 100 $\mu\varepsilon$ if the strain gradient were 2000 $\mu\varepsilon$/mm, such as in the vicinity of a shot-peened surface. Since samples vary in size and weight, not all the extreme requirements of accuracy and weight may be required at the same time, and a number of different translators may be needed. One fundamental requirement when designing a new instrument is to ensure that there is enough space for the sample tables and translators. Thus, the basic sample rotational table should be as low as possible below the beam height. Usually it is the reactor building floor that limits this distance. Whereas horizontal movement of the sample is relatively straight-forward, automated vertical movement calls for more careful design. All rotations and translations must be computer controlled. Stepper motors are often used to record angles and positions, but absolute encoders should be used to ensure perfect reproductibility. Some mounting and translation devices are shown in Figure 3.27a, b, and c, and the translation stage on

ENGIN-X can be seen in Figure 3.17b. The use of commercial robots, or multiaxis Stewart platforms shown in Figure 3.27d and e, are now being tried at some facilities as an alternative to conventional $xyz\Omega$ positioning devices [58].

In cases where the beam and instrument sample table cannot be easily viewed directly because of shielding, in particular on spallation source instruments, it is usual to have closed-circuit TV cameras trained on the sample environment so that the situation can be monitored. In all cases, the possibility of a collision between the moving sample and instrument hardware is an important consideration, since careful alignment may be destroyed. Software programs that simulate the situation and limit the range and sequence of positioning movements are important aids in avoiding collisions. An example is the software being developed for the ENGIN-X instrument at the ISIS facility [59]. During an extended set of measurements, it is advisable to repeat a measurement at a standard position in the sample at suitable intervals in order to check that no unwanted movement of the sample has occurred.

3.6.3.2   *Mounting the Sample*

As discussed above, the sample must be mounted accurately on the translation table, with its reference point or origin of coordinates at a known position relative to the instrument reference point at the center of the IGV, so that the position of strain measurement can be determined as the sample is translated and rotated. Ultimately, this entails locating the centroid of the SGV with respect to the centroid of the IGV for each measurement point, but usually a nominal sample position is positioned at the centroid of the IGV for the strain measurement, and then geometric and attenuation corrections made to determine the actual SGV centroid. The positioning of the sample relative to the centroid of the IGV is usually accomplished by use of at least two theodolites at ~90°. These are first set with the cross-wires focused on a pin mounted on the translator at the reference position, and then the sample is fixed to the sample table and moved to give a well-defined position, such as the sample reference point, at the same point as viewed by the theodolites. Lasers positioned on the optical benches and accurately aligned along the incoming and scattered beam paths may also be used. Theodolites accurately aligned on the optical benches can be used to view the sample through the apertures along each neutron beam path. Scanning a well-defined *edge* of the sample through the IGV and recording intensity versus position is probably the most accurate technique, as it uses the neutron beam itself, although it is time consuming and allowance must be made for absorption causing the intensity to vary with beam depth in the sample, as described in Section 6.5.2. If the sample is relatively small and of simple geometry, it can be positioned on the translator table relative to previously calibrated marks that define the instrumental gauge volume center.

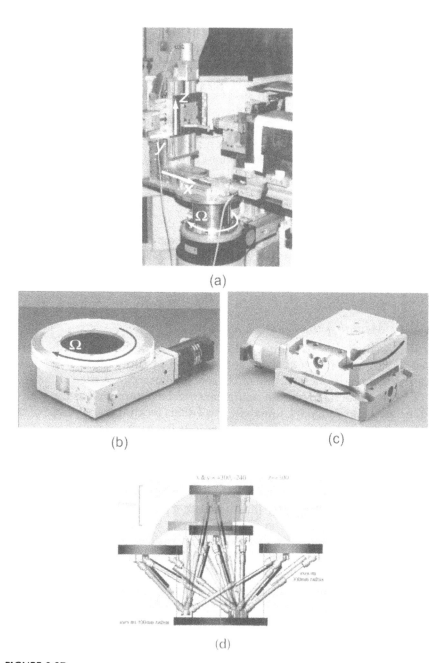

FIGURE 3.27

(a) Typical sample table components allowing x-y-z-Ω movement of the sample, on the TAS 8 instrument at Risø National Laboratory, Denmark. (b) A simple Ω rotation stage. (c) Two-axis tilting stage. (Courtesy of Grenoble Modular Instruments.) (d) An outline diagram of a prototype design of a hexapod (Stewart platform) installed at the new strain mapping instrument, SALSA at the ILL, which enables six-axis movement of the sample through the action of six rams. (e) A photograph of the hexapod [58].

(e)

FIGURE 3.27 (CONTINUED)

Since the shape and size of different samples varies considerably and is rarely of a perfect geometry, accurately mounting and aligning the sample is a time-consuming task. Methods of speeding up this procedure are therefore being developed. One such method involves the accurate off-line mounting of a sample onto a base plate, which then may be located onto a matching plate on the translator table of a diffractometer using accurately aligned bolt holes and spigots. Instruments that can characterize accurately the shape of a sample by coordinate measurement and locate it on a removable base plate are being developed [60]. Such coordinate measurement machines are normally used in workshops to make accurate geometric measurements of samples, in order to compare actual geometries with CAD models and to register components for accurate machining.

The accuracy to which the sample must be aligned to ensure that the measurement position is correct is illustrated by considering the case where it is to be moved horizontally a distance x_p during measurements. An error in alignment of α_e degrees will give a perpendicular error in position of $(x_p . \alpha_e . \pi/180)$. In the extreme case, over a 100 mm distance, α_e needs to be less than 0.006° to ensure that the error is less than 10 μm. However, although affecting the position of measurement in the sample, an angular error in alignment will not give rise to a serious error in strain if it is about a principal direction. For example, a misalignment relative to Q of 7.5° corresponds to an error in strain of only ~2% from a principal value. However, if a full angular variation of strain such as that using the $\sin^2 \psi$ method discussed in Section 5.1.3 is used then the alignment must be better than 1°.

The orientation of the sample is defined by the angles between the sample coordinate axes and the axes of the instrument; in particular the direction of Q and the vertical direction. In order to obtain the stress, the sample needs to be oriented about the measurement point so that at least three directions, usually the assumed principal strain axes, lie along Q (Figure 3.26). Whereas it is often possible to obtain two of the three principal orientations by rotation of the sample about a vertical axis, the third involves remounting the sample unless rotation about a horizontal axis is, exceptionally, available. One major consideration in mounting and orienting a sample is to minimize the total path length of the incident and diffracted beams through the sample. For cylindrical samples, it is the hoop strain component that usually involves the maximum path length.

As the strains measured in these directions are combined to determine the stress at a point in the sample, it is necessary that the gauge volume is as symmetric as possible, and that the positioning ensures that there is no displacement of the nominal point in the sample from the center of the IGV during rotation. This usually means that the scattering angle $\phi^s \sim 90°$ and that the aperture size in the horizontal plane is the same on the incoming and outgoing beams (i.e., ISW = DSW). The incident vertical aperture can only be relaxed if the strain variation in the vertical direction is small. Whereas this is often the case for the measurement of two strain components (e.g., axial and radial components of a cylindrical sample), it cannot also be so for the third, or hoop direction, which must be measured with a small vertical dimension of the gauge volume with consequent increase in measurement time.

An example of sample mounting, shown in Figure 3.26, is the orientation of a cylinder relative to Q to determine the three principal strains that are assumed to lie along the symmetry directions. Photographs of a cylindrical weldment in these three orientations on the TAS 8 diffractometer at Risø National Laboratory are shown in Figure 3.28.

3.6.4 Anomalous Angular Shifts Due to Surface Effects

It was apparent from the earliest attempts to measure strains near the surface of a sample, or at an interface between two materials, that anomalous shifts in Bragg peak angles may occur that could not be attributed to stresses and strains. These shifts arise from the fact (discussed in Sections 3.6.1 and 3.6.2) that when the sample partly fills the gauge volume, the position of the effective centroid of the SGV is not at the center of the IGV, usually the

(a) (b) (c)

FIGURE 3.28
Setup of a large cylindrical weldment on the TAS 8 instrument at Risø National Laboratory, Denmark, in order to measure (a) the radial, (b) the axial, and (c) the hoop strains with the direction of the neutrons indicated by arrows. The soller collimators and aperture holders, mounted on an optical bench, can be seen. In configuration (c), the incident beam passes into the cylinder through a hole cut in the cylinder, and the incident aperture is positioned inside the cylinder [61].

reference point of the instrument, due to geometric and attenuation effects. As a result, in addition to the fact that the effective position of measurement is not the nominal one, the strain is incorrectly measured prior to applying a correction for these effects. The term "surface effect" will be used here as a generic term for all possible errors introduced when measuring near-surfaces, either in the value of strain measured or the position of measurement, or both. To distinguish strain and position errors, we say that surface effects give rise to spurious near-surface strains and spurious near-surface locations, referred to more concisely as *spurious strains* and *spurious locations*. Indeed, we have seen that even if the IGV is filled by the sample, attenuation effects can shift the position of the SGV centroid (Figure 3.25), and thus can also give rise to anomalous angular shifts, which if combined with peak positions recorded for a correctly centered stress-free reference sample, will give rise to *spurious* values of strain unless accounted for. Note that spurious strains recorded near the surface for stress-free samples are sometimes called "apparent" strains or "pseudo strains," but we reserve the latter term to denote the anomalous strains arising from intergranular effects as discussed in Chapter 5. The term "surface effect" is also often used in the literature to describe just the occurrence of spurious strains.

The magnitude of the anomalous shift in measured angle, and thus the related spurious strain, is dependent on the configuration and type of the instrument used. The effect is most easily understood in the case of a simple monochromated, continuous-beam, two-axis diffractometer with a single detector, as in Figure 3.8. If one imagines the case of perfect resolution (i.e., no angular or wavelength spread in the incident or diffracted beam path), there would be no anomalous shift, as in, for example, an offset sample in reflection configuration at the corner of the IGV (Figure 3.23), since the detector would respond only at the correct Bragg scattering angle. Two effects can give rise to spurious strains from surface effects, the first arising because of a wavelength spread across the incident beam on the IGV, and the second because the instrument has imperfect angular resolution. The former means that the part of the IGV filled by the SGV determines the relevant wavelength profile, causing a shift in the mean wavelength. This is sometimes termed the *wavelength effect* (WE). The wavelength distribution is properly averaged only when the IGV is completely filled. The second effect, termed here the *positional discrimination effect* (PDE), means that the signal recorded by the detector is dependent on where within the IGV the diffracting grains are positioned, that is, on the position of the SGV in the IGV. If the scattering event occurs within the IGV at a position farthest from the monochromator, it will be measured as occurring at a *lower* angle of the detector than if it occurred within the IGV nearest the monochromator. Consequently, in the former case one obtains a negative anomalous shift in angle corresponding to a spurious tensile strain. By the same argument, a compressive spurious strain is recorded in the latter case. Examples of measurements are shown in Figures 3.30 and 3.32. This effect can be minimized by increasing the sample-detector distance and thus decreasing the positional discrimination relative to the angular discrimination.

The anomalous shifts in angle due to surface effects have been treated theoretically by simulation using ray-tracing techniques or Monte-Carlo codes [62,63] in order to identify the causes and characteristics of the problem, and to guide the selection of appropriate experimental configurations that can be used to minimize the effects. By way of an example, we discuss the results of Lorentzen [62] who carried out simulations for a simple, monochromatic continuous beam instrument in order to illustrate how the two different effects, the WE and PDE, contribute to the anomalous peak shifts. The calculations, which did not include attenuation of the rays in the sample, were made for various combinations of take-off angles from the monochromator and the sample. They show that the peak shifts are strongly correlated with the experimental configuration. The positional discrimination effect, referred to in his article as the "center of gravity" effect, is larger than the "wavelength effect," and as expected, reduces as the size of gauge volume is reduced. The two effects may add or subtract depending on the relative scattering angles at the monochromator and sample. In particular, it was found that the wavelength effect can be nullified by matching the two take-off angles, $|\phi^s| = |\phi^M|$. Thus, by using this focusing condition part of the problem can be overcome for the configurations considered, but the positional discrimination remains, causing peak shifts related to partial gauge filling. The surface-related spurious strains for a 90° scattering angle, when a plate sample is translated through the IGV in either the transmission or reflection configuration are shown in Figure 3.29a and b, respectively. These two situations arise when measuring the in-plane and out-of-plane strains in the plate, respectively. The consequent spurious surface shifts in strain have an antisymmetric profile as the sample is translated through the IGV,

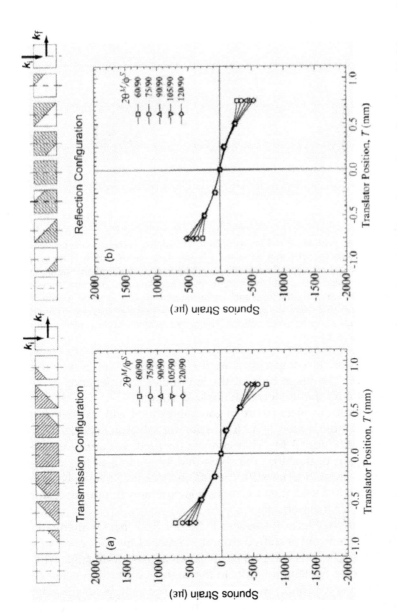

FIGURE 3.29

Illustration of spurious strains due to WE and PDE effects. Numerical results are shown for the simulation of a configuration with 0.5 mm wide apertures on both the incident and diffracted beams, situated at 5 mm distance from the centroid of the NGV, sample scattering angle $\phi^s = 90°$, and various scattering angles at the monochromator. The beam is incident from the top of the diagram, and the detector is to the right. A plate of thickness $1/\sqrt{2}$mm is translated through the NGV in (a) transmission, and (b) reflection configuration. The hatched areas indicate the sampled gauge volume in each case [62].

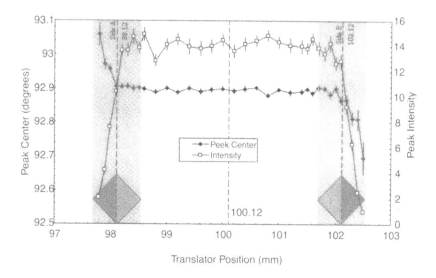

FIGURE 3.30

The 311 measured Bragg peak angle and intensity, from a strain-free nickel powder contained between aluminum sheets in cell 4 mm wide and 0.5 mm thick, as a function of translator position. The NGV was a $0.5 \times 0.5 \times 20$ mm³ with square horizontal cross-section. The sample was moved in 0.4-mm steps through the NGV from one side to the other in the transmission configuration used for measuring the 'in-plane' strain component. The variation reflects the anomalous angular shift [64]. Near the surfaces, the translator position does not accurately reflect the centroid of the sampled gauge volume. The correct locations of the centroid can be found using Equations (A.2.1a–d).

shifting from tension to compression as the centroid of the SGV moves toward the source. The magnitude of the spurious strain depends on the take-off angles at both the monochromator and sample.

The results of these simulations are in good agreement with experimental evidence obtained from a scan of a Ni powder, for which no elastic strains would be expected (Figure 3.30). They confirm the predicted antisymmetric nature of the angular peak shifts, and show that these can lead to values of the spurious strain exceeding 1.6×10^{-3}, or 1600 µε, even for a relatively small-gauge cross-section (0.5×0.5 mm²). Clearly, it is extremely important to minimize and correct for these effects.

In addition to surface effects, it was seen in Section 3.6.2 that for attenuating materials the SGV centroid is shifted from the center of the IGV even when the sample fills the IGV. This will give rise to anomalous shifts in angle which must be corrected for. Systematic shifts in the angle of diffraction peaks can also occur if there is a strong texture gradient in the sample as a function either of position or sample angle. There is then a variation of the strength of the scattering across the SGV that will bias the peak position to lower or higher angles. An example where this is important is in the measurement of strain components at positions across the radius of an extruded rod, where there is a texture gradient due to the variation of plastic deformation with depth.

The above discussion pertains to monochromatic beam continuous source instruments with a single detector. When a PSD is used, the situation is more difficult to analyze, as the IGV, SGV, and observed peak profile are interdependent, and geometric effects due to an off-center centroid of the SGV can be large [57]. If the divergence of the diffracted beam is larger than the angle that the aperture in the diffracted beam subtends at the reference point, shifts in the peak profile can occur due to asymmetric peak "clipping" [57]. The use of radial collimators helps to lessen this interdependence. Simulation using ray tracing is the best means of description, and an example is given in Section 6.5.2.

The use of pulsed source instruments with a PSD to measure near-surface strains requires further careful consideration. The channels of the radial collimator are associated with different elements of the detector and are combined to give the Bragg peak profile as a function of time of flight. A partly filled IGV will cause a change in the angular illumination of the detector elements, giving rise to a shift in the peak profile when the intensities are combined. This so-called "shadowing" effect is illustrated in Figure 3.31. The geometric spurious strains arising from these effects can, however, also be calculated using a ray tracing routine, and corrected for [29].

3.6.4.1 Corrections for Anomalous Angle Shifts

Model-Based Correction. One may model surface effects by ray tracing or Monte Carlo techniques, as mentioned above. However, this requires extremely accurate and reliable calculated predictions of the anomalous angle shifts, as the strain resolution sought in the measurement may amount to only a few percent of the magnitude of the anomalous shifts.

As an alternative to correcting anomalous angle shifts by simulation calculations, there are a number of ways in which the anomalous angle shifts from partially filled IGVs can be corrected or compensated for experimentally.

Exploiting Symmetry. In the case of relatively thin plate samples that have the same stress state on both surfaces, the symmetry of surface anomalies shown in Figure 3.29 can be exploited. Since the anomalous shifts in diffraction angle or time of flight on entry and exit of the sample are antisymmetric, the true strain profile can be deduced by taking the mean of the measured angular shifts on the entry and exit sides of the scan [64]. Clearly, this method requires the accurate location of the sample center. The principle of this technique, and an example actually taken from analogous measurements made using synchrotron x-rays rather than neutrons, is shown in Figure 3.32.

Sample Surface Reversal. The above method requires similarly treated surfaces and a thin sample. More generally, the measured strains

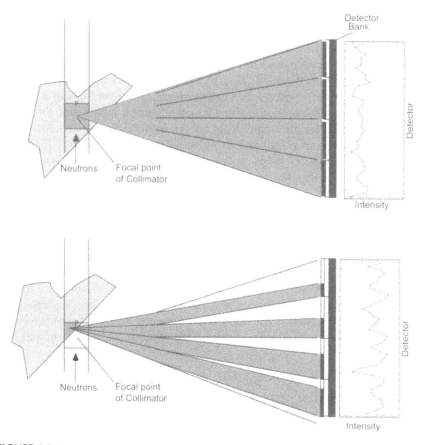

FIGURE 3.31
Schematic illustration of the 'shadowing' effect caused by five radial collimator foils when the normal instrument gauge volume is only partially filled on a pulsed source instrument. This causes a beating effect on the detector bank as shown on the right for the 41 foils of the ENGIN instrument (Figure 3.18). This causes the peak profiles for each detector to be shifted relative to each other, giving an overall anomalous peak angle when they are combined, which must be corrected [29].

near one surface of a plate sample may be corrected experimentally by first scanning the surface of the sample into the IGV, and then reversing it by rotation of 180° about the vertical axis and scanning the same surface out of the gauge volume. If absorption effects are small, equal and opposite geometric shifts are measured and may be averaged to give the true perpendicular strain as a function of depth. Unfortunately, such averaging approaches require one to subtract two large values of strain aimed at arriving at a very small value with high accuracy (Figure 3.32b). Such methods are vulnerable to large errors, either in the peak shift measurements or the exact location of the surface.

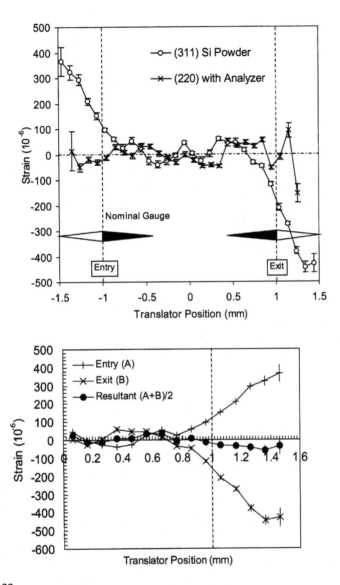

FIGURE 3.32

(a) Spurious lattice strain recorded from the 311 peak from a 2 mm thick Si powder sample, as it is scanned in the transmission configuration through the indicated nominal gauge volume (NGV) of the standard diffractometer 16.3 without an analyzer on the Synchrotron Radiation Source at Daresbury Laboratory (UK), and from the 220 peak on the BM16 instrument at the European Synchrotron Radiation Facility, Grenoble, France, with a perfect Si crystal analyzer, which gives excellent scattering angular resolution that minimizes positional discrimination effects. The vertical dashed lines indicate the translator readings when the surfaces of the powder are at the IGV centroid. The spurious strains are of geometric origin (PDE), and occur until the point at which the NGV is totally filled. (b) When a symmetric strain field is expected from edge to edge, the spurious surface effect can be removed by taking the mean of the 'entry' and 'exit' strains, as shown here for the data in (a) [64].

Use of an Identical Reference Sample. Another method is to use a strain free reference sample, such as a contained powder, having the exactly same geometry of the sample under strain measurement and, at each sample position, to subtract the two measured angles. As both include the same anomalous shifts, the corrected strain is obtained. Again, this requires very careful sample positioning.

3.6.4.2   Minimizing Anomalous Angle Shifts

It is clearly always important to minimize anomalous surface shifts due to partially filled IGVs by the experimental setup used. If the shifts can be drastically reduced, correction may not be necessary; minimizing these shifts will certainly help to reduce errors when using a calculated correction. The anomalous shifts may be minimized by using an experimental configuration that has the following characteristics:

- Reduces the gauge volume cross-section area in the scattering plane
- Reduced the angular divergence of the incident beam
- Increased the angular discrimination of the detector
- Minimal distance between the apertures defining the IGV and its centroid and therefore the sample surface
- Similar scattering angles ϕ^M and ϕ^s.

Figure 3.29 shows how such configurational changes can reduce the (calculated) spurious strains.

Use of an Analyzer Crystal. Also illustrated in Figure 3.32a is the effect of using a perfect Si crystal analyzer in a three-axis configuration, discussed in Section 3.7.1, set to diffract x-rays at the mean incident beam wavelength. A perfect crystal has a mosaic of less than a minute of arc, and so only scattered rays from the sample that are parallel to within this angle will be diffracted by the analyzer into the detector. In this way, the combination of a perfect crystal analyzer and detector set at its diffraction angle acts like the ideal collimator mentioned at the beginning of this section, and so essentially avoids positional discrimination. Such an arrangement can be used with effect on a synchrotron-source instrument with its extremely high incident flux of photons. However, neutron instruments have neutron fluxes of many orders of magnitude less, and so perfect Si or Ge crystals are not practical, as we have mentioned in Section 3.2.1. The use of a mosaic analyzer on a neutron diffractometer will not provide a marked increase in diffraction angle resolution. Since an analyzer crystal is energy selective it may help to minimize the WE-related shift in some cases.

Z-Scan Method. Experimental methods have been developed for near-surface measurements that avoid spurious near-surface strains, although they

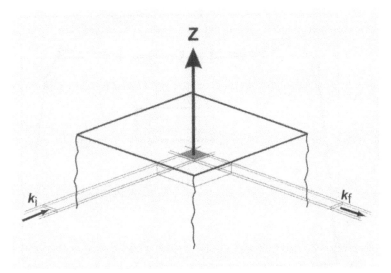

FIGURE 3.33
The Z-scan technique, where the sample is scanned in the z direction, gradually raising the sample surface into the gauge volume. The vertical extent of the related, here shown tightly defined, diffracted beam can be released to give more intensity in most cases [66].

do restrict somewhat the flexibility of the strain investigation. One such method is the *Z-scan* technique [65,66], where the sample surface is gradually raised into the IGV by scanning the sample in the vertical z-direction perpendicular to the horizontal scattering plane. The IGV is designed to have a very small vertical dimension, and there must be minimum vertical divergence of the incident beam, that is, no vertical focusing. Provided the incident beam fully defines the IGV in the vertical direction, the vertical aperture in the scattered beam may be relaxed to increase the detected intensity. The principle is illustrated in Figure 3.33. Clearly, the surface of the sample must be accurately aligned in the horizontal plane of the narrow incident aperture. Since the IGV is fully filled in the horizontal scattering plane, the centroid of the SGV is always on the reference axis, and consequently there are no geometric shifts.

The technique is restricted in flexibility in the sense that the gauge volume necessarily must be located near the sample's vertical edges; otherwise, the neutron beam path becomes prohibitively large and sample absorption must be low. On the other hand, it has the advantage that when the surface is gradually raised into the IGV, the scattered intensity increases with depth in a linear manner. This is in contrast to the severe rise in intensity occurring when translating the surface of a sample horizontally into an IGV with a square- or diamond-shaped cross-section. Near-surface measurements using this technique have been carried out to great accuracy as illustrated in Figure 3.34, where the steep gradient of the stress fields of a shot-peened surface have been measured with the SGV centroid approaching within 45 μm of the actual surface [67].

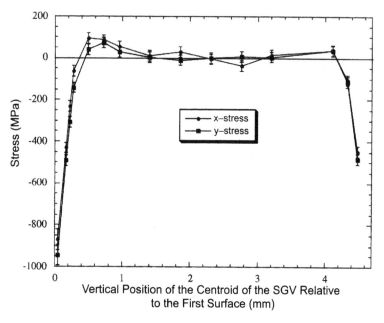

FIGURE 3.34

The near surface bi-axial stress profile from strains measured using the vertical Z-scan technique on a shot-peened nickel alloy (IN 718) plate sample. This sample was used for a round robin exercise in the VAMAS Technical Working Area 20 [67].

3.7 Specialized Instruments for Strain Measurement

3.7.1 Triple-Axis Spectrometer

The triple-axis spectrometer is a steady-state reactor source instrument used primarily for inelastic neutron scattering, in which the wavelength or energy, of the neutrons incident on, or scattered by, the sample may be scanned. The layout of the instrument is shown in Figure 3.35. The monochromator and analyzer crystals are set in the reflection condition, and their angle and that of corresponding diffracted beam direction are rotated in the ratio 1:2, which maintains the Bragg condition, to vary the wavelength. The detector may be a single detector or a PSD.

For strain measurement, the triple-axis instrument is normally set so that the analyzer diffracts neutrons with the same wavelength (energy) as those incident on the sample — that is, for elastic scattering from the sample. The analyzer-detector angles are held fixed as the scattering angle ϕ^s is scanned. While the third axis is an unnecessary complication for most strain measurements, it can prove advantageous if the sample is radioactive steel, when the Co^{60} γ-rays can be shielded from the detector as they pass through the analyzer

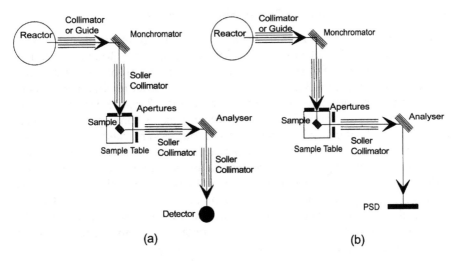

FIGURE 3.35

Schematic layout of a three-axis instrument with (a) a single detector and (b) a position-sensitive detector.

crystal that will only diffract the neutrons. Another use is to filter out unwanted neutrons that may be in the incident beam from the monochromator by setting the analyzer to only diffract the required wavelength. Whereas higher orders are not usually a problem, if the higher order of a monochromator is used as the main incident beam there may also be wavelengths of 2λ present in the beam. An example is when the 004 reflection planes from pyrolytic graphite (PG) monochromator are used to provide the desired incident wavelength, λ, giving better resolution than the 002 reflection because of its higher monochromator scattering angle ϕ^M. In this case, the 002 planes will usually give a more intense incident beam of 2λ, which may be incoherently scattered by the sample with a resulting increase in background. If the analyzer crystal PG 002 planes are set to detect only λ, the 2λ neutrons will not be detected.

Use of an analyzer can also help to improve the angular resolution in scattering angle of an instrument if the mosaic spread of the analyzer crystal is smaller than the collimation angle. However, as always, this improvement is at the expense of detected neutron intensity. The use of a perfect Si crystal analyzer on a synchrotron x-ray instrument was described above.

The conventional triple axis spectrometer can be used in two other modes in order to measure strain. The first is to scan both the wavelength incident on the sample and that diffracted from it, holding the scattering angle fixed at ϕ^s corresponding to the mean wavelength. In this configuration, the instrument is being used rather like a time-of-flight instrument. An alternative is to open up the resolution of either the monochromator or analyzer system so that the broad range of wavelengths in a scan is fully accepted, and to scan the other system with tight resolution. A specialized use of the conventional instrument with a PSD is described in the following section.

3.7.2   Use of Horizontally Curved Silicon Crystals

Attempts to improve strain measurement efficiency in terms of high detection rates of useful neutrons of a continuous beam diffractometer have led to the development of vertically bent extended monochromators and PSDs, as described in Sections 3.2.1 and 3.3. However, in order to obtain very good resolution, high monochromator scattering angles, $2\theta^M$, are necessary, with consequent loss of monochromated neutron flux. A compromise between resolution and intensity is always present, but as remarked in Section 3.4, one method of improving count rate is to relax unnecessarily tight resolution components. Since strain measurement requires only a small range of scattering angles ϕ^s to be detected and gauge volumes are usually small, this has led several groups to consider the use of monochromator crystals bent about a vertical axis to focus neutron in the horizontal plane onto a small IGV, with no intervening soller collimator. Such horizontally bent monochromators can give optimum resolution over a narrow range of ϕ^s.

Silicon and germanium crystals can be readily grown to large dimensions, but as remarked in Section 3.2.1, only with small mosaic angular spread. By bending the crystal, the effective mosaic can be increased for certain configurations of use, and the neutrons can be focused onto a small volume. Silicon has been used in this way, primarily by groups at the Nuclear Physics Institute Řež, near Prague, in the Czech Republic, in collaboration with Physikalisch–Techniche Bundesanstalt, Braunschweig, Germany [68–70], and at the University of Missouri [71]. Silicon single-crystal plates, typically of dimensions up to ~5 × 30 × 200 mm³ are cut from a large crystal in well-defined orientations and bent with a variable radius between ~10 m and infinity using a four-point mechanical bend [68,70,71]. Vacuum and pneumatic techniques have also been used to produce the bending [71]. The bent crystals may be stacked vertically at tilt angles to each other to give focusing also in the vertical plane. If the crystals are cut with a [01$\bar{1}$] axis vertical, various monochromator planes (*hkk*) may be brought into use by rotating the assembly about a vertical axis to give different wavelengths. A typical focusing silicon monochromator is shown in Figure 3.36.

As examples of the use of these crystals by the Řež group, a two-axis instrument [68] is shown in Figure 3.37, and a three-axis instrument [69] in Figure 3.38. These have very good angular resolution, enabling both microstrains from Bragg peak line widths and macrostrains from Bragg peak angular shifts to be measured. The two-axis instrument comprises the bent Si monochromator, with ϕ^s of 90° and a PSD. The PSD can be set to cover the required range of ϕ^s, and so there are no moving parts. The three-axis instrument utilizes two bent Si crystals as monochromator and analyzer in symmetrical geometry, and a conventional detector. The analyzer system is scanned in ϕ^s around the sample. The symmetric geometry is shown by theory to give the optimum resolution [69], so a choice of monochromator and analyzer plane is made to match that of the sample to give this symmetric scattering. However, this means that the scattering angle ϕ^s is ~50°,

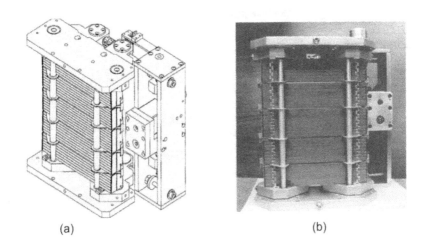

(a)   (b)

FIGURE 3.36
(a) Schematic and (b) photograph of the doubly bent Si monochromator made at the Missouri University Research Reactor for the new strain measurement instrument (SALSA) at the Institut Laue-Langevin. It comprises 39 perfect Si crystal pieces of $12 \times 165 \times 5$ mm dimensions. The fixed vertical curvature is realized by shaped posts, the variable horizontal curvature by a four-point bending mechanism. (Courtesy G. Bruno.)

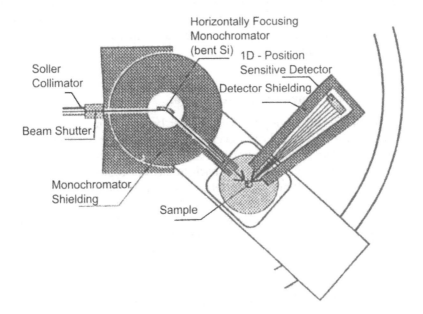

FIGURE 3.37
The layout of a high-resolution two-axis instrument strain measurement using a bent silicon crystal monochromator [68].

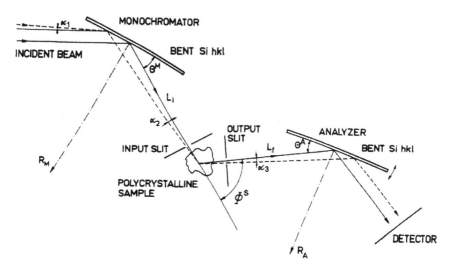

FIGURE 3.38
Schematic layout of the use of two bent silicon single crystals in a three-axis arrangement [69].

with consequent nonideal gauge volume. Bragg peak instrumental widths in ϕ can be as low as FWHM of ~6', corresponding to an instrument resolution of $\Delta d/d$ ~2×10^{-3}, which is essentially the closest that two peaks can be before they can no longer be distinguished. It can be seen from the Figure of Merit, expressed in Equation (4.17), that this can be beneficial for strain measurement. As a guide, the peak position can be measured to an accuracy of about 1/100th of this FWHM.

3.7.3   Specialized Time-of-Flight Techniques

A conventional neutron chopper in a reactor beam or time-pulsed source beam uses only a very small duty cycle, ~1% to 5%, albeit with a full range of wavelengths. The spallation neutron source compensates for this by the very high intensity, or brightness, of each pulse. For high-resolution work, this must be reduced further by shortening the pulse time or pulse rate. This is compared to a 100% duty cycle using just a small wavelength range in a conventional steady-beam instrument. Attempts to improve the duty cycle of a time-pulsed reactor beam have been made by using choppers with "open" and "closed" segments that give a sequence of intensity pulses of different duration with time. The sequence may be either "pseudo random" or "sinusoidal," and in each case cross-correlation techniques are used to extract the useful signal at the detector, and the duty cycle rises to ~25%. One disadvantage of this method is systematic background levels arising from the analysis algorithms, which can give rise to problems in detecting small Bragg peaks. However, for strain measurement, a strong peak can be selected and the technique can be used to give high resolution.

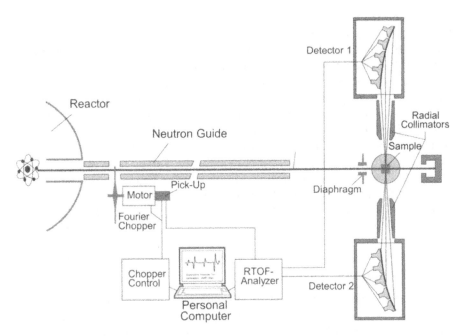

FIGURE 3.39
Schematic layout of the reverse-time-of-flight diffractometer FSS at GKSS. The detectors are positioned accurately to give time focusing [33].

Researchers at the GKSS Research Centre (Geesthacht, Germany) have developed the technique for strain measurement by using the principle of Fourier analysis in which the beam is sinusoidally modulated, and the diffraction spectrum is analyzed in terms of sine and cosine components. The instrument, called the Fourier Strain Spectrometer (FSS) [33], uses the reverse tof (RTOF) method [72,73], in which the timing trigger is set by the detected neutron rather than the source pulse or rotor position. This allows a range of rotation frequencies to be used. The detected neutron intensity and phase relative to the chopper modulation is measured as a function of frequency. The chopper, shown in Figure 3.14c, is in the form of a disk with a comb-like pattern of alternating absorbing gadolinium and transmitting aluminum segments around the periphery, which rotates about a horizontal axis in front of a similar stator. The detector banks of ^{6}Li-glass scintillators are arranged in time-focusing geometry at ϕ^s ~90° on either side of the beam incident on the sample [73]. This time focusing must be done by hardware arrangement, as software corrections cannot be made in RTOF. The IGV is defined by apertures in both incident and diffracted beams, and the instrument layout is shown in Figure 3.39. The analysis of raw data is carried out by online computers to give line shapes that are close to the derivative of Gaussian shapes, so that the point of crossing the axis defines the center [72,73]. The resolution is better than $\Delta d/d = 10^{-4}$.

A method using frame overlap to increase the duty cycle of a conventional tof instrument on a continuous flux neutron source, to 5.8%, has been developed by Stuhr at SINQ [74,75]. The POLDI instrument uses a disk chopper with eight apertures cut at random angular spacing around the edge over a 90° sector; these are accurately repeated over the other quadrants to give short (8 μs) time pulses. The neutron counts from the eight overlapping frames are collected from a time-focused He³ PSD detector over a 30° range of scattering angle. The overlapping tof diffraction patterns are deconvolved by identifying their time origin from their time intervals, and from the slope of peak intensity contours in a plot of intensity versus time of arrival and angle ϕ^s. Using neutron optics to define the beam, a resolution of $\Delta d/d \sim 2 \times 10^{-3}$ FWHM can be attained. As a large range of ϕ^s is required the instrument is best used for strain measurement along a principal strain direction. It has advantages for measurements in large grained samples.

3.7.4   Transmission Techniques

The sharp increase in transmission through a sample as the wavelength is increased and no longer fulfills the Bragg condition for each lattice plane, described in Section 2.3.3, can be used to measure strain. The edges occur at wavelengths $\lambda_e^{hkl} = 2d_{hkl}$ for each lattice plane of spacing d_{hkl}. As each edge corresponds to a back-scattering angle $\phi^s = 180°$, the resolution is high, and the scattering vector and measured strain direction lie along the incident beam. There is a Bragg edge in transmission corresponding to each Bragg peak in the normal diffraction pattern, and the step change in transmission at each edge is analogous to the height of the Bragg peak at $\phi^s = 180°$ and the steepness of the edge is related to the Bragg peak width. The edge profile may be parameterized and a Rietveld-type fit may be made to the spectrum of edges as shown in Figure 3.40 or to the differentiated spectrum [77].

Following early tests made at the Harwell LINAC in the 1980s, the technique known as "strain radiography," has recently been developed as a method of strain measurement at GKSS and LANSCE, and at ISIS [76]. As the intensity in transmission is very high, it is possible to observe the Bragg edge pattern after a single pulse of neutrons, so that transient effects during loading or relaxation can be studied on a millisecond or even 100 μs time scale [77,78].

The technique is best applied to plate samples exhibiting plane stress, so that the through-thickness stress component normal to the surfaces may be taken as zero. If an area detector is used to measure the transmitted neutron intensity and pixels are grouped in fine spatial regions, a transmission image may be obtained. Such pixellated detectors allow maps of the normal strain (edge shifts), textural components (edge shapes and strength), or crystallographic phase analysis (relative strengths of edges corresponding to different phases) over the area of the sample illuminated by the incident beam that is imaged at one setting [76,77,79]. An example of such a map of strain is

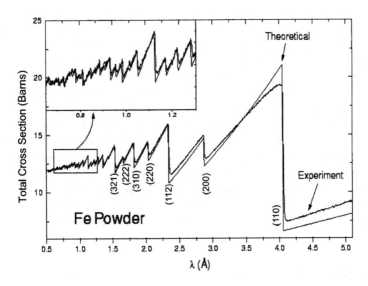

FIGURE 3.40
The normalized powder transmission spectrum from a ferritic steel sample observed on the ISIS transmission facility. The profile of the Bragg edges is fitted by the theoretical attenuation cross-section [76].

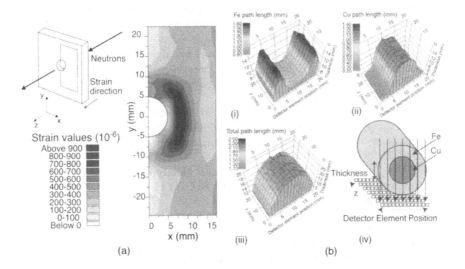

FIGURE 3.41
Illustrating results of the use of the transmission imaging technique using a pixelated detector. (a) Variation of the, through-thickness average, out-of-plane elastic strain component around a cold expanded hole in a 12 mm thick steel plate [79]. (b) Phase selective imaging of neutron path lengths through the test sample shown in (iv). (i) In Fe, (ii) in Cu, (iii) total path length [82].

shown in Figure 3.41a. The strain components in the plane of the sample may be deduced by tilting the sample about the two perpendicular directions as shown in Figure 4.22. In this way, it resembles the $\sin^2\psi$ technique used in x-ray diffraction stress measurement, which uses the fact that the normal stress at the surface is zero to obtain a zero-stress reference. As discussed in Section 4.6.3, another application is as a means of determining d_{hkl}^0, the zero-stress reference lattice spacing of a material, by making a number of measurements at different tilt angles [76,80].

The relative strengths of the transmitted Bragg edges in the pattern from a two phase material can be determined from a two phase Rietveld-like profile fit. The transmitted intensity from each phase recorded on a two dimensional pixellated detector can then be used to build up a two dimensional phase sensitive radiograph, with intensities from each phase related to their path length as a function of position. The results of a demonstration experiment on a cylindrical sample of copper surrounded by iron are shown in Figure 3.41b. The beam is incident perpendicular to the axis of the sample. This technique opens up the possibility of phase selective tomographic imaging [82].

References

1. P. Ageron, Neutrons in the investigation of matter: neutron beam reactors, *Neutron News*, 9, 14–16, 1998.
2. J.M. Carpenter and W.B. Yelon, Neutron sources, in *Methods of Experimental Physics*, Vol. 23A, *Neutron Scattering*, K. Sköld and D.L. Price, Eds., Academic Press, New York, 1986, 99–196.
3. S.J. Cocking and F.J. Webb, Neutron sources and detectors, in *Thermal Neutron Scattering*, P.A. Egelstaff, Ed., Academic Press, New York, 1965, chapter 4.
4. R. Kahn and A. Menelle, Improvement of neutron guides at LLB, *J. Neutron Res.*, 4, 175–181, 1996.
5. A. Menelle, Upgrade of a curved multichannel neutron guide using supermirror coatings at the Laboratoire Léon Brillouin, *Neutron News*, 6, 10–13, 1995.
6. K. Kakurai and Y. Morii, Neutron scattering activities at JRR-3M, *Neutron News*, 9, 15–22, 1998.
7. M. Ono, Y. Waku, K. Habicht, and T. Keller, Neutron Larmor diffraction measurement of strain in a ductile composite $Al_2O_3/Y_3Al_5O_{12}$ (YAG), *Appl. Phys. A*, 74A, S73–S75, 2002.
8. D. Mildner, H. Chen, G. Downing, and V. Sharov, Focused neutrons a point to be made, *J. Neutron Res.*, 1, 1–11, 1993.
9. M.R. Daymond and M.W. Johnson, An experimental test of a neutron silicon lens, *Nucl. Instr. Methods Phys. Res. A*, 485, 606–614, 2002.
10. B.E. Allman, A. Cimmino, S.L. Griffin, A.G. Klein, K.A. Nugent, I.S. Anderson, and P. Høghøj, Novel optics for conditioning neutron beams II: focusing neutrons with a 'lobster-eye' optic. *Neutron News*, 10, 20–23, 1999.c,
11. C.G. Windsor, *Pulsed Neutron Scattering*, Taylor & Francis, London, 1981.
12. C.C. Wilson, A guided tour of ISIS — the UK spallation neutron source, *Neutron News*, 6, 27–34, 1995.
13. R. Eccleston and C.C. Wilson, A guided tour of ISIS: 2004 update, *Neutron News*, 15, 15–18, 2004.

14. A Second Target Station at ISIS, CCLRC, Rutherford Appleton Laboratory, Didcot, Oxon, UK, RAL-TR-2000-032, 2000.
15. K. Crawford, The intense pulsed neutron source, *Neutron News*, 1, 9–15, 1990.
16. J.A. Roberts, The Los Alamos neutron science center (LANSCE), *Neutron News*, 10, 11–14, 1999.
17. M. Furasaka, KENS neutron scattering facility, *Neutron News*, 3, 8–14, 1992.
18. P. Allenspach, The continuous spallation neutron source SINQ, *Neutron News*, 11, 15–18, 2000.
19. M.R. Daymond, personal communication, 2003.
20. A.J. Allen, C. Andreani, M.T. Hutchings, and C.G. Windsor, Measurement of internal stress within bulk materials using neutron diffraction, *NDT Int.*, 15, 249–254, 1981.
21. A.J. Allen, M.T. Hutchings, C.G. Windsor, and C. Andreani, Neutron diffraction methods for the study of residual stress fields, *Adv. Phys.*, 34, 445–473, 1985.
22. R.E. Lechner, R.V. Wallpach, H.A. Graf, F.-J. Kasper, and L. Mokrani, A monochromator with variable horizontal and vertical curvatures for focusing in real and reciprocal space, *Nucl. Instr. Methods*, A338, 65–70, 1994.
23. T. Vogt, L. Passell, S. Cheung, and J.D. Axe, Using wafer stacks as neutron monochromators, *Nucl. Instr. Methods*, A338, 71–77, 1994.
24. J.D. Axe, S. Cheung, D.E. Cox, L. Passell, T. Vogt, and S. Bar-Ziv, Composite germanium monochromators for high resolution neutron powder diffraction applications. *J. Neutron Res.*, 2, 85–94, 1994.
25. B. Lebech, K. Theodor, B. Breiting, P.G. Kealy, B. Haubeck, J. Lebech, S.A. Sørensen, and K.N. Clausen, A facility for plastic deformation of germanium single-crystal wafers, *Physica B*, 241–243, 204–206, 1998.
26. T.C. Hsu, F. Marsiglio, J.H. Root, and T.M. Holden, Effects of multiple scattering and wavelength-dependent attenuation on strain measurements by neutron scattering, *J. Neutron Res.*, 3, 27–39, 1995.
27. L. Edwards, D.Q. Wang, M.W. Johnson, J.S. Wright, H.G. Priesmeyer, F. Rustichelli, G. Albertini, P.J. Withers, and I.B. Harris, Precise measurement of internal stresses within materials using pulsed neutrons (PREMIS). In Proceedings, 4th International Conference on Residual Stresses, Society for Experimental Mechanics, Baltimore, MD, 1994, 220–229.
28. T. Pirling, A new high precision strain scanner at the ILL, *Mater. Sci. Forum*, 321–324, 206–211, 2000.
29. P.J. Withers, M.W. Johnson, and J.S. Wright, Neutron strain scanning using a radially collimated diffracted beam, *Physica B*, 292, 273–285, 2000.
30. P.C. Brand and H.J. Prask, New methods for the alignment of instrumentation for residual-stress measurements by means of neutron-diffraction, *J. Appl. Crystallogr.*, 27, 164–176, 1994.
31. G.C. Stirling, Experimental techniques, in *Chemical Applications of Thermal Neutron Scattering*, B.T.M. Willis, Ed., Oxford University Press, Oxford, 1973, 31–48.
32. C.G. Windsor, Experimental techniques, in *Methods of Experimental Physics*, vol. 23A, *Neutron Scattering*, K. Sköld and D.L. Price, Eds., Academic Press, New York, 1986, 197–257.
33. J. Schröder, V.A. Kudryashev, J.M. Keuter, H.G. Priesmeyer, J. Larson, and A. Tiitta, FSS — A novel RTOF-diffractometer optimized for residual stress investigations, *J. Neutron Res.*, 2, 129–141, 1994.
34. M.W. Johnson, L. Edwards, and P.J. Withers, ENGIN — A new instrument for engineers, *Physica B*, 234, 1141–1143, 1997.

35. L.E. Edwards, P.J. Withers, and M.R. Daymond, ENGIN-X: A neutron stress diffractometer for the 21st century, in Proceedings, 6th International Conference on Residual Stress, Oxford, 2000, IOM Communications, Ltd., London, vol. 2, 2000, 1116–1123.

36. J.A. Dann, M.R. Daymond, J.A.J. James, J.R. Santisteban, and L. Edwards, ENGIN-X: A new diffractometer optimised for stress measurement, in Proceedings, International Colloquium on Advanced Neutron Sources XVI, Forschungszentrum Jülich GMBH, Eds. G. Mank and H. Conrad, Dusseldorf–Neuss, 2003, 231–238.

37. M.R. Daymond and L. Edwards, ENGIN-X: A fully refined diffractometer designed specifically for measurement of stress, *Neutron News*, 15, 24–29, 2004.

38. M.A.M. Bourke, D.C. Dunand, and E. Üstündag, A spectrometer for strain measurement in engineering materials, *Appl. Phys.*, 75A, 1–3, 2002.

39. M.R. Daymond and M.W. Johnson, Optimisation of the design of a neutron diffractometer for strain measurement, in Proceedings of International Colloquium on Advanced Neutron Sources XV, S. Itoh and J. Suzuki, Eds., KEK Report, Tsukuba, Japan, 2000–22, 499–503.

40. M.W. Johnson and M.R. Daymond, An optimum design for a neutron diffractometer for measuring engineering stresses, *J. Appl. Crystallogr.*, 35, 49–57, 2002.

41. M.R. Daymond, M.A.M. Bourke, R.B. Von Dreele, B. Clausen, and T. Lorentzen, Use of Rietveld refinement for elastic macrostrain determination and for the evaluation of plastic strain history from different spectra, *J. Appl. Phys.*, 82, 1554–1556, 1997.

42. H.M. Rietveld, A profile refinement method for nuclear and magnetic structures, *J. Appl. Crystallogr.*, 2, 65–71, 1969.

43. G.S. Pawley, Unit-cell refinement for powder diffraction scans, *J Appl. Crystallogr.*, 14, 357–361, 1981.

44. N.J. Rhodes, M.W. Johnson, and C.W.E. Eijk, The future of scintillation detectors in neutron scattering instrumentation, *J. Neutron Res.*, 4, 129–133, 1996.

45. J.R. Santisteban, L. Edwards, M.E. Fitzpatrick, A. Steuwer, P.J. Withers, M.R. Daymond, M.W. Johnson, N. Rhodes, and E.M. Schooneveld, Strain imaging by Bragg edge neutron transmission, *Nucl. Instr. Methods Phys. Res. A*, 481, 765–768, 2002.

46. M.C. Crawford, Gas detectors for neutrons, *J. Neutron Res.*, 4, 97–107, 1996.

47. K.D. Müller, R. Reinartz, R. Engels, P. Reinhart, J. Schelton, W. Schäfer, E. Jansen, and G. Will, Development of position-sensitive neutron detectors and asscociated electronics, *J. Neutron Res.*, 4, 135–140, 1996.

48. D.A. Witte, A.D. Krawitz, R.A. Winholz, R.R. Berliner, and M. Popovici, An instrument for stress measurement using neutron diffraction, *J. Neutron Res.*, 6, 217–232, 1998.

49. B. Guérard, Latest Microstrip Gas-Counter Developments, Institut Laue-Langevin Annual Report, Grenoble, 1998.

50. G. Caglioti, Conventional and three axis neutron powder diffraction, in *Thermal Neutron Diffraction*, B.T.M. Willis, Ed., Oxford University Press, Oxford, 1970, 14–33.

51. G. Caglioti, A. Paoletti and F.P. Ricci, Choice of collimators for a crystal spectrometer for neutron diffraction, *Nucl. Instr.*, 223–228, 1958.

52. F.M.A. Margaça, Optimised geometry for a stress measurement two-axis diffractometer at a reactor, in*Measurement of Residual and Applied Stress Using Neutron Diffraction*, M.T. Hutchings and A.D. Krawitz, Eds., NATO ASI Series E: Applied Sciences, Vol. 216, Kluwer: Dordrecht, 1992, 301–311.

53. M.J. Cooper and R. Nathans, The resolution function in neutron diffractometry, II. The resolution function of a conventional two-crystal neutron diffractometer for elastic scattering, *Acta Crystallogr.*, A24, 481–484, 1968.

54. M.J. Cooper and R. Nathans, The resolution function in neutron diffractometry, III. Experimental determination and properties of the 'elastic two-crystal' resolution function, *Acta Crystallogr.*, A24, 619–624, 1968.

55. G.A. Webster, Ed., Polycrystalline Materials — Determinations of Residual Stresses by Neutron Diffraction, ISO/TTA3 Technology Trends Assessment, Geneva, 2001.

56. M.T. Hutchings and C.G. Windsor, personal communication, 2001.

57. P.J. Webster, G. Mills, X.D. Wang, W.P. Kang, and T.M. Holden, Impediments to efficient through-surface strain scanning, *J. Neutron Res.*, 3, 223–240, 1996.

58. G. Bruno, T. Pirling, S. Rowe, W. Hutt, P.J. Withers, and C. Carlile. A look ahead in residual stress analysis: the strain imager at the ILL, in Proceedings, Advances in Neutron Scattering Instrumentation, 47th International Society for Optical Engineering (SPIE) Annual Meeting, Seattle, 2002, 64–74.

59. J.A. James, J.R. Santisteban, L. Edwards, and M.R. Daymond, A virtual laboratory for neutron and synchrotron strain scanning, Physica B: Condensed Matter, 350, E743–E746, 2004.

60. K.T.W. Tan, G. Johnson, N.J. Avis, P.J. Withers, and J. Kelleher, The engineering body scanner concept, *J. Neutron Res.*, 11, 245–251, 2004.

61. P.J. Bouchard, M.T. Hutchings, and P.J. Withers. Characterisation of the residual stress state in a double 'V' stainless steel cylindrical weldment using neutron diffraction and finite element simulation, in Proceedings, 6th International Conference on Residual stress, Oxford, 2000, IOM Communication Ltd., London, vol. 2, 2000, 1333–1340.

62. T. Lorentzen, Numerical analysis of instrumental resolution effects on strain measurements by diffraction near surface and interfaces, *J. Neutron Res.*, 5, 167–180, 1997.

63. X.L. Wang, S. Spooner, and C.R. Hubbard, Theory of peak shift anomoly due to partial burial of the sampling volume in neutron diffraction residual stress measurements, *J. Appl. Crystallogr.*, 31, 52–59, 1998.

64. P.A. Browne, Determination of *Residual Stress in Engineering Components using Diffraction Techniques*, Ph.D. thesis, University of Salford, 2001.

65. M.T. Hutchings and C.G. Windsor, Variation of the hoop strain away from the edge of two WC/Co composite discs, AEA Technology Report, AEA-InTec-0864, 1992.

66. D. Wang, I.B. Harris, P.J. Withers, and L. Edwards. Near-surface strain measurement using neutron diffraction, in Proceedings, 4th European Conference on Residual Stress, Soc. Francaise de Met. et de Mater., vol. 1, 1996, 69–78.

67. L. Edwards, Near-surface stress measurement using neutron diffraction, in *Analysis of Residual Stress by Diffraction using Neutron and Synchrotron Radiation*, A. Lodini and M.E. Fitzpatrick, Eds., Taylor and Francis, London, 2003, 233–248.

68. P. Mikula, M. Vrana, P. Lukas, J. Saroun, H.J. Ullrich, and V. Wagner, Neutron diffractometer exploiting Bragg diffraction optics: a high resolution strain scanner, in Proceedings, Fifth International Conference on Residual Stresses, Linköping, 1997, 721–725.
69. M. Vrana, P. Lukas, P. Mikula, and J. Kulda, Bragg diffraction optics in high resolution strain measurements, *Nucl. Instr. Methods*, A338, 125–131, 1994.
70. V. Wagner, P. Mikula, and P. Lukas, A doubly bent silicon monochromator, *Nucl. Inst. Methods*, A338, 53–59, 1993.
71. M. Popovici and W.B. Yelon, A high performance focusing silicon monochromator, *J. Neutron Res.*, 5, 181–200, 1997.
72. V.A. Kudryashev, H.G. Priesmeyer, J.M. Keuter, J. Schröder, and R. Wagner, On the shape of the diffraction peaks measured by Fourier reverse time-of-flight spectrometry, *Nucl. Instr. Methods*, B101, 484–492, 1995.
73. V.A. Kudryashev, H.G. Priesmeyer, J.S., J.M. Keuter, R. Wagner, and V.A. Trounov, Optimisation of detectors in time-focusing geometry for RTOF neutron diffractometers, *Nucl. Instr. Methods*, B93, 355–361, 1994.
74. U. Stuhr, Concept and performance of POLDI — the new tof-strain-scanner at PSI, in Proc. International Colloquium on Advanced Neutron Sources XVI, Forschungscentrum Jülich GMBH, 2003, G. Mank and H. Conrad, Eds., Düsseldorf-Neuss, 2003, 257–265.
75. U. Stuhr, POLDI — the new time of flight diffractometer at PSI for strain field measurements, *J. Neutron Res.*, 9, 423–429, 2001.
76. A. Steuwer, P.J. Withers, J.R. Santisteban, L. Edwards, M.E. Fitzpatrick, M.R. Daymond, M.W. Johnson and G. Bruno, Neutron transmission spectroscopy: a solution to the d^0 problem?, in Proceedings, 6th International Conference on Residual Stress, Oxford 2000, IOM Communications Ltd., London, vol. 2, 2000, 1239–1246.
77. H.G. Priesmeyer, M. Stalder, S. Vogel, K. Meggers, and W. Trela, Progress in single-shot transmission diffraction, in Proceedings, International Conference Neutrons in Research and Industry, SPIE Proceedings Series, vol. 2867, 1997, 164–167.
78. R. Pynn, Neutron scattering methods for materials science, in Proceedings, Materials Research Society Symposium, 1990, 15–26.
79. J.R. Santisteban, L. Edwards, A. Steuwer, and P.J. Withers, Time-of-flight neutron transmission diffraction, *J. Appl. Crystallogr.*, 34, 289–297, 2001.
80. A. Steuwer, P.J. Withers, and R. Burguete, The d^0 distribution in Al-2024 friction stir welds, *J. Neutron Res.*, 9, 345–350, 2001.
81. M.R. Daymond, personal communication, 2000.
82. A. Steuwer, P.J. Withers, J.R. Santisteban, and L. Edwards, Using pulsed neutron transmission analysis for crystalline phase imaging and analysis, *J. Appl. Phys.* (submitted 2004).

4

Practical Aspects of Strain Measurement
Using Neutron Diffraction

The principles of neutron diffraction and the instrumentation needed to undertake experiments are described in Chapters 2 and 3. In this chapter, the practical aspects of diffraction peak measurement and the analysis of the raw data required to determine accurate lattice strain are described. Appendix 3 provides a checklist of points for consideration when embarking on a strain measurement that builds on many of the aspects discussed in this and the previous chapter.

4.1 Introduction

In this chapter, considerations crucial to the acquisition of reliable strain measurement are discussed. These include the instrument setup, the practical choice of suitable diffraction peaks to evaluate lattice strain, the optimum choice of parameters defining the scan of the peak, and the method of peak profile analysis. The main points for consideration and a sequence for procedure are summarized in Appendix 3.

During the 1997–2000 period, a committee comprising of scientists and engineers from around the world involved in stress measurement using neutron diffraction, met and discussed the procedures currently used with a view to setting up a standardized approach. This committee worked within the framework of VAMAS, the Versailles Agreement on Advanced Materials and Standards, as Technical Working Area (TWA) 20, "Measurement of Residual Stress." Their draft report was published in 2000, and contains details of their recommended standard procedures [1]. The topics discussed in this chapter amplify and build on some of the VAMAS committee's work and enlarge on aspects touched on in Chapter 3. In this discussion, it is necessary to anticipate important aspects of the conversion of strain to stress described fully in Chapter 5, and in particular the effects of elastic and plastic anisotropy and intergranular stresses.

The equation for lattice strain, ε^{hkl} (Equation (3.8)),

$$\varepsilon^{hkl} = \left(d_{hkl} - d_{hkl}^0 \right) / d_{hkl}^0 = \frac{\Delta d_{hkl}}{d_{hkl}} = -\left[\cot(\theta_{hkl}^{B0}) \right]\left(\theta_{hkl}^B - \theta_{hkl}^{B0} \right) \qquad (4.1)$$

is very similar to that for the continuum engineering strain. However, it is important to remember that while diffraction does allow one to probe the atomic lattice spacings with great precision, it has quite different characteristics from a conventional strain gauge. For this reason, the analysis of diffraction data is far from straightforward. In fact, a great deal of information of importance to engineers and materials scientists can be obtained from the behavior of a diffraction peak profile from a sample under load, including its shape, center, width, and height. The interpretation, from both the engineering and materials science perspectives, of the lattice strain data in terms of stress using the appropriate diffraction elastic constants is the main topic of Chapter 5.

It is essential for both the measurement and interpretation of strain data from a polycrystalline sample, especially in the process of converting Bragg peak angles into strains and thence calculating stresses, to appreciate the highly selective nature of the diffraction technique. As has been mentioned in Chapters 2 and 3, using the powder method, a single *hkl* diffraction peak is inherently associated with a subset, or family, of the order of several thousand grains within the sampled gauge volume of, typically, a few cubic millimeters. This subset consists of grains with *hkl* plane-normal oriented in the direction of the scattering vector Q, and the technique thus provides a very selective measurement. As described in Section 1.4.1, strain variations occurring over a characteristic distance l_0 greater than the corresponding sampling gauge volume dimension are recorded as shifts in the angle or wavelength at which a diffraction peak is measured, whereas those having a characteristic length shorter than the corresponding sampling gauge dimension, often termed *microstrains*, are evidenced by changes in the profile of the diffraction peak in width or shape. As a result, the angle representative of the center of the peak provides the lattice strain in this subset averaged over the gauge volume. While it should always be remembered that the measured lattice strain is an average, it is often termed simply the *lattice strain* and will for reasons of brevity often be referred to as such in this book. That the lattice strain is proportional to the macroscopically applied stress, at least below the proportional (elastic) limit, is evident in Figure 4.1. The consequence of the selective nature of the individual different reflections is also clear from the figure, which shows their different average lattice strain response from that of a conventional strain gauge to applied loading, even at low loads. The different slopes of the strain-stress curves before macroscopic plastic yielding in Figure 4.1 is called *elastic anisotropy*, and arises from the different elastic properties of the various crystal lattice planes. Clearly,

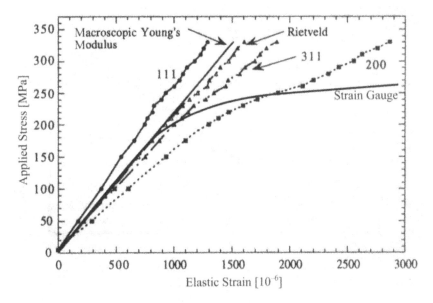

FIGURE 4.1

The elastic lattice strain response for austenitic stainless steel parallel to uniaxial loading as determined from individual peak analysis and by Rietveld analysis of the whole diffraction profile. The latter is in good correspondence with the macroscopic strain gauge in the elastic regime, and is linear with the applied stress in the plastic regime. It is clear that the departure from linearity in the macroscopic strain response (same abscissa scale) occurred at around 200 MPa, and the 0.2% yield stress occurs at around 260 MPa [2].

this elastic anisotropy must be taken into account when interpreting diffraction data, and is discussed fully in Section 5.5.

It is also clear from Figure 4.1 that unlike a conventional strain gauge, first-order lattice strains — which by their nature must be elastic — are not sensitive to macroscopic plastic strain. This is because plastic strain occurs by slip processes and the passage of dislocations through the crystalline lattice, and does not give rise to any increase in the lattice spacing *per se*. However, plastic strain can give rise to intergranular misfits leading to Type II intergranular stresses. As a result, lattice strain gives a measurement of the stress even in the plastic regime. As to whether this stress is representative of the bulk stress depends on whether there are intergranular stresses formed that cause the elastic lattice strain for that particular reflection to differ from the bulk elastic strain. Intergranular strains generated by plastic straining give rise to the nonlinearities in the applied stress versus lattice strain curves seen in Figure 4.1. These nonlinearities are termed *plastic anisotropy* and also need to be accounted for when interpreting strain data in terms of stress. The complete understanding of intergranular stresses is an important and ongoing challenge, and the current situation is discussed in Section 5.6.

4.2   Measurement of Diffraction Bragg Peak Profile

In this section, a number of factors affecting the approach to making a measurement of strain are enumerated and discussed. The questions to be asked when embarking on a series of stress measurements are summarize-summarized in a logical sequence in Appendix 3.

4.2.1   Choosing an Appropriate Gauge Volume

The factors that must be considered when deciding on an appropriate gauge volume have been introduced in Section 3.5. On the one hand, as large a gauge volume as possible is desirable to maximize the data acquisition rate, but on the other, restricting the gauge volume in certain directions to maximize spatial resolution is also desirable, such as in directions of high strain gradient. At the present time, a practical lower limit of the sampled gauge volume, V_v, for neutron diffraction is a minimum dimension of ~0.33 mm, and a sampled volume of ~1 mm^3. As mentioned in Chapter 3, one way of finding a balance between these competing demands is to use a "matchstick" shape gauge volume where possible, aligned with the long axis parallel to a direction in which the strain field is not expected to vary steeply — that is, it has a large characteristic length l_0 (Section 1.4.1). It is also important to consider the type of stress of interest. In general, neutron diffractometer gauge volumes tend to measure macroscopic Type I stress variations for monolithic materials and phase-averaged Type II stresses for multiphase materials, that is, $V_0^I > V_v > V_0^{II}$.

The grain size in the sample material may place a limit on the minimum gauge volume possible if the conventional powder method is to be employed. The number of grains in the gauge volume is important in determining the quality of the diffraction pattern. Whether a sufficient number of grains lies within the gauge volume to achieve powder diffraction can be assessed by recording the diffraction pattern on a two-dimensional positional sensitive detector, as seen in Figure 4.2 for the low scattering angles characteristic of high-energy synchrotron x-ray data when uneven cone intensity becomes apparent. The same effect would be seen for neutron data; however, at the present time it is rare that two-dimensional data are collected. An effective way to determine if the grain size is sufficiently small is to monitor the variation in integrated intensity as the sample is translated through the gauge volume, as shown in Figure 4.2c and d. The desired outcome is a diffraction peak intensity variation that is within counting statistical uncertainty (see Section 4.4.3.3). Alternatively, the sample may be rotated in the beam over 10 to 15° about the vertical Ω axis, and a peak intensity monitored every 0.5°. Grains that are too large will produce large intensity variations, that is, variations which exceed those expected from counting statistics [1]. Texture, on the other hand, causes a gradual change of peak intensity over

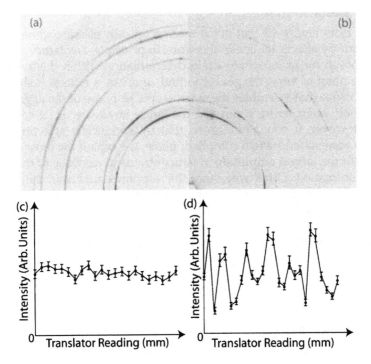

FIGURE 4.2

Typical quarter diffraction patterns recorded using synchrotron x-ray diffraction for Al characteristic of (a) an instrumental gauge volume (IGV) large enough to meet the powder criterion, and (b) an IGV smaller than that necessary for a powder pattern. In (c) and (d), respectively, schematic intensity profiles representative of cases (a) and (b) on translating the sample through these IGVs.

small ranges of sample rotation. In broad terms, a point-to-point integrated intensity variation of more than 25% is suggestive of an insufficient number of grains in the diffracting condition, and a larger gauge volume should be tried. Alternatively, if feasible, data from a large number of equivalent positions could be averaged.

4.2.2 Choosing an Appropriate Reflection

In order to obtain a high-intensity peak representative of a large number of grains, it is usually advisable to choose a peak with a high structure factor and multiplicity (see Section 2.3.5). While these are clearly important factors, the effects of texture, elastic anisotropy, and plastic anisotropy due to intergranular stresses mean that the choice of Bragg reflection to measure strain must be made with great care, since these affect the subsequent conversion to stress, as discussed in Section 5.2. Whereas the influence of elastic anisotropy of polycrystals, discussed in Section 5.5, is understood quantitatively, that of plastic anisotropy is only qualitatively understood for most materials (Section 5.6).

4.2.2.1 Elastic Regime

It is clear from Figure 4.1 that in the macroscopically elastic regime, crystal-line anisotropy effects are linear; therefore, in principle, any lattice reflection can be chosen for macroscopic strain determination within this region [1]. From one point of view, the goal is to find, and use, a reflection showing a linear behavior that resembles the macroscopic response of the aggregate so that the bulk engineering strain and the lattice strain are in 1:1 correspon-dence. However, it could be argued that by specifically selecting an *hkl* reflection corresponding to a compliant plane, we exploit the most sensitive gauge with the largest amplitude in lattice strains in response to the macro-scopic deformation. Either way, since the responses are linear, the choice is not critical in the elastic regime as long as it is known that the sample has not previously undergone plastic deformation. However, in the conversion from strain to stress, we must address the inherent elastic anisotropy and use the correct *diffraction elastic constants*. These elastic constants may be determined by a calibration experiment in which a test sample of the material is subjected to known elastic loading *in situ* on a diffractometer, or they may be calculated as discussed in Section 5.5.

4.2.2.2 Plastic Regime

As discussed in detail in Section 5.6, there is at present insufficient under-standing of the nature of intergranular stresses to accurately account for and to separate out analytically their contribution to the measured lattice strain in the plastic regime. Thus, the safest approach to relate the measured lattice strain to the macrostress in the plastic regime is to utilize as an internal lattice strain gauge one of those lattice planes that have been calibrated experimen-tally and known to develop the smallest residual intergranular stresses. In other words, use a reflection that is essentially insensitive to plastic defor-mation in that it provides an essentially linear lattice strain–stress response in both the elastic and the plastic regime. However, linearity in the lattice strain parallel to the applied load direction does not always mean that there is also linearity in the lattice strain measured perpendicular to the load direction. The former is usually taken to be sufficient. Advice on the most appropriate lattice planes to use for different crystal structures is given in the draft VAMAS standard [1]. Examples of the suitable reflections that are now known to exhibit good linearity for different materials, and some which do not, are summarized in Table 4.1. It should be pointed out that unless it is known for certain that the sample has not undergone plastic deformation, only these reflections should be used to determine macrostress in an engi-neering component.

One indication that a sample might have undergone plastic deformation is the presence of texture due to reorientation of the grains. The presence of *texture* in a sample influences the choice of Bragg reflection, because it may be necessary to use different reflections to obtain sufficient intensity to measure strains in different directions. Texture may also influence the

TABLE 4.1

Lattice Planes Weakly and Strongly Affected by Intergranular Strains

Material	Recommended Planes (Weakly Affected by Intergranular Strains)	Problematic Planes (Strongly Affected by Intergranular Strains)
fcc (Ni [43], Fe [44], Cu [45])	(111), (311), (422)	(200)
fcc (Al [46], [47])	(311), (422), (220)	(200)
bcc (Fe [47])	(110), (211)	(200)
hcp Ti [31])	Pyramidal $(10\bar{1}2)$, $(10\bar{1}3)$	Basal (0002) and prism $(10\bar{1}0)$, $(1\,\bar{2}10)$
hcp (Be [49])	Second-order pyramidal $(20\bar{2}1)$, $(11\bar{2}2)$	Basal, prism, and first-order pyramidal $(10\bar{1}2)$, $(10\bar{1}3)$

Note: Only those that are weakly affected, and thus show good linearity of strain under at least uniaxial loading, are recommended for stress measurement. Note that *hcp* materials may be strongly affected by texture [1].

determination of stress from measured strains in both the elastic and plastic regimes, since it restricts the subset of grains contributing to a reflection and affects the averaging in the calculation of elastic constants.

4.2.2.3 Avoidance of Anomalous Shifts of Angle Due to Bragg Edges

The occurrence of a rapid change in transmission of neutrons at a Bragg edge, described in Section 2.3.3, can give serious anomalies in the conventional method of neutron strain measurement. This occurs when the neutron path through the sample is long and the scattering cross-section is large, as in the case of iron. Suppose that the wavelength selected for the measurement is 2.86Å, as might be a natural choice of wavelength for measurements in *bcc* steel components since the 110 reflection is then at $\phi^s = 90°$. Unfortunately, as can be seen from Figure 3.40, this wavelength coincides exactly with the Bragg edge for the 002 reflection. If the wavelength distribution in the incident beam has a spread of about 0.05 Å neutrons on the short wavelength side are removed from the beam relative to those on the long wavelength side because of the higher cross-section on the short wavelength side. Thus, the average wavelength of the incident beam varies with path length through the material. In principle, this effect occurs at all wavelengths since the cross-section is never flat, but it is very marked at Bragg edges. If the nominal wavelength, which might well have been carefully measured in a calibration experiment with a standard powder such as silicon, is used to derive the lattice spacing and the strain from the peak positions, then a serious systematic error will occur. Although the effect may lead to peak asymmetry, it is unlikely that the experimenter would be able to differentiate it from other sources of asymmetry and the resulting effect on peak position could go unnoticed.

An experiment demonstrating these effects for body-centered cubic steel was performed by Hsu et al. [3]. The average apparent strain was measured for three wavelengths, one exactly at the Bragg edge for the 112 peak, 2.34 Å as is shown in Figure 3.40, one just above at 2.50 Å and one below the cut-off at 2.28 Å. The steel sample was strain-free and was arranged in reflection geometry with a path length of 34 mm. An anomalous shift in strain of more than 1000×10^{-6} was found when the wavelength corresponds exactly to the Bragg cut-off. The negative sign of the anomalous shift was consistent with the fact that the average wavelength was shifted to a higher value as the beam traversed the long beam path, although the apparent compressive shift was larger than expected. Anomalous shifts will be found for samples in both reflection and transmission geometries. As a general rule, the wavelength chosen must not lie within a range $\Delta\lambda = \pm 0.02\ \lambda_e^{hkl}$, where $\lambda_e^{hkl} = 2\ d_{hkl}$ corresponds to a Bragg edge [3].

Figure 4.3 shows the variation of the integrated intensity of the 110 peak in the steel sample as a function of wavelength. This result indicates a simple strategy for improving the penetration into thick samples, or alternatively improving the intensity for a given path length, by choosing the wavelength for the experiment for which the total cross-section is smallest. This will be close to a minimum in the cross-section versus wavelength (Figure 3.40), but will avoid the actual Bragg edge. It is clear from Figure 4.3 that the penetration may be improved by 50% simply by choosing the wavelength appropriately.

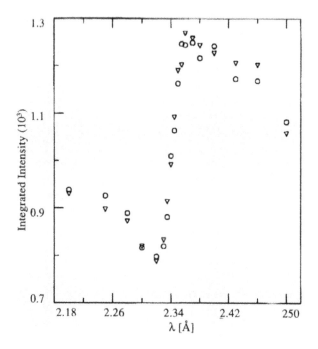

FIGURE 4.3
Integrated intensity of the 110 diffraction peak for polycrystalline iron as a function of wavelength near the 112 Bragg cut edge at 2.34Å (see Figure 3.40). The total path length through the sample was 13 mm. The open circles and triangles represent two different scans though the wavelength range [3].

4.2.2.4   Measurement of Many Peaks in a Diffraction Pattern

In contrast to the single peak measurement described above, there are some advantages if several peaks are measured together in the form of a diffraction pattern profile. Such a profile could be the result of successive recording of a number of peaks by scanning the neutron wavelength or diffraction angle at a powder diffractometer on a steady-state neutron source. More commonly, however, it is the result of a *time-of-flight* experiment from a pulsed source, where the experimental data are always in the form of a diffraction profile including many Bragg peaks. These can be used individually, or more usually together for determining the lattice strain. A Rietveld profile analysis may be carried out on the entire pattern recorded, as described in Section 3.2.2, with further details given in Section 4.5 below. This provides average lattice parameters over the range of peaks, with smaller uncertainties per given measurement time than on a continuous source instrument.

In general, if it is thought that intergranular stresses are important, the collection of data from a number of peaks is recommended whether the measurement is made using an instrument on a pulsed or continuous neutron source. Other more practical factors also determine the choice of reflection as described in Appendix 3.

4.2.3   Scanning the Diffraction Peak

In order to scan a single Bragg reflection on a simple diffractometer efficiently, one must make a choice of the range of angles covered, the size of angular steps, and the duration of the count at each angle — all with a view to determining the peak center to the desired accuracy. If a position-sensitive detector is used that covers an adequate range of angles, the step size is defined by the pixel resolution, although these may be later binned together, and one can determine the count time by observing the build up of counts in the peak and consequent accuracy in determining its center. The same is true for most instruments on a pulsed neutron beam. The intensity and width of the peak, and the intensity of the background are determining factors. However, the overall time that needs to be allocated to the measurements and the consequent cost are usually the overriding considerations. Optimum use of that time is therefore paramount, and is discussed in the next section.

4.3   Analyzing Bragg Peak Profiles

The following sections discuss the analysis of diffraction peak profiles, and using the theory of statistical fitting how one can decide on the optimum scan parameters. Of course, in many cases it is sufficient to proceed empirically by trial and error, acquiring continuously and repeatedly fitting

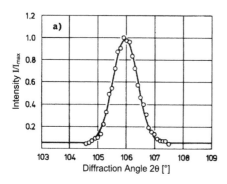
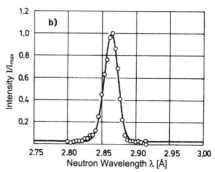

FIGURE 4.4

Typical diffraction peaks measured for the 002 peak of austenitic steel on a diffractometer operating on a steady-state neutron source, (a) by a $\phi^s = 2\theta$ scan at a wavelength of 2.86Å, and (b) by a λ-scan at an angle of ϕ^s =106°. The solid lines are Gaussian functions fitted to the profile.

the diffraction peaks until the required accuracy is achieved. However, it is often useful to be able to determine *a priori* the expected level of accuracy for a given signal, either to plan experiments ahead of time, to compare the capabilities of different instruments or to design new ones, or to check the performance and reliability of a chosen peak-fitting routine. Simple analytical expressions are available for predicting the precision of peak center in terms of the diffracted intensity, the peak width, and the peak-to-background signal [4]. The expressions are derived using standard procedures [5,6] applied to Gaussian peak shapes, but appear to be broadly applicable to other similar peak shapes.

The diffraction peak can be measured as a function of angle or wavelength using steady-state or pulsed neutron sources. Whichever the approach, the outcome is a diffraction peak, or a diffraction profile over a series of peaks, for which the number of neutrons detected varies as a function of diffraction angle or wavelength. Typical peaks measured by scanning ϕ^s and λ using a continuous source instrument are illustrated in Figure 4.4.

In this section the task of obtaining a measure of strain from the recorded profiles over one diffraction peak, or over several peaks, is considered. Although, as discussed in Section 5.7, the diffraction peak contains a lot more information about the state of strain within the sampled gauge volume than just the average value, the task of determining a representative peak center is emphasized here, since it is the shift in the peak center from that of a stress-free sample that is most simply related to a representative lattice strain. This is true whether the strain ε^{hkl} has been recorded as a shift, from that of a stress-free sample; in angle, θ^B; wavelength, λ; or time-of-flight, t, for the *hkl* reflection (see Equations (3.8) and (3.12)):

$$\varepsilon^{hkl} = -\cot\theta_0^B \Delta\theta^B \;;\; \varepsilon^{hkl} = \frac{\Delta\lambda_{hkl}}{\lambda_{hkl}} \;;\; \text{or}\; \varepsilon^{hkl} = \frac{\Delta t_{hkl}}{t_{hkl}} \tag{4.2}$$

In much of the following we shall focus on the standard continuous source diffractometer as the main example for discussion.

4.3.1 Expressions for Peak Profiles

4.3.1.1 Symmetric Peaks from Continuous Source Diffractometers

Having obtained the raw experimental data in the form of diffraction peaks measured in terms of counts at successive angular positions of a detector, the task of obtaining the best estimate of the peak center is best achieved through a curve-fitting procedure. This involves the selection of an appropriate profile function to be fitted to the data.

For powder diffractometers at steady-state sources, most of the elements defining the instrument resolution function can adequately be described by symmetric functions, as discussed in Sections 3.2.1 and 3.4.1. Consequently, it is usually adequate to select a symmetric fitting function, the most common choice being a Gaussian-shaped function:

$$F(x_i) = H_0 e^{-\frac{(x_i - x_0)^2}{2u_x^2}} + B_0 \tag{4.3}$$

where H_0 is the fitted peak height, u_x the standard deviation, x_o the derived peak center (in angle, time, or wavelength), and B_0 the fitted flat background at each point. In practice, the peak profile is measured at angular, time or wavelength intervals, x_i, stepped in n increments Δx. This profile has a full width at half maximum (FWHM) of $[2\sqrt{(2\ln 2)}]u_x = 2.355\ u_x$. The integrated area, I, under the peak itself is given by

$$I = \Delta x \sum_{i=1}^{n} \left[F(x_i) - B_0 \right] = H_0 u_x \sqrt{2\pi} \tag{4.4}$$

However, of more significance is the total number of counts in the measured peak, as this dictates the accuracy of measurement, defined here as the summed or integrated count, I_C,

$$I_C = \sum_{i=1}^{n} \left[F(x_i) - B_0 \right] = \frac{H_0 u_x \sqrt{2\pi}}{\Delta x} \tag{4.5}$$

In the above expression, the integrated count has been calculated from the best-fit curve; it could also be calculated directly from the actual counts. Measured peak profiles sometimes appear to have tail sections more closely resembling a Lorentzian-shaped curve, as in

$$F(x_i) = \frac{H_0 u_x^2}{\left[u_x^2 + (x_i - x_0)^2 \right]} + B_0 \qquad (4.6)$$

where the symbols, apart from u_x have the same meanings as above. For this profile the FWHM $= 2u_x$, but the integrated intensity is not well defined because of the infinite tails. Typical Gaussian and Lorentzian peak profiles are shown in Figure 4.5.

Often in the elastic regime, the peak profile is well represented by a Gaussian function, but in the plastic regime the profile may become more Lorentzian in shape. Combinations of these two functions are often used, most commonly the "Voigt" or the so-called "Pearson Type VII" functions; both provide a scheme for a gradual variation in shape from a Lorentzian to a Gaussian. The Voigt function is a convolution of a Gaussian and a Lorentzian function, and because it involves a convolution it is rather more difficult to implement in curve-fitting routines. A typical formulation of the "Pearson Type VII" function is given by

$$F(x_i) = H_0 \left[1 + \frac{(x_i - x_0)^2}{m u_x^2} \right]^{-m} + B_0 \qquad (4.7)$$

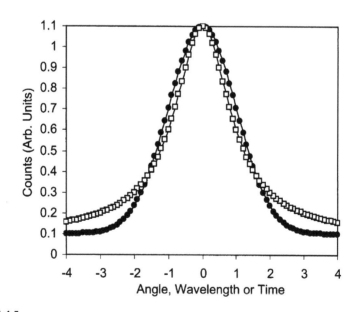

FIGURE 4.5
Gaussian (filled circles) and Lorentzen (open squares) curves according to Equations (4.3) and (4.6), with $H_0 = 1$, $u_x = 1$ and $x_0 = 0$ and $B_0 = 0.1$ with $\Delta x = 0.1$.

where m is the shape factor governing the balance between a Gaussian ($m = \infty$) and a Lorentzian ($m = 1$) shape. For further details, see Snyder [7] and Hall et al. [8].

The common statistical term for describing the width of a distribution is the standard deviation, u_x. For diffraction peak analysis, however, it is common to use the FWHM. In all cases as exemplified above, simple relations can be established between the standard deviation and the FWHM, but is crucial that the quantity quoted is made quite clear at all times when describing uncertainty in results.

4.3.1.2 Asymmetric Peaks from Time-Pulsed Source Instruments

In contrast to symmetric peak functions which are normally adequate for describing data originating from steady-state sources as shown in Figure 4.4, diffraction peaks from pulsed sources tend to be asymmetric in character due to the moderation process, as mentioned in Section 3.2.2, necessitating a more complicated function. A typical example of a pulsed-source single-peak profile from Si powder is shown in Figure 4.6. As fitting the whole diffraction pattern profile simultaneously using a Rietveld type approach is often required, the wavelength-dependent evolution of the peak profile must also be taken into account.

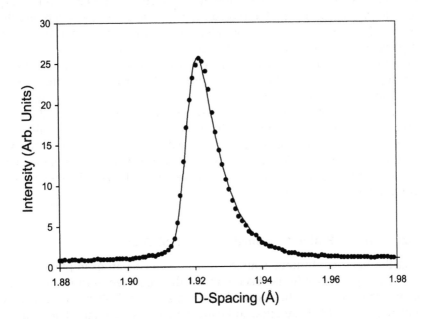

FIGURE 4.6
A Si 220 Bragg peak recorded on the ENGIN beamline at the ISIS time-of-flight source. (Courtesy of M.R. Daymond.) The solid line is the best fit of the function given by Carpenter et al. [11].

Incorporating the basic physics behind the asymmetry, a function consisting of leading and trailing exponentials joined back to back convolved with a shifted symmetric Gaussian, was the first peak shape function that enjoyed widespread use [9,10]. This approach is particularly useful for heavily poly-ethylene-moderated sources, such as some of those at Los Alamos National Laboratory. To achieve better fits to the data on high-resolution instruments across the full range of wavelengths and characteristics of different types of moderator, more complicated fitting functions have been developed. The function developed by Carpenter et al. [11] fits the peak profile by a function consisting of a sharp rise followed by the sum of two exponential decays, convolved with a shifted Gaussian. This function is shown as the solid line fit to the data in Figure 4.6, and is currently used to fit much of the data collected on ENGIN-X. The approach by Ikeda and Carpenter [12] has been developed to include a more physical and detailed account of the incident wavelength spectrum resulting from a short-wavelength epithermal contribution and a longer-wavelength moderated contribution. The function consists of the sum of a delta function and a decaying exponential with arbitrary relative areas, convolved with a "slowing down" function, and involves four wavelength-dependent parameters. This function is better for undermoderated sources, which are common at ISIS. These three functions are discussed by Carpenter and Yelon [13], and are illustrated schematically in Figure 4.7. A more comprehensive list of common fitting functions is given in Young [14].

It is clearly very important to record carefully the parameters defining the peak profile used in the analysis of pulsed source diffraction data. With such complicated functions, it is not appropriate to define a "center" for each profile but rather a well-characterized feature such as the channel corresponding to the steepest slope of the leading edge or the peak intensity. As long as this characteristic feature is the same for corresponding peaks from the sample and reference, it can be used to determine shifts in the time channel, and hence macrostrain, with very good accuracy. If the absolute lattice spacings, or intrinsic widths of the peaks, are required, it is necessary to fit the same peak profile function with the same instrumental parameters to all corresponding peaks from the calibration sample, sample under study, and the reference sample.

4.3.2   Fitting a Profile to a Single Diffraction Peak

Using the basic continuous source diffractometer as an example, once the form of the peak profile has been chosen, the next step is to fit this to the data in order to extract the essential peak parameters. This is a well-defined problem in the field of statistics, and usually involves the method of *nonlinear least squares* or *linear regression*, described in many textbooks on the statistical analysis of experimental data. Most software codes are based on the so-called Levenberg–Marquardt method [15], which has become the standard in the field. A detailed description is found in *Numerical Recipes* [6], which also

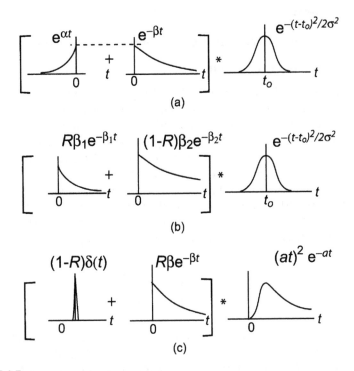

FIGURE 4.7
Schematic representation of three functions used to fit diffraction peak profiles on a spallation source diffraction instrument. The asterisk denotes a convolution: (a) Jorgensen [10]; (b) Carpenter et al. [11]; and (c) Carpenter and Ikeda [12]. (Based on Carpenter and Yelon [13].)

includes written procedures and instructions for incorporating them into a curve-fitting computer code.

In fitting the theoretical profile $F(x_i)$ to the measured counts $C(x_i)$ at each of n angles x_i, the function χ^2, defined by

$$\chi^2 = \sum_{i=1}^{n} \frac{\left[F(x_i) - C(x_i)\right]^2}{[\Delta C(x_i)]^2} \tag{4.8}$$

is usually minimized by varying the parameters in the function $F(x_i)$ in order to obtain their optimum values. $\Delta C(x_i)$ is the uncertainty in the measured value $C(x_i)$. Clearly, if the fit is such that each calculated intensity is equal to the measured value within the expected uncertainty, the numerator and denominator at each point will on average be equal to unity, and the summation will equal the number of points on the scan, $\chi^2 = n$. Indeed, χ^2 is often defined with a prefactor $1/n$ multiplying the summation, so that $\chi^2 = 1$ for the statistically best fit. An alternative prefactor sometimes used is $1/(n - m)$, where m is the number of parameters fitted.

4.4 Accuracy of Diffraction Peak Center Determination

4.4.1 Accuracy of Fitted Parameters

Following the above discussion of profile fitting, we can now consider uncertainty in the fitted parameters, and the optimization of a scan through a peak to give a desired accuracy. We first consider the minimum total number of counts required for determination of the peak center to a desired accuracy. This makes possible estimation of the required counting time, t_M, and thus prevents counting for longer than necessary.

As we have seen, in a wide range of cases a Gaussian profile superimposed on a flat background is often an acceptable approximation for a continuous source diffractometer. We therefore describe the theoretical intensity count at an angle x_i as given by Equation (4.3), which we may write in the form

$$F(x_i) = \frac{I_c \Delta x}{u_x \sqrt{2\pi}} \exp\left(-\frac{(x_i - x_0)^2}{2u_x^2}\right) + B_0 \tag{4.9}$$

This function has the advantage that the peak can be defined by four parameters: height $H_0 = I_c \Delta x / [u_x (2\pi)^{1/2}]$, standard deviation (u_x), peak center (x_0), and background (B_0). Also of importance is the measurement time to acquire the peak profile, t_M; the integrated count, I_C; the number of points measured, n; and their spacing, Δx.

4.4.1.1 Uncertainty in Case of Negligible Background

Defining a "goodness-of-fit" function χ^2 as in Equation (4.8), the maximum probability that is the "best fit" is obtained when χ^2 is a minimum, that is when $\partial \chi^2 / \partial x_0 = 0$, and strictly also $\partial \chi^2 / \partial I = 0$, $\partial \chi^2 / \partial u_x = 0$, and $\partial \chi^2 / \partial B_0 = 0$. The uncertainty in the best fit can be determined by considering the spread of probability around this optimal point. The approach is given in full in Withers et al. [4]; here we give only the basic results. Provided the Gaussian peak shape model is appropriate, they found that the center position of the peak is not strongly correlated with the width. Three assumptions were made: (a) the detected signal counts follow Poisson-type statistics, that is, the uncertainty of a recorded count is equal to the square root of its magnitude, true for fully statistical counting events; (b) there are sufficient points n present within the peak to allow the summation in Equation (4.5) to be approximated by an integral; and (c) that the background is negligible compared to the signal ($H_0 \gg B_0$). In this case, we obtain the expression for the uncertainty $\Delta_u(x_0)$ in fitted peak position x_0:

$$\Delta_u(x_0) \approx \frac{u_x}{\sqrt{I_c}} \tag{4.10}$$

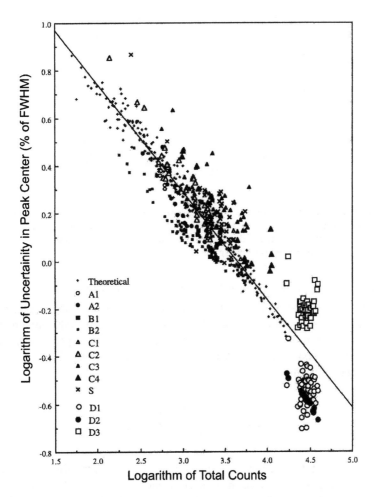

FIGURE 4.8

Collection of data from many different published strain measurements, showing the logarithm of the ratio of uncertainty in the peak position to the peak width *versus* the logarithm of the total number of counts in a diffraction peak [16].

This expression has been used in defining a figure of merit for instrument design and optimization, discussed in Section 4.4.2. It is interesting to note that an empirical approximate form of this relation has also been determined by Webster and Kang [16], who plotted the uncertainty in measured peak position, as a percentage of the FWHM, against peak intensity, for a large body of data, as shown in Figure 4.8. Their graphical results can be fitted by the relationship:

$$\Delta_u(x_0) = 1.05 \; u_x/(I_C)^{0.45} \qquad (4.11)$$

where I_C is the integrated count in the peak, a result that is in very close agreement with the analytical form.

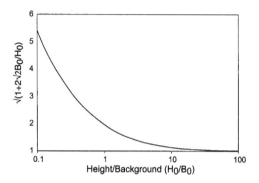

FIGURE 4.9
The increase in peak position uncertainty, $\Delta_u(x_0)$, predicted by Equation (4.12), normalized by that expected on the basis of zero background level, given by Equation (4.10), as a function of height-to-background ratio. The ordinate expresses the penalty factor incurred as increased uncertainty for increasing levels of background.

4.4.1.2 Effects of Background on Strain Uncertainty

In the above discussion, the background B_0 was assumed negligible. However, when a significant background is present, one commonly must acquire a larger signal count to achieve the same uncertainty in peak position as when there is no background. Withers et al. [4] have shown that in this case, the uncertainty is modified:

$$\left[\Delta_u(x_0)\right]^2 \approx \frac{u_x^2}{I_C}\left(1+2\sqrt{2}\,\frac{B_0}{H_0}\right) \tag{4.12}$$

It is clear that the accuracy expected on the basis of the signal alone (Equation (4.10)) is overly optimistic by a factor of $(1 + 2\sqrt{2}\,B_0/H_0)^{1/2}$. This penalty of greater uncertainty in peak position with decreasing peak height to background is shown in Figure 4.9. To achieve a prescribed uncertainty in the peak position, one must counteract this with longer count times to increase I_C. For example, a signal-to-background ratio H_0/B_0 of 1 gives approximately twice the uncertainty of that with no background, and would require an acquisition time four times longer to achieve the same accuracy. Given the rapidly increasing uncertainty in peak position for a given acquisition time, measurements for which $H_0/B_0 < 1$ can normally be regarded as unfeasible.

As part of an empirical study, Stoica et al. [17] independently developed a very similar formulation:

$$\left[\Delta_u(x_0)\right]^2 \approx \frac{u_x^2}{I_C}\left(1+3\,\frac{B_0}{H_0}\right) \tag{4.13}$$

This is in very good agreement with the analytical approach. Similar analytical expressions have been derived for the uncertainty in peak integral counts, $\Delta_u(I_C)$, and in peak width, $\Delta_u(u_x)$, taking a scan width of $\pm 2.5u_x$ [4]:

$$\left[\Delta_u(I_C)\right]^2 \approx I_C\left(1+\sqrt{2}\,\frac{B_0}{H_0}\right) \tag{4.14}$$

$$\left[\Delta_u(u_x)\right]^2 \approx \frac{u_x^2}{3I_C}\left(1+4\sqrt{2}\,\frac{B_0}{H_0}\right) \tag{4.15}$$

The above analytical formulations express succinctly the uncertainty penalty of fitted parameters that is paid when the background level becomes significant compared to the peak height. The analytical approach is only strictly valid as B_0/H_0 approaches 0 or ∞. For more general ratios, Monte Carlo simulations have therefore been used; these study the effect of background level on the uncertainties in fitting a peak profile of fixed total counts I_C [14]. The background level was varied systematically for a simulated peak profile with amplitude $H_0 = 1 \times 10^5$ counts and width given by $u_x = 1.0$, $\Delta x_i = 0.1$, so that $I_C = 2.5 \times 10^6$ counts. Four Gaussian profile parameters (x_0, H_0, B_0, u_x) were fitted to the simulated counts, $C(x_i)$, which varied with Poisson uncertainties at each angle x_i. The results are shown in Figure 4.10, where the determined uncertainty in peak position, intensity, and standard deviation is divided by that calculated from Equations (4.12), (4.14), and (4.15), and plotted against the height-to-background ratio (H_0/B_0). The deviations from the dashed lines in Figure 4.10 show how the analytic expressions are in error by only 8% (at $H_0/B_0 \approx 5$), 3% (at $H_0/B_0 \approx 2$), and 14% (at $H_0/B_0 \approx 9$) for the peak position, intensity, and standard deviation, respectively.

The expressions given above cannot be expected to be correct if the peak profile assumed is not an accurate description of the data, since some element of the uncertainty will then also arise from the incorrect profile [6]. Nevertheless, using the simple Gaussian formulation, Equation (4.9), good agreement is observed between the expression given in Equation (4.13) and data from a round-robin exercise on strain measurement for a ceramic composite. This is seen in Figure 4.11, where the effect of the background "penalty factor" is clear at low signal-to-background ratios. It is important to note that this increase in uncertainty in peak position arises because of the uncertainty in background count at *each point*, even if the average background level is well known. In other words, increasing the width of the scan to include a large section of the background profile will not improve peak position uncertainty. Even if the profile is not Gaussian, it appears that as long as the "correct" peak profile is used in the fitting procedure, the above expressions are still likely to be useful. This has been shown to be true for an arbitrary, not necessarily symmetric, peak profile in the case of negligible background [18].

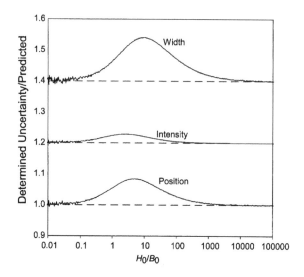

FIGURE 4.10
The results of a four-parameter fit of a Gaussian profile to a large series of Monte Carlo peak simulations made with constant $I_c = 2.5 \times 10^6$, with $H_0 = 1 \times 10^5$, $u_x = 1.0$, and step size $\Delta x_i = 0.1$, and various background levels B_0. The resulting uncertainty in peak angle, normalized to that calculated from Equation (4.12), is plotted as a solid line as a function of signal to background ratio (H_0/B_0). Also shown are the variations in expected intensity and peak width normalized respectively by Equations (4.14) and (4.15), displaced upward by 0.2 and 0.4 [4].

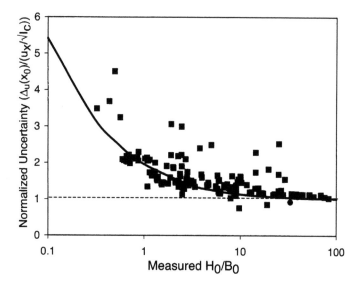

FIGURE 4.11
The measured uncertainty in the determination of peak angle, or peak position in time, normalized by $(u_x/\sqrt{I_c})$ for the two phases of an Al_2O_3/SiC nanocomposite sample plotted against measured peak height to background. The data points were acquired from many neutron facilities around the world as part of the VAMAS TWA20 standardization exercise. As in Figure 4.9, the curve represents the function $(1 + 2\sqrt{2}B_0/H_0)$ [4].

4.4.1.3 Systematic Uncertainties

A vital part of any exercise in fitting a theoretical profile to experimental data, whether it is a single peak or a diffraction pattern, is to take great care to judge the validity of the results obtained. Even if it is felt that there is an acceptably small uncertainty in the fitted peak centers, the strain measurement may not necessarily be accurate, since many possible systematic effects, discussed in Chapter 3, may affect the interpretation. These include instrument resolution, nonhomogeneous materials properties, an off-center centroid of the sampled gauge volume due to attenuation or texture effects, or a partially filled instrument gauge volume and uncertainties in positioning of the sample — especially in the case of steep strain gradients. It is therefore advisable to test for systematic uncertainties by empirical means, measuring the strain at the same position using different configurations of the sample and instrument.

4.4.2 Time Required to Achieve a Given Accuracy in Peak Center Measurement

Results of the above discussion of strain measurement accuracy can be used to assess the time required, t_M, to achieve a specified uncertainty, $\Delta_u(x_0)$, in the derived peak center position, x_0. From Equation (4.12),

$$t_M = \frac{u_x^2}{\left(\Delta_u(x_0)\right)^2 I_C}\left(1 + 2\sqrt{2}B_0 / H_0\right) \tag{4.16}$$

where I_C, H_0, and B_0 are the rates at which total count, peak height, and background are acquired, respectively; that is, $I_C = I_C t_M$, $H_0 = H_0 t_M$, and $B_0 = B_0 t_M$. This gives the well-known result that the acquisition time is inversely proportional to the square of the required uncertainty. In other words, it takes four times as long to achieve twice the accuracy in angle or strain. The equation also highlights the need to minimize the background, especially when the signal intensity is low, as for example, when making measurements at positions deep within a component [19].

In order to compare the performance for strain measurement of different instruments, it is useful to define a Figure of Merit (FoM), as the "inverse of the time taken to measure a lattice spacing to a given accuracy" [18]. From Equation (4.16),

$$FoM = 1 / t_M = \frac{I_C\left[\Delta_u(x_0)\right]^2}{u_x^2\left(1 + 2\sqrt{2}B_0 / H_0\right)} \tag{4.17}$$

This is of more widespread applicability than the simpler form given in Section 3.4.1, which is only valid for low background levels compared to the peak height. This FoM should be maximized when designing an instrument, and is especially important when considering an instrument for making measurements of strain at positions very deep within a sample, because at large depths the background level will be comparable to the signal. In such cases, an instrument with inherently low background will have a high FoM. For more general peak shapes, a proportionality constant can be introduced in the denominator. Of course, the peak width has both an instrument- and sample-dependent contribution, but this formulation reminds us that an efficient instrument for strain measurement should combine high detected count rate (I_C), narrow instrument peak width (u_x), and low background count rate (B_0). A high detected count rate requires a high incident flux combined with a large solid angle of high-efficiency detectors.

4.4.2.1 Measuring at Positions Deep within a Sample

As the path length through the sample increases, such as when making measurements at increasing depths in reflection geometry, beam attenuation causes an exponential decrease in the diffracted intensity (see Figure 2.11). This means that the time to acquire a given integrated peak count (I_C) increases exponentially with depth. As a result, when making measurements at depth it is normally assumed that as the effective position sampled moves deeper into the sample, the measurement time must be increased exponentially to offset the increase in attenuation. However, this is oversimplistic since it does not take into account the background signal (B_o) that Equation (4.16) shows must also be considered when estimating the acquisition time required to achieve a given strain uncertainty. For measurements made at positions deep in the sample where the signal per unit time may be low, the penalty factor $(1 + 2\sqrt{2}\,B_0/H_0)$ becomes increasingly large. Consequently, it is often the background level that limits the depth to which practical measurements can be made and, as mentioned above, characterizes a good instrument for measuring strain.

This practical limitation to the depth of the measurement position is now considered further with reference to measurements made on a 4.5-mm thick Ni powder sample at increasing depths, as it is translated through a small instrumental gauge volume. Figure 4.12 shows the variation in the log of the count rate for the integrated peak counts (I_C), peak height (H_0), and background level (B_0) as the sample is moved through the instrument gauge volume in reflection geometry. As expected, there is an exponential decay of count rate with increasing depth. In this case, the background count rate also decreases essentially exponentially. This is often the case, as noted by Withers [19], who has modeled the variation in background level as comprising: a constant instrument component B_{inst}, which arises even when there is no sample present, and several sample dependent contributions B_{sample}. Taking into account the decrease in both the diffraction peak count rate and

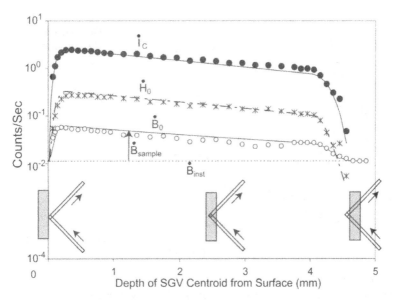

FIGURE 4.12
The variation, with depth of the sampled position, of the count rate for the integrated Ni (311) count, I_c, the peak height, H_0, and the background, B_0, as a 4.5-mm thick Ni powder sample is translated in reflection geometry through a $0.5 \times 0.5 \times 5\text{mm}^3$ instrumental gauge volume. Corrections have been made for surface effects as described in Section 3.6. The measurements were made using the Chalk River, Canada, E3 strain instrument, with $\phi^s = 92.9°$. (Courtesy of P.J. Webster and P. Browne.) Instrumental and sample contributions to the total background level B_0 are distinguished (based on Withers [19]).

the background count rate with increasing depth it is possible to use Equation (4.16) to calculate the time required to achieve a stated strain measurement accuracy as a function of depth. For the data shown in Figure 4.12, the time required to attain a strain accuracy of 10^{-4} as a function of depth is shown in Figure 4.13. In the latter case the sampled gauge volume considered is $1 \times 1 \times 40 \times = 40 \text{ mm}^3$. As one would expect, in the absence of background, the time would increase exponentially. On the other hand, if the background level were to remain constant, equal to that encountered near the surface, the time required would soon become prohibitive. However, the time for a decreasing background, which is commonly observed experimentally, is parallel to the zero background curve for positions near the surface because the background rate falls off almost exponentially in proportion to the counts in the peak (Figure 4.12). However, at large depths the flat background component dominates; consequently, the response becomes parallel to the constant background curve.

It is clear from Figure 4.13 that at some point it becomes economically impractical to probe any deeper into the sample. Few scientists or engineers would consider it feasible to count for more than 12 h on a single point, and from Figure 4.13 this places the maximum depth of the position of measurement at about 9.8 mm for a gauge volume of 40 mm³. This introduces the

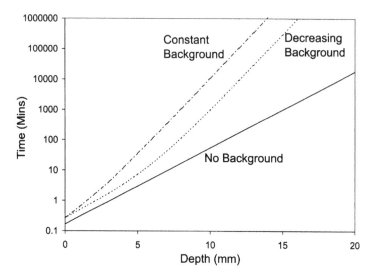

FIGURE 4.13

The time needed to attain a 10^{-4} strain accuracy at Chalk River for Ni measured in reflection geometry using data extracted from Figure 4.12, but with a $1 \times 1 \times 40$ mm³ sampled gauge volume. Predictions are shown for the cases of no background, a background level equal to that encountered near the surface, and one that decreases with depth.

concept of the *economically feasible path length* of the neutron in the material, including both incident and diffracted beams, which here is ~28 mm, as a useful benchmark. As the path length for $\phi_s = 90°$ is $2\sqrt{2}$ times the depth for the reflection configuration, this governs the maximum depth at which meaningful measurements can be made on a given instrument for a given material and scattering angle. If times longer than 1 h are unacceptable, then the maximum depth would be 7.5 mm for this gauge volume. Withers [19] has suggested that a peak height-to-background ratio (H_0/B_0) of 1 might be a more appropriate criterion than a time-based one, because beyond this ratio the penalty factor in Equation (4.16) becomes prohibitive. Based on the same incident flux and instrument setup as that used to acquire the data for Ni powder in Figure 4.12, it is possible to estimate the count rate for the diffraction peak and background with increasing depth for any material [20]. The diffraction peak height and background count rates for three engineering materials are shown in Figure 4.14 for a scattering angle 90°. Note that for the Chalk River instrument used, the exponentially falling sample-dependent component of the background is not very important in determining the maximum feasible depth, that is, the point at which $B_0/H_0 = 1$. In such a case, the depth capability of the instrument could be improved by lowering the instrument background, although in this example only with difficulty as the instrument is saturated on a beam tube which passes through the reactor core.

Using the above maximum allowable time and peak-to-background ratio criteria, some economically feasible path lengths for 1–8Å neutrons in a range

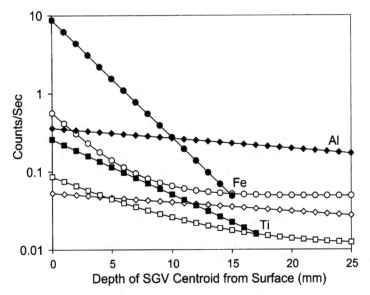

FIGURE 4.14

The predicted variation in count rates for the diffraction peak height (solid symbols) and the background level (open symbols) with increasing depth penetration of the sampled gauge volume for a range of engineering materials, based on Withers [20]. The predictions are based on $\phi^s = 90°$, and a sampling gauge of $1 \times 1 \times 40$ mm³ for the Al(311), Ti(100), and Fe(211) peaks, with incident flux and instrumental background representative of the Chalk River, Canada, E3 instrument for strain measurement.

TABLE 4.2

Maximum Economically Feasible Path Lengths for Selected Common Engineering Materials (mm)

Criterion	Al	Ti	Fe	Ni	Cu
$H_0/B_0 = 1$	185	24	43	21	39
12 h	297	48	52	27	51
1 h	160	22	40	20	36
Webster	237	23	37	21	40

Note: These are based on a gauge volume of 40mm³, 90° scattering angle, 10^{-4} accuracy, and are calculated either for a height to background ratio of 1, or for a 12-h or 1-h measurement time [20]. These are compared with Webster's rule-of-thumb estimate [21].

of engineering materials are given in Table 4.2 [19]. Those for Fe, Al, and Ti can be read off the graph in Figure 4.14 using the B_0/H_0 criterion. The 1 hour values in Table 4.2 are in good agreement with a largely empirical relation for the economically feasible path length (x) obtained from the expression,

$$\bar{b}^2 e^{-\mu_a x}\left[\sigma_{coh}/\left(\sigma_{coh}+\sigma_{incoh}\right)\right]=1$$

derived by Webster et al. [21]. Here μ_a is the linear attenuation length. Note that the high incoherent scattering component characteristic of Ti compromises the maximum depth at 1 h because of the large background rate, but not at the larger depths characteristic of the 12-h and B_0/H_0 criteria, because the sample dependent background has decayed such that it is only a minor contribution to the total background count at such depths. While in practice, the maximum path length will depend on the instrument, maximum allowable gauge volume, required strain accuracy, and a number of other factors, the exponential decrease in intensity with depth means that their effects are relatively weak, and so the values in Table 4.2 provide a useful indication of what is currently achievable by neutron diffraction.

4.4.3   Analytical and Empirical Studies of Scan Optimization

From the previous sections, it is clear that determining the best estimate of a peak's angular center from the fitting procedure and hence deriving the lattice strain are not always trivial matters. It is important to evaluate the sensitivity of the entire fitting procedure to the choices made, and the extent to which this sensitivity is of concern when the data are converted to lattice strains. Windsor [22], with the use of an analytical approach, and Brand [23], by using numerical methods, have considered the dependence of strain accuracy obtained by fitting a peak center on the main scan parameters for the case when a fixed time has been allocated for performing a peak scan. Webster and Kang [16] have studied the problem by trial-fitting techniques. We here summarize the latter authors' results.

4.4.3.1   Range of Scan

Webster and Kang [16], assuming a Gaussian peak profile, recommend that for good determination of a peak center the peak should be measured over a range of around seven times the standard deviation, u_x, that is a measurement range of $\pm 3.5 \, u_x$ about the center of the diffraction peak or, equivalently, about three times the peak FWHM. They also suggest that around 20 points should be measured over this range. For a single detector that is scanned in angle, the larger the range the longer the acquisition time, so that there is some advantage in reducing the range somewhat, provided that the data are sufficient to confirm that a flat background is present. The fractions of the total integrated intensity, I, under a Gaussian profile within various ranges of angle x_i are summarized in Table 4.3. Because a significant fraction of the signal lies within $\pm 2.5 u_x$, a range of this size may be sufficient, provided that the background is a simple flat level. For detectors that record the whole diffraction peak at once, such as position-sensitive detectors using monochromatic sources and time-sensitive detectors using polychromatic beams, there is no time penalty for recording over a larger range.

TABLE 4.3

Fraction of Total Signal Intensity I in Gaussian Profile Lying Within Various Angular Measurement Ranges

Range	$\pm 1u_x$	$\pm 1.5u_x$	$\pm 2u_x$	$\pm 2.5u_x$	$\pm 3u_x$	$\pm 3.5u_x$
Fraction within range	0.682	0.866	0.954	0.988	0.997	0.999

TABLE 4.4

Numerical Results of Fitting Gaussian Profile Function with Various Background Functions to Measured Bragg Peak Superimposed on Nonuniform Background Shown in Figure 4.15

Background Function	Peak Center (deg.)	Uncertainty (deg.)
Flat	95.6794	±0.0491
Linear	95.6726	±0.0281
Parabola	95.6510	±0.0181

4.4.3.2 Fitting a Sloping Background

The sensitivity of the fitted peak center to a nonuniform background is exemplified by fitting a Gaussian with a range of plausible background functions to the same experimental data set, as shown in Figure 4.15. Note the large changes in fitted peak position given in Table 4.4 and the associated uncertainties. Visually, from the graphs presented in Figure 4.15, all choices appear to render acceptable fits to the data, but some marked differences are observed in the fitted peak centers with uncertainties (Table 4.4). It is important to bear in mind that the calculated uncertainties in the peak centers reflect how well the fitting function resembles the experimental data, as well as the statistical quality of the data set. Clearly, the strains reported from a diffraction experiment are never more accurate than the statistical accuracy by which one is able to calculate characteristic features of the peaks such as peak center angles. From the numerical results of Table 4.4, we note that the uncertainties, representing one standard deviation, do in fact overlap, so that normal statistical analysis would claim a rather high likelihood that the three positions found are correct within the experimental accuracy. However, the scientist analyzing the data will have to pick one approach to describe the background. The differences in the fitted angles may not seem large, but note that at $\phi^s \approx 90°$ an absolute difference of more than $0.03°$ in ϕ^s corresponds to a difference in strain on the order of 250×10^{-6}. This far exceeds the typical ideal accuracy aimed for, which is of the order of $\pm 100 \times 10^{-6}$.

The correlation between various fitting parameters and individual uncertainties are equally important. These correlations are calculated from the off-diagonal elements of the inverse variance–covariance matrix, and in general, high correlations will simply add to the uncertainties in the parameters concerned. For example, correlations between two parameters will give a

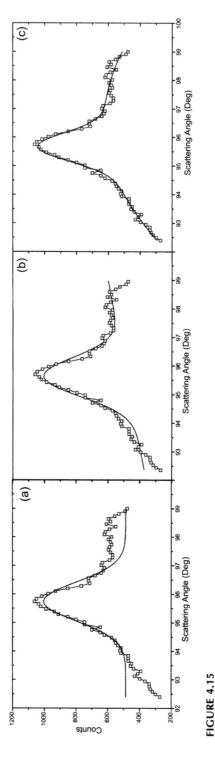

FIGURE 4.15

Examples of fitting a measured peak profile with a Gaussian function using different descriptions of the observed irregular background: (a) flat, (b) linear, and (c) parabolic.

range of possible values that may be represented by an ellipse in parameter space.

As an example of the effect of correlated parameters, consider the cases in Figure 4.15 with either a flat or linear sloping background. It is clear that the latter will gives a better fit to the data. However, in that case, the uncertainty in the peak center is strongly linked to the uncertainty in the slope of the background. This is because as the slope of the background is altered, the peak center angle adjusts proportionately to obtain the best fit. This means that if the background is poorly known, it may increase the scatter in the fitted peak center. Similarly, if the peak is slightly asymmetric, and the background not sufficiently defined outside the range of the peak, changes in asymmetry may be fitted by changes in the slope of the background. This can lead to a low statistical uncertainty in the peak position, but larger point-to-point variations in the inferred strain due to the combined effect of coupled background gradient and peak positional errors. In such cases, a flat background may be preferable since there is then no coupling between the peak center and the background parameters, and so the variations in the determined peak center will be less affected by the background. As a result, the point-to-point scatter in the strain may in fact be less, despite a larger quoted statistical uncertainty in the peak position. An alternative strategy to counter this problem is to fit all data within a scan using a sloping background of fixed gradient.

An alternative approach to improving the fit to the profile would be to fit more than one peak while keeping the background gradient fixed. However, it is important to remember that the aim is not to achieve the best fit *per se*, since this could be achieved by using tens of fitting parameters, but to obtain the most reliable estimates of strain, or less commonly, peak width or integrated intensity. It is rare for an accurate representation of the background to be an end in itself. The art is to get an acceptable fit using the least number of fitted parameters. It should also be kept in mind that strain requires the subtraction of two peak angle positions. Thus, if both the sample under examination and the stress-free sample are fitted in the same manner, the shift in angle can be determined to greater accuracy than the individual peak angles.

A general rule of thumb, which under normal conditions has proven to render good-quality strain measurements, is to aim that uncertainties in the ϕ^s are of the order of 0.01° or better, for the situation when $\phi^s \sim 90°$. This corresponds to $\sim 100 \times 10^{-6}$ uncertainty in strain. Using this guideline, none of the profile fits in these examples has sufficient statistical accuracy, and the measurement should probably be repeated using a longer acquisition period to improve the statistics!

4.4.3.3 *Weighting*

In order to find the correct statistical fit to the data, it is very important to use the correct uncertainties that weight each data point term in Equation (4.8).

Many commercially available peak fitting routines automatically give each data point equal weighting irrespective of the number of counts. The correct statistical error on a count of $C(x_i)$ is $\Delta C(x_i) = \sqrt{C(x_i)}$. Failure to correctly assign these individual weights can result in calculated uncertainties in peak position that are significantly below the theoretical limit given by Equation (4.12). In judging the validity of reported uncertainties in derived peak positions, and hence strain, the algorithm used to define the weights must therefore be carefully assessed.

4.4.4 VAMAS Recommendations on Peak Fitting

The VAMAS Technology Trends Assessment document [1] offers some practical advice and warnings in situations where fitting a profile to the measured peak is difficult. These are summarized here to supplement to the points already made in the previous sections. The following cases were considered:

  Poor Peak to Background Ratio: If the ratio of the peak count, with background subtracted, to the background is less than two ($H_0/B_0 < 2$), it may prove difficult to determine the diffraction peak position (Figure 4.9). Note that in Section 4.4.2, we suggested that in most cases prohibitively long acquisition times are required to obtain good strain accuracy for $H_0/B_0 < 1$.

  Overlapping Peaks: Fitting overlapping peak profiles should be avoided wherever possible. However, in the study of multiphase materials, overlapping peaks are sometimes unavoidable. Within the VAMAS program, a study was undertaken to look at the effect of a contaminant peak (Peak 2) on a major Gaussian peak (Peak 1) having center $\phi^s_{01} = 2\theta_{01} = 90°$, $FWHM_1 = 0.5°$, $H_{01} = 1000$, $B_{01} = 200$, and $\Delta x = 0.1°$. Three different fitting strategies, described below, were examined. In order to follow this analysis, two cases are considered in which the contaminant (Gaussian) peak has $H_{02}/H_{01} = 1$ and $FWHM_2/FWHM_1 = 1$.

  Case I: The contaminant peak is centered at $\phi^s_{02} = 2\theta_{02}$, at $1.8 \times FWHM_1$ from the major peak center, as plotted in Figure 4.16a and Figure 4.17a.

  Case II: The contaminant peak is centered at $\phi^s_{02} = 2\theta_{02}$ at $1.3 \times FWHM_1$ from the major peak center as plotted in Figure 4.16b and Figure 4.17b.

   (i) *Using a single Gaussian peak to the whole profile of Peak* 1. Taking an uncertainty in diffraction peak position corresponding to no more than 100×10^{-6} in strain as acceptable, it is possible to evaluate the conditions under which a satisfactory peak fit is obtained. The condition under which this is achieved for the

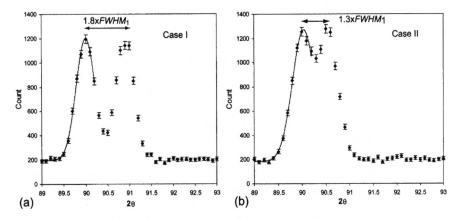

FIGURE 4.16
Truncated peak fits to the data conforming to (a) Case I and (b) Case II based on the primary peak parameters given in the text. In both cases, $H_{02}/H_{01} = 1$, and $FWHM_2/FWHM_1 = 1$; Case II lies outside the acceptable limit defined in Table 4.5.

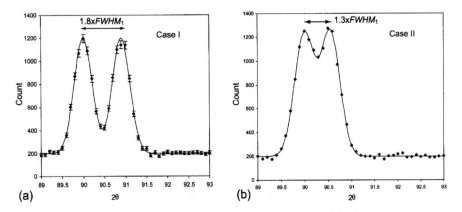

FIGURE 4.17
Two Gaussian peak fits to the data conforming to (a) Case I and (b) Case II based on the primary peak parameters given in the text. In both cases, $H_{02}/H_{01} = 1$, and $FWHM_2/FWHM_1 = 1$.

above conditions is almost independent of H_{02}/H_{01} (in the range 0.1 to 1.0) and can be approximated by

$$(2\theta_{02} - 2\theta_{01}) > (FWHM_1 + 0.8\ FWHM_2) \qquad (4.18)$$

This means that an error of around 100×10^{-6} is expected for a separation of $1.8 \times FWHM_1$, making Case I just acceptable, with a much larger uncertainty in Case II, of $\sim 1000 \times 10^{-6}$. This analysis would suggest that using a single peak is only a good strategy when the two peaks have almost no overlap.

(*ii*) *Truncated Peak Fit.* The effect of an overlapping peak can be minimized by performing a single peak fit to an incomplete profile of the major peak of interest, in which the side that is influenced by the nuisance peak is truncated, as shown in Figure 4.16. A study looking at the effect of the contaminant peak (Peak 2) on the primary peak (Peak 1) having $\phi^s_{01} = 2\theta_{01}$ has indicated that the optimum "cut-off" point for a truncated peak is $0.5 \times FWHM_1$ from the Peak 1 center position on the contaminated side. The feasibility of this method depends on the peak separation, width and intensity ratios of the two peaks. Table 4.5 provides a guideline for the case where a fit to a truncated peak gives strains within 100×10^{-6} of the "true" values, and indicates that Case I is acceptable and Case II unacceptable. In Figure 4.16a, the uncertainty in the fitted ϕ^s_{01} is $\sim 80 \times 10^{-6}$, while in Figure 4.16b, it is $\sim 130 \times 10^{-6}$.

(*iii*) *Double Peak Fit.* A profile comprising two peaks is fitted to the whole range of measurement over the two overlapping peaks as shown in Figure 4.17. Use of such a profile leads to a large number of parameters to be fitted, and a consequent large increase in their collective uncertainties. In this case, peak fitting can achieve a satisfactory strain error for peak position for both Figure 4.17a of $\sim 40 \times 10^{-6}$, and Figure 4.17b of $\sim 70 \times 10^{-6}$. However, since the errors in peak position are coupled to each other and to other fit parameters, such uncertainty can only be achieved if all the other peak fit parameters are held constant. In general, uncertainties become large for peak ratios $H_{02}/H_{01} < 1$ when $(2\theta_{02} - 2\theta_{01})/FWHM_1 < 1$. In certain cases, some of the fitted parameters can be fixed or related to each other. For instance, if the volume fraction of a two-phase material is known, the intensity ratio of the two peaks can be calculated. It may also be valid to apply constraints to some of the parameters; this can significantly improve the stability of the fits.

Asymmetric Peak Profiles. The asymmetric peaks characteristic of spallation source instruments are discussed in Section 4.3.1. However, asymmetry in a peak profile may also be found on a continuous source instrument. It may arise due to sample-dependent effects such as nonhomogeneities and stacking faults, as well as instrument effects. Unless the origin and behavior of this asymmetry is fully understood, the use of asymmetric peaks should be avoided for strain measurements because the degree of asymmetry and the peak position are strongly coupled.

TABLE 4.5

Guideline to *Limits of Applicability* of Truncated Peak Fit to Major Peak (Peak 1), Contaminated by Minor Peak (Peak 2)

H_{02}/H_{01}	$FWHM_2/$ $FWHM_1$	$(2\theta_{02} - 2\theta_{01})/$ $FWHM_1$	H_{02}/H_{01}	$FWHM_2/$ $FWHM_1$	$(2\theta_{02} - 2\theta_{01})/$ $FWHM_1$
0.1	0.1	0.5	0.3	0.1	0.5
0.1	0.5	0.5	0.3	0.5	0.5
0.1	1.0	0.5	0.3	1.0	0.5
0.5	0.1	0.5	1.0	0.1	0.7
0.5	0.5	1.0	1.0	0.5	1.0
0.5	1.0	1.5	1.0	1.0	1.5

Note: The parameters fitted are discussed in the text: H_{0i}, $FWHM_i$ and $\phi^s_{0i} = 2\theta_{0i}$, denote the peak amplitude, width, and peak center position, respectively, with subscripts 1 and 2 referring to the two peaks. For height ratios H_{02}/H_{01} smaller than those given, or peak separation $(2\theta_{02} - 2\theta_{01})/FWHM_1$ closer than given, the fitted Peak 1 angle deviates from the true value by more than an amount equivalent to a strain of 100×10^{-6}. (Courtesy J.W.L. Pang and P.J. Withers.)

4.4.4.1 Strategy of Peak Fitting

It must be emphasized that it is crucial to consider carefully the strategy to be adopted *prior* to undertaking lengthy fitting analysis. While the actual peak centers deduced may differ according to the approach taken, if the strategy is sensible and is the same for both sample and stress-free peaks, then the strains obtained should be correct within an acceptable error. Naturally, for well-defined diffraction peaks on a regular background, the issue is less important than for weaker peaks on a poorly defined background. In developing a strategy, the following activities are helpful:

- First, undertake trial fitting procedures on a limited data set using a range of fitting strategies, and then keep to a consistent plan throughout the whole analysis.

- Perform a sensitivity analysis to ascertain how sensitive the derived peak centers are to the selection of the fitting function and its free parameters.

- Use the same fitting function and number of free parameters for all diffraction peaks of a data set, including the strain-free reference sample, as far as possible.

- Aim to obtain calculated uncertainties in the fitted peak centers of ~0.01° or better, for $\phi^s = 90°$. This corresponds to strain uncertainty of ~100×10^{-6}.

- Validate uncertainties in peak centers obtained by checking the reproducibility of measuring and fitting a selected peak.

- Visually inspect all peak fits and objectively evaluate how well the fits resemble the data. Note any sudden or progressive alteration in the uncertainties on all freely fitted parameters. This should help identify if, when, and why the fits to peaks become less trustworthy.

- Check the correlation matrix off-diagonal components, which if they approach unity, indicate a high correlation between fitted parameters. If a parameter correlates strongly with the peak center position, consider using a fixed value for this parameter for all fits.

4.4.5   Sources of Background Counts

In many practical cases, appreciable background contributions to the measured counts are unavoidable, especially when undertaking measurements at positions deep within the sample. It should be noted that there will always be a continuous background originating from incoherent elastic scattering and inelastic scattering from the sample, and for elements with large coherent cross-sections, multiple scattering as the beam passes through the sample will contribute (Section 2.6). In addition, there will be a thermal and fast neutron background from the instrument and surroundings, and air scattering from the neutron beam path. If the discrimination in the detector is not perfect γ-rays will be detected, and there may be inherent electrical noise from the amplifiers. As we have seen in the previous section, there is always the possibility of anomalous shifts in Bragg angles from uneven background contributions, stray beams, or stray background, and such effects should always be tested for.

4.5   Analysis of Complete Diffraction Profiles for Strain

In the discussions of accuracy above, we have concentrated on the analysis of single diffraction peaks on a continuous source instrument. When a more complete diffraction pattern has been measured, for example, on a spallation source time of flight instrument, it is more usual to fit the whole diffraction pattern using a Rietveld or Pawley–Rietveld approach mentioned in Section 3.2.2. Consequently, there has not been much work carried out on the accuracy of individual time-of-flight peaks. It is possible to fit the instrument-dependent parameters — such as the leading and trailing moderator-dependent exponentials in Carpenter's [9] expression — globally for all peaks, and the plane-specific ones such as peak position expressed in time of flight and if necessary peak intensity, on a peak-by-peak basis.

Several computer codes for performing Rietveld refinement of diffraction patterns are available free of charge. Among the most common ones are the Generalized Structure Analysis System (GSAS) by Larson and Von Dreele [25], the DBWS code by Wiles and Young [26], and the RIETAN code by Izumi et al. [27]. A more comprehensive list is found in a survey by the IUCr

(Commission on Powder Diffraction) by Smith et al. [28]. For further reading on the Rietveld refinement method, see, for instance, Young [29].

Just as when fitting a fitting a single peak, performing the Rietveld refinement on a diffraction pattern is potentially a delicate matter, where the choices made can have a strong impact on the stresses and strains deduced from the data. The software codes have numerical criteria for finalizing the fitting procedure, but to cite Young [14]: "Numerical criteria are important, but numbers are blind; it is imperative to use graphical criteria also." This, of course, is universally true for any fitting procedure; the experimenter should always visually compare the result of the fitting procedure to the actual data. As stated in Section 4.3.2, the calculated uncertainties in the fitted parameters merely tell one part of the story; to quote Prince [30], "If the fit of the assumed model is not adequate, the precision and accuracy of the parameters cannot be validly assessed by statistical methods."

For the purpose of residual stress determination from a Rietveld analysis, a number of strategies can be chosen. The refinement can be constrained to the known crystal structure, such as face-centered cubic, so that the method combines all the peak information to achieve the most representative average lattice parameter or parameters, such as a for a cubic system. Alternatively, it can be allowed to accommodate the anisotropic variation with hkl which might be expected from any elastic or plastic anisotropy, discussed in Sections 5.5 and 5.6. For cubic materials the average lattice parameter obtained from a Rietveld or Pawley–Rietveld analysis has been shown empirically to be largely independent of intergranular effects, even to several percent plastic strain; that is, it is linear as a function of stress as seen from Figure 4.1. Since the Rietveld fit produces a weighted average of the strain determined from all peaks in the diffraction pattern, one might expect it to be similar to the bulk macroscopic strain measured in that direction. In practice, the elastic modulus thus determined is usually found to be close to the macroscopic Young's modulus. For noncubic materials, it is necessary to identify a suitable parameter that is linear in strain. For example, for hexagonal materials an appropriate linear parameter is $(2\varepsilon_a + \varepsilon_c)/3$, where ε_a and ε_c are the strains determined from the a and c lattice parameters, respectively [31]. The interpretation of such strain data is described further in Sections 5.6.6 and 6.6.1.

4.6   Strain-Free Reference

4.6.1   Requirement

All diffraction-based measurements of internal strain, and hence absolute residual stress, are based on a comparison of a lattice plane spacing measured in the sample, d_{hkl}, with a reference value d_{ref} (d^0_{hkl}), measured in a "stress-free" sample of identical material, as given in Equation (4.1). In the case of a Rietveld refinement of a diffraction pattern, average lattice constants a, b, and c are often measured from many reflections, and these are compared with the "stress-free" values a^0, b^0, and c^0 to give a bulk strain.

Although the reference is usually denoted a "stress-free" d-spacing, this only corresponds to "strain-free" in the full triaxial case, when all three principal values of the macroscopic stress and strain tensors are zero, and the reference value is needed to establish the absolute strain and stress tensors. There are cases when one component of the stress tensor may be zero and "stress-free," but this does not necessarily relate to a corresponding "strain-free" component, as in the case of the out-of-plane strain component at a free surface of a biaxial in-plane stress state (plane stress). The case of plane-strain represents the converse example.

In certain cases, an actual measurement of the reference d-spacing on a zero-stress sample may be unnecessary. If it is known that a component of stress is strictly zero, this may be used to obviate the need for an accurate d^0_{hkl} value. Such a case is the boundary condition that the normal stress component at a free surface must be zero, a condition that is used extensively in laboratory-based x-ray investigations of near-surface stress fields, using the so-called $\sin^2\psi$ technique described in Section 5.1.3. A less strict condition is that the normal stress in a thin plate is approximately zero, as mentioned above. In the general triaxial case, if only the deviatory stress is required without the hydrostatic component, any reference d-spacing can be used. The average of the three diagonal strain components (ε_{11}, ε_{22}, ε_{33}) measured relative to this arbitrary reference can then be subtracted from the diagonal elements of the measured strain tensor to yield the traceless deviatory strain tensor. This is described further in Section 5.1.1.

4.6.2   Other Factors That Can Affect Lattice Spacing

There are many factors that can give rise to a change in the lattice spacing of a material that are not stress related whether on a local or overall scale. Care must be taken that these do not mask or affect the required measurement of true strain in either the sample under investigation or the reference sample. These factors include the following:

Compositional Changes. Changes in composition brought about by the precipitation or dissolution of second phases during heat treatment, such as during welding, or by segregation during casting. These can cause variation in d-spacing with position, which can be an important problem in many practical cases, especially for heat-treatable alloys discussed in Section 6.2.3. An example of the scale of this problem is the effect of copper content on the d-spacing of an aluminum alloy shown in Figure 4.18a. Compared to typical strain-induced changes to the d-spacing, the changes due to varying alloy content are dramatic, and would be detrimental to any accurate

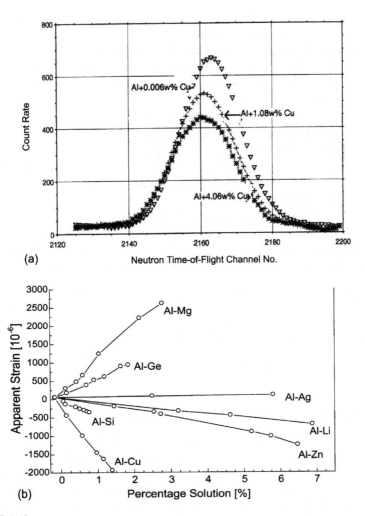

FIGURE 4.18

(a) Typical shifts of the Al 311 Bragg peak in aluminum copper alloys with increasing copper content, indicating that the aluminum lattice shrinks as Cu is added. Measurements were made using the time-of-flight diffractometer at the FRG-I reactor, GKSS Research Center, Geesthacht, Germany [32]. (b) The variation (expressed in microstrain) of the unstrained lattice parameter with solute content for different solutes in aluminum [33].

derivation of strains from measures of *d*-spacing. By way of examples, a change of 0.005 wt% in carbon solute content in austenitic steel induces a change in *a* equivalent to 600 με, while the effects of various solute elements on the lattice spacing of aluminum are summarized in Figure 4.18b.

Phase Transformations. Phase transformations can occur when a sample has undergone large temperature changes, such as during welding, which may cause compositional changes in the form of the presence of different phases that are uneven across the sample. In this case, a global reference sample d^0-spacing cannot be assumed. Reference samples fabricated from a range of positions in a similar weldment may be necessary as outlined below. In other cases of uniform heat treatment, it is essential to ensure that the reference sample has undergone the same treatment.

Changes in Temperature. Any variation in temperature can give rise to an apparent strain, and, if rapid, peak broadening. Typically for metals, a change in temperature of $1°$ gives a fractional change in lattice spacing, or apparent strain, of order 10×10^{-6} or $10 \ \mu\varepsilon$. It is therefore important that sample being investigated and the reference sample are at the same temperature.

Geometrical Effects. The geometric effects discussed in Section 3.6 can give rise to anomalous shifts in measured diffraction angles, and thus give spurious strains. These are particularly severe for cases where the sample partly fills the instrument gauge volume.

Intergranular Strains. Intergranular strains caused by prior plastic deformation are discussed in Section 5.6. If not avoided or corrected for, they can give rise to "pseudo-strains," which mask the true macroscopic strain. They can exist and vary on a length scale approximately equal to the grain size and so can reside even in the relatively small blocks of material used as "stress-free" samples.

4.6.3 Measurement of Reference d^0-Spacing

The optimum method of determining d^0_{hkl}, or d_{ref}, for each reflection used to measure strain will depend on the particular application under consideration. Since the value of the reference d-spacing occurs in the calculation of every strain, it is essential to spend adequate time on its measurement in order to determine its value with sufficient accuracy. It has been suggested that ten times the time used to acquire the other data points should be allocated to its measurement, so that it has three times smaller uncertainty (see Equation (4.16)). Appropriate measurement methods include:

- Measuring a far-field value in the sample where it is expected that the stress is likely to be very small
- Measurement on stress-free powders or filings
- Cutting stress-free reference cubes or combs from the sample
- Application of force/moment balance.

These methods and some practical applications are described below, and further examples are provided in Chapter 6. Careful measurements of a standard sample, such as Si powder, to accurately calibrate the wavelength

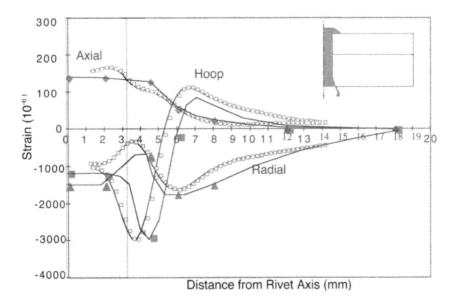

FIGURE 4.19

The axial, radial, and hoop strain components (solid symbols) measured 3 mm below the surface at positions radially from the axis of a 3 mm radius 2124 Al rivet joining two 7010 Al plates (see inset). The rivet spacing was 30 mm and the strain-free d^0 value was measured laterally towards the edge of the plate, far from the line of rivets. The closely spaced open symbols represent the predictions from a finite element model smeared to match the spatial resolution of the sampling gauge. The good agreement with the data validates the use of the far field measurement of d^0. (Courtesy of P.J. Withers and R.V. Preston.)

and zero encoder angle of ϕ^s so that measured d-spacings are absolute, can enable a value of d^0 measured on one instrument to be used on another.

4.6.3.1 Far-Field Reference

A common approach is to use a *far-field* value of d-spacing as a reference. Typically, this reference value is measured in a region of the sample that is considered unlikely to be affected by the process from which the residual stress field in question originated. A typical example is the strain field around a rivet holding together two plates. Because the field is mechanically induced and would be expected to fall off rapidly with distance from the rivet, a representative *far-field* d-value can be obtained as far away from the rivet as possible, and the reference level is established by averaging the d-spacing measured through thickness, or by averaging over a number of orientations. The example in Figure 4.19 shows the strains measured radially from a 2124 Al rivet holding together two 7010 Al plates. Strictly, stress balance should be applied to ensure that the far-field region is indeed in a state of low stress. That this is the case here is evident from the good agreement with the finite element (FE) model, which necessarily obeys stress balance. Note that this method of globally fixing d^0 can be very inappropriate in cases where there maybe localized heating, such as near a weld, due to possible local compositional changes (discussed above).

4.6.3.2   Powders, Cubes, and Combs

In this approach, pieces of the material freed from the constraint of the surrounding macroscopic stress field, as shown in Figure 4.20, are used to measure d_{ref}. Several methods of obtaining such samples can be used, including powders or filings, small reference cubes, or "combs" prepared from an identical material to that under measurement or, destructively, the same sample. As discussed in Chapter 1, the shorter the range of the stress field, the finer the sample must be sectioned to relieve the stress.

Use of Powder Reference Samples. The powder approach is based on the assumption that small particles of a powder are unable to sustain any macroscopic stress state — that is, they are stress-free. However, care must be taken to ensure that the mechanical process of producing the powder or filings in itself does not give rise to residual stresses due to plastic deformation, especially short-range intergranular stresses, which even in fine powders may not be fully relaxed. The powder approach is attractive in the sense that powders are easy to handle when contained in vanadium or quartz containers. However, they must completely fill the IGV, or be very accurately centered at the instrument reference point to avoid any geometrical shift in measured diffraction angle. Any absorption in the powder may move the effective centroid of the powder sample away from the reference point.

Use of Small Cuboid Reference Samples. An alternative is to cut, using electro-discharge machining (EDM) for example, small reference cuboids from the part of the sample or an identical one, at positions where measurements have been made to measure d^0. Generally, cubes are used, and the smaller the better as long as the grain size is sufficiently small. However, compared to the powder approach, small cubes are more difficult to handle and to accurately position at the instrument reference point. If there is a variation in composition in the sample, the reference strain-free lattice spacings, d^0, must be measured at exactly the locations of the strain data points themselves. This calls for an extensive process of extracting small reference cubes throughout, for example, weld region, and subsequently measuring them individually. A practical example of this cumbersome method is given by Krawitz et al. [34], who showed from experimental measurements that it was necessary to obtain individual reference values for each location in question.

Another example of the technique was applied to a mock-up double-"V" weldment, shown in Figure 4.21a, between two ~781 mm, 16 mm thick, Type 316 stainless steel tubes. Depending on their proximity to the weld, different regions had been subjected to a different heat treatment. The measurement of triaxial strain near the weld line was made using the 111 reflection on the

(a)

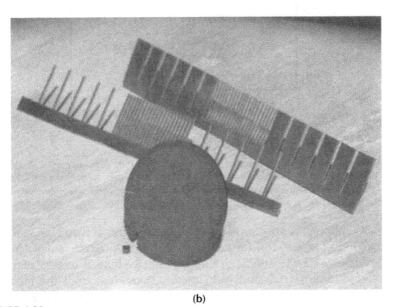

(b)

FIGURE 4.20

Powders, cubes, and combs for the determination of *strain-free* reference values. (a) Powder along with powder sample can; stress-free cubes and a typical tensile specimen. (b) A comb and its "negative" cut at mid-plate thickness with the teeth parallel to the welding direction of an Al friction-stir weldment (see Figure 6.14), and a disc-like structure cut from a steel automotive crank shaft from which a series of small stress-free reference cubes are being cut at different radial distances.

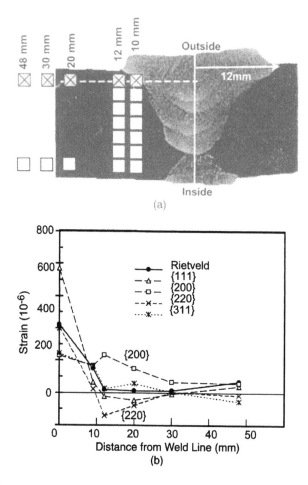

FIGURE 4.21
(a) A double "V" butt weld joining two 16-mm thick AISI type 316L austenitic steel pipes. Squares denote positions at which "matchstick" 2×2×16mm³ samples were taken. (b) Pseudo-strains measured in the "matchstick" samples taken from positions (crossed squares in (a)) away from the weld along a line 2 mm below the outer surface [35].

TAS 8 conventional, continuous-source, diffractometer at Risø National Laboratory [35]. In order to test for variation of d^0 with position, small "matchstick" samples, $2 \times 2 \times 16$ mm³ with a long axis in the hoop direction, of parent heat affected zone (HAZ) and weld material were cut from the mock-up weldment by a "stress-free" electrodischarge machining technique at positions 2 mm from the outer surface. Lattice parameter variations in planes oriented at 45° to the cylinder radial-axial directions were measured for each matchstick on the ENGIN instrument. Both a full Rietveld analysis using all the diffraction peaks and separate analyses of four individual diffraction peaks (111, 200, 022 and 311) were carried out for each matchstick. Any residual "strains" in the matchstick samples were determined by reference to measurements of d^0 made at a far-field point at 48 mm from the weld center line, assumed to be at zero strain. Strains at this location were assumed to be unaffected by local compositional variations and by plastic strain-induced residual microstresses. The results of a hardness survey confirmed that no significant plastic strain had been induced this far from the weld.

The measured strains for the matchsticks are shown in Figure 4.21b. Except in the weld itself, at a distances below ~12 mm from the weld line, where compositionally induced changes in d^0 are likely, the strains determined from a Rietveld analysis are negligible. However, in the heat-affected zone at distances from 12 to 30 mm, considerable intergranular strains are evident as different strains as a function of hkl. These are most probably due to plastic anisotropy, discussed in Section 5.6, and are most marked for the 200 reflection and least for the 111 reflection. This is characteristic of plastic anisotropy in stainless steel, with 200 exhibiting the largest intergranular strains and 220 smaller strains in the opposite sense. It is seen that the variations in strain from these peaks bound the Rietveld curve in both the weld and heat-affected zone. As noted in Table 4.1, use of 111 or 311, and not 200, is recommended for macrostrain measurement in plastically strained stainless steel.

It is clear from these results that if single diffraction peak measurements of d^0 from small-cut reference samples are used as the "zero strain" reference, they will incorporate a contribution from residual intergranular strains, pseudo strains discussed in Section 5.6, if plasticity has occurred. Provided that the extent of plasticity is exactly the same as that in the original sample, they will have essentially the same intergranular strain. As a result, the use of d^0 measured in the same locations will give the correct elastic lattice strains in the sample, since the intergranular strains — as indeed will any compositional variations — in sample and reference will cancel out. However, if a far-field zero strain reference d^0 is used, any possible intergranular strains in the measurements on the sample must be corrected for in the analysis.

Use of Comb Reference Samples. An elegant alternative approach, which is suitable in appropriate circumstances, is to manufacture so-called reference "combs" from a similar sample, examples of which are shown in Figure 4.20b. The comb may be cut in an appropriate manner corresponding to the expected stress state. Small regions of the material are thus made almost free from the constraint of the surroundings, but left in registry through a small mechanical connection to the base material. Such structures circumvent the problems associated with handling small reference cubes, and at the same time retain the positional relationship between each "tooth." The comb design is effective when it is anticipated that no steep variation in d^0 is expected in the direction chosen to be the "long" axis of the "teeth," as, for example, parallel to the welding direction for a weldment. In such cases, because the stress changes only over relatively long distances, the long axis of the tooth does not inhibit stress relaxation. In Figure 4.20b, the teeth are more closely spaced in the near-weld region for better spatial resolution. As with the use of cubes of material, compositional variation and the effect of intergranular strains will be corrected.

4.6.3.3 d^0 Measurement Using X-Rays

At the low-surface penetrations typical of laboratory x-rays, for example, about ~23 μm into Ni, an in-plane biaxial stress field ($\sigma_{33}= 0$) can be assumed. The $\sin^2\psi$ technique, described in Section 5.1.3 where the notation is given, is commonly used to measure the principal in-plane stresses σ_{11} and σ_{22}. However, as well as the in-plane principal stress values, d^0 can also be determined from the relation [33]

$$d^0 = \left(\frac{v}{1+v} \right) \frac{d_{\phi\psi} + d'_{\phi\psi} - 2d_\perp}{\sin^2 \psi} + d_\perp \tag{4.19}$$

where $d_{\phi\psi}$ and $d_{\phi'\psi}$ are the d-spacings measured at a polar angle ψ from the surface normal and at azimuthal angles ϕ and $\phi'= \phi + 90°$ to the in-plane x-axis. Note that this equation, which follows from Equation (5.14a) and the invariance of the trace of the biaxial stress tensor under rotation of axes so that $[\sigma_\phi + \sigma_{\phi'}] = [\sigma_{11}+ \sigma_{22}]$, is independent of the actual azimuthal ϕ angle chosen, and thus requires no knowledge of the principal in-plane stress directions [33,36]. At first glance, it would appear that Poisson's ratio v is generally not known with sufficient accuracy to apply Equation (4.19) for the precise measurement of d^0; however, the first term acts as a small correction to the second, and so the method can normally give a value of d^0 that is accurate enough to be used to determine the strain.

4.6.3.4 d^0 Measurement by Neutron Transmission

Since the normal stress component is zero ($\sigma_{33} =0$) for the case of plane stress, such as for a thin slice or plate sample, the $\sin^2\psi$ technique can be applied in such cases using neutron techniques. This method is particularly well suited to the examination of thin slices using the neutron transmission technique described in Section 3.7.4. The real potential of the transmission geometry for the determination of the unstressed lattice parameter is revealed when the technique is used in combination with an array of detectors to produce a spatial radiographic-like "image" of d^0. Measurements are made at different tilts ψ, and at each of two angles ϕ and $\phi + 90$. As an example of the neutron transmission technique the experimental setup used on the ENGIN instrument at ISIS to determine d^0 is shown in Figure 4.22, where the corresponding angles ϕ and ψ are indicated. Measured hkl Bragg edges at each angle can be fitted individually to enable the corresponding d^0_{hkl} values in Equation (4.19) to be determined. Alternatively, a Rietveld-type multiple peak fit can be made to determine an average stress free constant a^0 from an equation corresponding to Equation (4.19) in which $a_{\phi\psi}$ replaces $d_{\phi\psi}$.

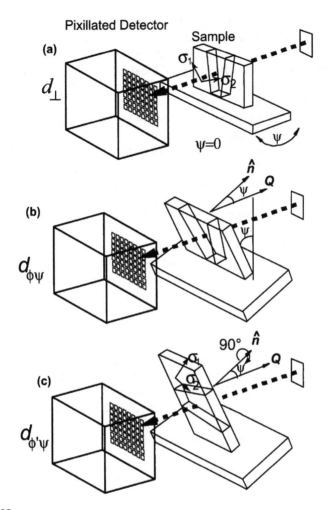

FIGURE 4.22

Schematic showing the set up for a two-dimensional full-field measurement of d^0 on the weld sample in Figure 4.23 by transmission using the $sin^2\psi$ method on the ENGIN instrument at ISIS. Two orthogonal axes must be used for the $sin^2\psi$ tilts as shown by rotating the sample 90° about $\hat{n}$, but since it is not necessary for these to have a special relationship to the in-plane principal axes, it is possible to map d^0 with no prior knowledge of these, even when the in-plane stress field is complex [37].

This method is best illustrated by an example, this is provided by the mapping of d^0 for a slice taken from a ferritic weldment, used as a round-robin sample in the VAMAS TWA 20 program [1]. A number of samples were cut from a weld laid down at TWI Ltd. (United Kingdom), in a 6 mm wide, 8.5 mm deep, U-groove machined in the middle of a BS 4360 50D ferritic steel plate $1000 \times 150 \times 12.5$ mm³ in size. The groove was filled by a 12-pass, mechanized, tungsten inert gas (TIG) weld. A 3 mm thick, 40 mm long, slice was electrodischarge machined from the 12.5 mm thick plate and weldment for the transmission measurements. This slice was placed in front of a two-dimensional pixellated detector comprising a 10×10 array of 2×2 mm^2 scintillation detectors on a 2.5 mm pitch, and tilted around the transverse (TD) and short transverse (STD) directions of the original plate, as shown in Figure 4.22. In this way, each pixel recorded a transmission Bragg edge spectrum which was refined to yield an average $a_{\phi\psi}$ at each angle ψ, corresponding to a region not smaller than $2 \times 2 \times 3$ mm^3 of a sampled gauge volume. The area examined in the experiment is shown as a white rectangle in Figure 4.23a, superimposed on the etched macrograph of the weld. Not all available pixels were used in the experiment due to the smaller dimension of the sample in the STD direction. ψ angles of 0°, 18.4°, 24.4°, and 28.4° were used for the definition of the slopes of the $a_{\phi\psi}$ versus sin$^2\psi$ graph. Relatively low angles were chosen in order to minimize excessive distortion of the mapped area. Since the distortion is proportional to distance from the rotation axis, pixels located more than 10 mm from the axis of rotation were not included in the map of a^0. In principle, it would be possible to correct for the spatial distortion by interpolation of the results taken at each ψ angle.

The map of the unstressed lattice parameter a^0 is shown in Figure 4.23b; this is expressed as spurious strain relative to the value of a^0 measured far away from the weld. This map has been interpolated from an original grid of seven by four points. The map resembles the shape of the weld, but is asymmetric relative to the weld center presumably due to the welding sequence. The extent of the lattice parameter variation found was equivalent to 600 $\mu\varepsilon$. Neglecting this difference and assuming a single constant value for a^0 would introduce an error of 130 MPa in the residual stress, which is of the same order of magnitude as typical stresses found in practice.

In the special case of a uniaxial stress field, Equation (4.19) can be simplified:

$$d^0 = d_\perp + \frac{\nu}{1+\nu} \frac{d_\psi - d_\perp}{\sin^2 \psi} \qquad (4.20)$$

In this case, the principal axes are known and the sample need only be tilted to one angle. An important aspect of this approach is the angle ψ^* for which

$$\sin^2 \psi^* = \frac{\nu}{1+\nu} \qquad (4.21)$$

The angle ψ^* is called the *point of invariance* [38,39], and at this point d_{ψ^*} is equal to d^0. In the case of a Rietveld-type fit to a transmission spectrum, d may again be replaced by the average lattice parameter a.

The validity of Equation (4.20) has been demonstrated by determining the stress-free lattice parameter of an austenitic stainless steel sample while subjected to uniaxial loading to 131 MPa and 210 MPa [37]. Measurements of ten Bragg edges were made on the ENGIN instrument at ISIS, and the average

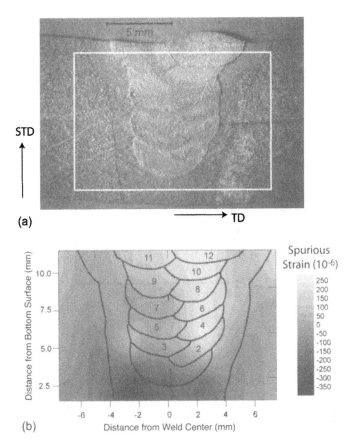

FIGURE 4.23
(a) Etched macrograph of the 12-pass ferritic steel weld. The dotted rectangle indicates the area studied. (b) Interpolated map of the unstressed lattice parameter a^0 for the area indicated in (a). The change in a^0 is expressed in term of a spurious strain relative to the value far away from the weld, in the parent material. The numbers and solid lines in (b) indicate the weld pass sequence [37].

lattice constants a_ψ, determined. From the slopes of the lines fitted to a_ψ versus $\sin^2\psi$ in Figure 4.24, and using an effective Poisson's ratio for stainless steel of about $v = 0.3$, values of a^0 determined from Equation (4.20) under the two different loading conditions are 3.60891Å and 3.60889 ± 0.0001. These are equal within an equivalent strain of 5με, and are close to the value obtained directly by measurement on the sample with no load applied, $a^0 = 3.60873$Å, differing by an amount equivalent to ~45 με. The point of invariance, as given by Equation (4.21), for $v = 0.3$ is $\sin^2\psi^* = 0.3/1.3$ ~0.23, and is marked by the dashed line in Figure 4.24. Strictly speaking, the $\sin^2\psi$ line should rotate about this pivot point at ψ^*, with the intercept of the fitted line at $\sin^2\psi = 0$ increasing as the slope decreases. The vertical line denotes the point of invariance, giving the reference a^0 values at the ordinate. In principle, these should be the same but in actual fact they differ by an equivalent strain of about 5 με. The

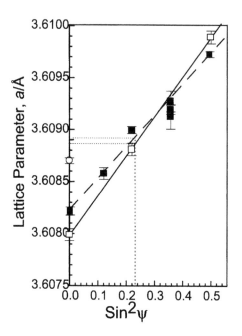

FIGURE 4.24
The average lattice parameter a_ψ, measured from a Pawley refinement of the lowest ten Bragg edges, plotted against $sin^2\psi$, from a stainless steel sample uniaxially loaded to 131 MPa (solid squares) and 210 MPa (open squares). The full and long broken lines are the best linear fit to the data. The vertical line denotes the point of invariance, giving the reference values at the ordinate. The value of the reference a^0 determined from a single measurement at zero load at $\psi = 0$ is denoted by the open circle [42].

estimates of a^0 determined from the point of invariance are also in close agreement with that determined with no stress applied.

4.6.3.5   *Imposing Stress Balance*

An approach to the determination of reference d-spacings, which precludes the need for their measurement, is to adopt the fundamental continuum mechanics–based requirement that force and moment balance across one or more selected cross-sections of the sample. The approach is to measure the required field of d-spacings, or diffraction angles, in the sample, and, using a nominal d_{ref}, or angle, to calculate the triaxial strain and stress in the sample. Subsequently, the reference value is varied iteratively in order to find the true reference value, that is, the value that renders a stress field in which force and moment balance occur. The principle, shown in Figure 4.25, is to imagine cutting the sample along a certain plane and replacing the actions of the cut-away part on the remaining part by tractions and moments in this plane. It is a necessary requirement for mechanical equilibrium of a stationary object that force and moment are balanced across such a plane section. In adopting this approach, great care must be taken in selecting appropriate cross-sections

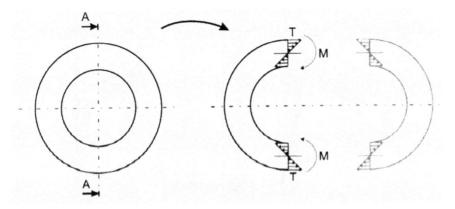

FIGURE 4.25
Schematic sectioning of a tubular structure across the section AA, over which tractions, T, and moments, M, are required to balance in order to retain motion stability.

over which to require forces and moments to balance. It must be ensured that the experimental data set covers the entire cross-section, and not just a part of it, and also that a single value of d_{ref} is appropriate throughout.

This method is best held in reserve to check the validity of the d^0 value obtained from one of the primary methods described above. Indeed, it should always be used, when applicable, to check the validity of a measured stress field [40].

4.7 Reproducibility Tests

As part of the initial phase of the Versailles Project on Advanced Materials and Standards (VAMAS) Technical Working Area 20, before the final establishment of a protocol for measurement, a round-robin study was undertaken [41]. The aim of the study was to assess the efficacy of the various approaches to strain measurement taken at different facilities, and hence the reproducibility and reliability of the data. Eighteen facilities worldwide participated in the study. It involved the measurement of the strain and stress distribution in a shrink-fit aluminum ring and plug sample in which the elastic strain and stress can be analytically predicted. The sample comprised a ring of outer and inner diameters of 50 and 25 mm respectively, into which a plug, oversized by 0.015 mm, had been cooled and inserted. A range of reflections, 111, 002, 311, and so on, were used to determine lattice strains, and the instrumental gauge volume used at the different facilities ranged from 2 mm^3 to 480 mm^3. The averaged hoop strain measurements are shown in Figure 4.26a. The best estimate of the three principal stresses shown on Figure 4.26b were obtained from a Bayesian best fit to all the data. The stresses were obtained

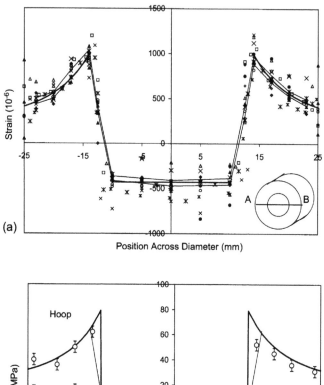

FIGURE 4.26
(a) The measurements (points) of the hoop strain component, obtained at 18 facilities worldwide on two identical shrink fit Al plug samples, across a mid-plane diameter (A-B in inset). The bold line shows the averaged strain variation, and the other lines the uncertainties of one standard deviation, about 75 με from the mean. Note how surface effects lead to larger errors at the edges of the ring. (b) The best estimate of the three principal stresses (points) obtained from the measured principal strains, compared with results of an analytical calculation (bold and broken lines) assuming an interference between ring and plug of 21.5 μm. The sloping lines at the interface represent smoothing of the calculated stress variation by the instrumental gauge volume [41].

from the principal strains using Equation (5.8) and appropriate values of E_{hkl} and v_{hkl}. The overall standard error of the results was 75 με in strain and 7 MPa in stress, giving confidence that if the points discussed in this chapter are followed, reliable data can be obtained. Accurate results rely not just on accurate strain measurement, but also on accurate sample location, especially at surfaces and regions of steep strain gradients as discussed in Section 3.6.3. The positioning error in this study was found to have a standard deviation of 90 μm if data from four laboratories were excluded, and 230 μm if results of the whole group were considered.

4.7.1 Recording Measurement Details

When carrying out neutron diffraction strain and stress measurements, it is extremely important to record the instrument setup used in every detail, and to note exactly how each measurement was made. The extensive round-robin exercise described above illustrates this point. Some facilities have aspects of this record-keeping formalized as part of a quality assurance (QA) system, while others leave it entirely to the user! The key rule is that records should be sufficient to enable another user, who is intelligent but inexperienced, to repeat the measurement in the same manner so that the same results are obtained within statistical uncertainties. It is often the case that "obvious" points in the procedure, or setups that are "standard" go unrecorded, only to cause great problems when repeating or continuing measurements after a period of time. Thus, the rule to follow is — "record everything." The advent of digital photography enables visual records to be readily kept of the instrument, sample, and setup. The VAMAS TWA20 group has formally listed full details of exactly what should be recorded [1].

References

1. G.A. Webster, Ed., Polycrystalline Materials — Determinations of Residual Stresses by Neutron Diffraction, ISO/TTA3 Technology Trends Assessment, Geneva, 2001.
2. M.R. Daymond, M.A.M. Bourke, R.B. Von Dreele, B. Clausen, and T. Lorentzen, Use of Rietveld refinement for elastic macrostrain determination and for the evaluation of plastic strain history from different spectra, *J. Appl. Phys.*, 82, 1554–1562, 1997.
3. T.C. Hsu, F. Marsiglio, J.H. Root, and T.M. Holden, Effects of multiple scattering and wavelength-dependent attenuation on strain measurements by neutron scattering, *J. Neutron Res.*, 3, 27–39, 1995.
4. P.J. Withers, M.R. Daymond, and M.W. Johnson, The precision of diffraction peak location, *J. Appl. Crystallogr.*, 34, 737–743, 2001.
5. D.S. Sivia, *Data Analysis — A Bayesian Tutorial*, Oxford University Press, Oxford, 1996.
6. W.H. Press, S.A. Teukolsky, W.T. Vetterling, and B.P. Flannery, *Numerical Recipes*, Cambridge University Press, Cambridge, 1986.

7. R.L. Snyder, Analytical profile of x-ray powder diffraction profiles in Rietveld analysis, in *The Rietveld Method*, IUCr Monographs on Crystallography, R.A. Young, Ed., Oxford University Press, Oxford, 1993, 111–131.

8. M.M. Hall, V.G. Veeraghavan, H. Rubin, and P.G. Winchell, The approximation of symmetric x-ray peaks by Pearson type-VII distributions, *J. Appl. Crystallogr.*, 10, 66–68, 1977.

9. J.M. Carpenter, H.M. Mueller, R.A. Beyerlein, T.G. Worlton, J.D. Jorgensen, T.O. Brun, K. Sköld, C.A. Pelizzari, S.W. Peterson, N. Watanabe, M. Kimura, and J.E. Gunning. Neutron diffraction measurements on powder samples using the ZING-P pulsed neutron source at Argonne, in Proceedings, Neutron Diffraction Conference, 1975, Petten, Reactor Centrum Nederland, 192–208.

10. J.D. Jorgensen, Compression mechanisms in α-quartz structures — SiO_2 and GeO_2, *J. Appl. Phys.*, 49, 5473–5478, 1978.

11. J.M. Carpenter, R.A. Robinson, A.D. Taylor, and D.J. Picton, Measurement and fitting of spectrum and pulse shapes of a liquid methane moderator at IPNS, *Nucl. Inst. Methods Phys. Res.*, A234, 542–551, 1985.

12. S. Ikeda and J.M. Carpenter, Wide-energy-range, high resolution measurements of neutron pulse shapes of polyethylene moderators, *Nucl. Inst. Methods Phys. Res.*, A239, 536–544, 1985.

13. J.M. Carpenter and W.B. Yelon, Neutron sources, in *Methods of Experimental Physics*, vol. 23A, *Neutron Scattering*, K. Sköld and D.L. Price, Eds., Academic Press, New York, 1986, 99–196.

14. R.A. Young, Introduction to the Rietveld method, in *The Rietveld Method*, IUCr Monographs on Crystallography, R.A. Young, Ed., Oxford University Press, Oxford, 1993, 1–38.

15. P.J. Huber, *Robust Statistics*, Wiley & Sons, New York, 1981.

16. P.J. Webster and W. Kang, Optimisation of data collection and processing for efficient strain scanning, *J. Neutron Res.*, 10, 93–110, 2002.

17. A.D. Stoica, M. Popovici, C.R. Hubbard, and S. Spooner, Neutron monochromators for residual stress mapping at the new HB-2 beamport, Oak Ridge National Laboratory, 1999, available at: www.osti.gov/bridge, 2002.

18. M.W. Johnson and M.R. Daymond, An optimum design for a neutron diffractometer for measuring engineering stresses, *J Appl. Crystallogr.*, 35, 49–57, 2002.

19. P.J. Withers, Depth capabilities of neutron and synchrotron diffraction strain measurement instruments: Part I — The maximum feasible path length, *J. Appl. Crystallogr.*, 37, 596–606, 2004.

20. P.J. Withers, Depth capabilities of neutron and synchrotron diffraction strain measurement instruments: Part II — Practical implications, *J. Appl. Crystallogr.*, 37, 607–612, 2004.

21. P.J. Webster, X.D. Wang, and G. Mills, Strain scanning using neutrons and synchrotron radiation, in Proceedings European Conference on Residual Stress IV, 4, 1996, 127–134.

22. C.G. Windsor, The precision of peak position determination in diffraction measurements of stress, in *Measurement of Residual and Applied Stress Using Neutron Diffraction*, M.T. Hutchings and A.D. Krawitz, Eds., NATO ASI Series E: *Applied Science*, vol. 216, Kluwer, Dordrecht, 1992, 285–296.

23. P.C. Brand, *Stress Measurements by Means of Neutron Diffraction*, Ph.D. thesis, University of Twente, 1991.

24. G.E. Bacon, *Neutron Diffraction*, 3rd ed., Clarendon Press, Oxford, 1975.

25. A.C. Larson and R.B. Von Dreele, GSAS: Generalized structure analysis system, Los Alamos National Laboratory, Los Alamos, NM, 1985.
26. D.B. Wiles and R.A. Young, A new computer program for Rietveld analysis of x-ray powder diffraction patterns, *J. Appl. Crystallogr.*, 14, 149–151, 1981.
27. F. Izumi, H. Asano, H. Murata, and N. Watanabe, Rietveld analysis of powder patterns obtained by TOF neutron diffraction using cold neutron source, *J. Appl. Crystallogr.*, 20, 411–418, 1987.
28. D.K. Smith and S. Gorter, Powder diffraction program information, *J. Appl. Crystallogr.*, 24, 369–402, 1991.
29. R.A. Young, Ed., *The Rietveld Method*, IUCr Monographs on Crystallography, vol. 5, Oxford University Press, Oxford, 1993.
30. E. Prince, Comparison of profile and integrated intensity methods in powder refinement, *J. Appl. Crystallogr.*, 14, 157–159, 1981.
31. M.R. Daymond, personal communication, 2000.
32. A. Mohr and H.G. Priesmeyer, Experimental correlation between the elemental composition and the lattice dimensions in alloys, using Al-Cu as a model substance, Proceedings, International Conference on Neutrons and their Applications, Crete, 1994, Soc. of Photo-optical Instrumentation Engineers, vol. 2339, 350–355, 1994.
33. C.S. Barrett and T.B. Massalski, *Structure of Metals*, Pergamon, Oxford, 1966.
34. A.D. Krawitz and R.A. Winholtz, Use of position-dependent stress-free standards for diffraction stress measurements, *Mat. Sci. Eng. A*, 185, 123–130, 1994.
35. P.J. Bouchard, M.T. Hutchings, and P.J. Withers, Characterisation of the residual stress state in a double 'V' stainless steel cylindrical weldment using neutron diffraction and finite element simulation, in Proceedings, 6th International Conference on Residual Stress, Oxford, 2000, IOM Communications Ltd., London, vol. 2, 2000, 1333–1340.
36. C.S. Barrett and M. Gensamer, *Physics*, 7, 1, 1963.
37. J.R. Santisteban, A. Steuwer, L. Edwards, P.J. Withers, and M.E. Fitzpatrick, Mapping of unstressed lattice parameters using pulsed neutron transmission diffraction, *J. Appl. Crystallogr.*, 35, 497–504, 2002.
38. B. Eigenmann and E. Macherauch, Röntgenographische untersuchung von spannungszuständen, werkstoffen I-IV, *Matwiss. Werkstofftech*, 26, 148–160, 1995.
39. V. Hauk, *Structural and Residual Stress Analysis by Non-Destructive Methods*, Elsevier Science, Oxford, 1997.
40. M. Preuss, J.W.L. Pang, G.J. Baxter, and P.J. Withers, Inertia welding nickel-based superalloy, part II: residual stress development, *Metall. Mater. Trans.*, 33A, 3127–3134, 2002.
41. G.A. Webster, Neutron diffraction measurements of residual stress in ring and plug, Versailles Project on Advanced Materials and Structures (VAMAS), TWA20, Tech Report No. 38, 2000.
42. A. Steuwer, J.R. Santisteban, P.J. Withers, L. Edwards, and M.E. Fitzpatrick, In-situ stress determination by pulsed neutron transmission, *J. Appl. Crystallogr.*, 35, 1159–1168, 2003.
43. T.M. Holden, A.P. Clarke, and R.A. Holt, Neutron diffraction measurements of intergranular strains in Monel 400, *Metall. Mater. Trans*, 28A, 2565–2576, 1997.
44. B. Clausen, T. Lorentzen, M.A.M. Bourke, and M.R. Daymond, Lattice strain evolution during tensile loading of stainless steel, *Mater. Sci. Eng.*, A259, 17–24, 1999.

45. B. Clausen, *Characterisation of Polycrystal Deformation by Numerical Modeling and Neutron Diffraction Measurements*, Ph.D. thesis, Technical University of Denmark, Risø National Laboratory, 1997.
46. J.W.L. Pang, T.M. Holden, and T.E. Mason, In situ generation of intergranular strains in an Al7050 alloy, *Acta Mater.*, 46, 1503–1518, 1998.
47. J.W.L. Pang, T.M. Holden, and T.E. Mason, The development of intergranular strains in a high-strength steel, *J. Strain Anal.*, 33, 373–383, 1998.
48. J.W.L. Pang, T.M. Holden, and T.E. Mason. In-situ generation of intergranular strains in zircaloy under uniaxial loading, in Proceedings, 5th International Conference on Residual Stresses, Linköping, Sweden, 1997, 610–615.
49. M.R. Daymond, M.A.M. Bourke, and R.B. Von Dreele, Use of Rietveld refinement to fit a hexagonal crystal structure in the presence of elastic and plastic anisotropy, *J. Appl. Phys.*, 85, 739–747, 1999.

5

Interpretation and Analysis of Lattice Strain Data

Neutron diffraction lattice strain measurement provides much more information than a conventional strain gauge. By sampling a well-defined subset of grains, each diffraction peak provides insight into the Type I, II, and III elastic strains within the sample. While of special interest to the materials scientist, this level of detail presents a complication to the engineer wishing to interpret strain data in terms of macrostress. In this chapter, the interpretation of measured strains in terms of the macrostress and microstress in the sample is described. Although the determination of stress from strain using a continuum model is straightforward, the appropriate elastic constants must be used. These are discussed in terms of basic micromechanics and models that enable reliable interpretation of data in both the elastic and plastic regimes.

5.1 Inferring Stresses from Lattice Strains

5.1.1 Basic Continuum Relationship between Strain and Stress

Stress ($\underline{\underline{\sigma}}$) and strain ($\underline{\underline{\varepsilon}}$) are tensor quantities related to one another by the elastic stiffness tensor $\overline{\overline{C}}$, and the elastic compliance tensor $\underline{\underline{S}}$:

$$\sigma_{ij} = \sum_{kl} C_{ijkl}\varepsilon_{kl} \quad , \quad \text{and} \quad \varepsilon_{ij} = \sum_{kl} S_{ijkl}\sigma_{kl} \tag{5.1}$$

where $\underline{\underline{\sigma}}$ and $\underline{\underline{\varepsilon}}$ have 3×3 components, 6 of which are independent, and $\underline{\underline{C}}$ and $\underline{\underline{S}}$ have $3 \times 3 \times 3 \times 3$ components, of which as many as 36 can be independent [1]. As a result, the conversion of measured strain components to stress is an inherently difficult task, requiring measurement of the strain in many directions at a point in a sample before the stress at that point can be inferred. Systematic and statistical errors in the individual strain measurements combine to reduce the accuracy of the inferred stresses. Furthermore, for reasons of sample texture, it may not be possible to measure strain in sufficient directions using a single reflection, while the presence of intergranular stresses may complicate the interpretation of the measured lattice strain. In some cases, the measured strains can be used directly. For example, the validation of a finite element model is clearly best undertaken in strain rather than in stress. In many circumstances, however, it is the macrostress in a component that is of ultimate importance, and it is necessary to infer this stress from measured strains. This necessitates a full understanding of the effects of elastic and plastic anisotropy in crystals, and how these are averaged in relating strain to stress in polycrystalline samples.

Essentially, most engineering investigations are based on isotropic continuum mechanics. In this case, $\underline{C}$ can be written in terms of just two independent elastic components, such as Young's modulus, E, and Poisson's ratio, v. Consequently, the relationship between stress and strain can be expressed using the generalized Hooke's law equations as

$$\sigma_{ij} = \frac{E}{(1+v)}\left[\varepsilon_{ij} + \frac{v}{(1-2v)}\left(\varepsilon_{11}+\varepsilon_{22}+\varepsilon_{33}\right)\right] \tag{5.2}$$

where $i, j = 1,2,3$ indicate the components relative to chosen axes.

From the above expression, it is clear that in order to derive the stress tensor, the strain tensor must first be evaluated. In the general case, this can be done by relating a specific strain component, $\varepsilon(lmn)$, measured in the direction for which l, m, and n are the direction cosines to orthogonal sample coordinate axes Ox, Oy, and Oz, to the six components of the strain tensor ε_{ij} referred to these axes by [2]:

$$\varepsilon(lmn) = l^2.\varepsilon_{11} + m^2.\varepsilon_{22} + n^2.\varepsilon_{33} + 2lm.\varepsilon_{12} + 2mn.\varepsilon_{23} + 2nl.\varepsilon_{31} \tag{5.3}$$

Hence, the measurement of strain components $\varepsilon(l_{d'}, m_{d'}, n_{d'})$ in six different directions, $d' = 1,6$, leads to a system of six linear equations in the six unknown components of the strain tensor, from which the latter can be determined:

$$
\begin{bmatrix}
\varepsilon(l_1,m_1,n_1)\\
\varepsilon(l_2,m_2,n_2)\\
\varepsilon(l_3,m_3,n_3)\\
\varepsilon(l_4,m_4,n_4)\\
\varepsilon(l_5,m_5,n_5)\\
\varepsilon(l_6,m_6,n_6)
\end{bmatrix}
=
\begin{bmatrix}
l_1^2 & m_1^2 & n_1^2 & 2l_1m_1 & 2m_1n_1 & 2n_1l_1\\
l_2^2 & m_2^2 & n_2^2 & 2l_2m_2 & 2m_2n_2 & 2n_2l_2\\
l_3^2 & m_3^2 & n_3^2 & 2l_3m_3 & 2m_3n_3 & 2n_3l_3\\
l_4^2 & m_4^2 & n_4^2 & 2l_4m_4 & 2m_4n_4 & 2n_4l_4\\
l_5^2 & m_5^2 & n_5^2 & 2l_5m_5 & 2m_5n_5 & 2n_5l_5\\
l_6^2 & m_6^2 & n_6^2 & 2l_6m_6 & 2m_6n_6 & 2n_6l_6
\end{bmatrix}
\bullet
\begin{bmatrix}
\varepsilon_{11}\\
\varepsilon_{22}\\
\varepsilon_{33}\\
\varepsilon_{12}\\
\varepsilon_{23}\\
\varepsilon_{13}
\end{bmatrix}
\tag{5.4}
$$

Note that the off-diagonal ($i \neq j$), or shear, components of the strain tensor are sometimes written in terms of an 'engineering shear strain' γ_{ij}, with $\gamma_{ij} = 2\varepsilon_{ij}$. Similarly, the shear stress σ_{ij} is written as τ_{ij} and the shear modulus, μ, defined by $\gamma_{ij} = 2\varepsilon_{ij} = \tau_{ij}/\mu$. For isotropic materials, $2\mu = E/(1 + v)$.

For subsequent determination of the stress tensor great care must be taken in selecting those directions ($l_{d'}$, $m_{d'}$, $n_{d'}$) that give truly independent strain components, and which render the highest accuracy of the stress tensor. Considering the inherent measurement uncertainty associated with individual strain measurements, it is generally best to overdetermine the problem by measuring more components than theoretically needed to solve the mathematical problem. Eight to ten carefully selected strain directions typically render reasonable accuracy in the calculation of the stress tensor [3].

From this strain tensor determination, components of the stress tensor using Equation (5.2) can be calculated. However, in many cases symmetry considerations enable assumptions about the principal stress and strain directions to be made, and then the elaborate and time-consuming process of measuring the full strain tensor based on six or more measurements is avoided. In this case, because the stress tensor is diagonal, only the three orthogonal principal strain components need be measured. If this is not possible, the full strain tensor, $\underline{\varepsilon}$, determined from Equation (5.4) must be diagonalized to find the principal axes Ox_D, Oy_D, and Oz_D. The expression for the calculation of the principal stress components follows from Equation (5.2), written in terms of three orthogonal principal strain components as

$$\sigma_{x_D x_D} = \frac{E}{(1+v)}\left[\varepsilon_{x_D x_D} + \frac{v}{(1-2v)}\left(\varepsilon_{x_D x_D} + \varepsilon_{y_D y_D} + \varepsilon_{z_D z_D}\right)\right] \tag{5.5a}$$

$$\sigma_{y_D y_D} = \frac{E}{(1+v)}\left[\varepsilon_{y_D y_D} + \frac{v}{(1-2v)}\left(\varepsilon_{x_D x_D} + \varepsilon_{y_D y_D} + \varepsilon_{z_D z_D}\right)\right] \tag{5.5b}$$

$$\sigma_{z_D z_D} = \frac{E}{(1+v)}\left[\varepsilon_{z_D z_D} + \frac{v}{(1-2v)}\left(\varepsilon_{x_D x_D} + \varepsilon_{y_D y_D} + \varepsilon_{z_D z_D}\right)\right] \tag{5.5c}$$

It should be noted that a consequence of the form of Equation (5.2), and the invariance of the trace of a tensor under axes rotation, is that any three measured orthogonal normal strain components, ε_{11}, ε_{22}, and ε_{33}, can form the basis for calculating the corresponding three orthogonal stress components without any knowledge of the shear (off-diagonal) strain components. Thus, these three orthogonal stress components are correctly determined from the orthogonal strain components with respect to any axes even when they are *not* the principal ones. Of course, the *principal* stresses can only be determined by a full stress tensor determination as described above without simplifying assumptions.

The strain and stress tensor may be written as the sum of two parts: a traceless part, termed the *deviatory* or *deviatoric* strain or stress, obtained by subtraction of the mean of the diagonal terms of the full tensor, and an isotropic part equal to this mean of the diagonal terms, termed the *hydrostatic* strain or stress. Since the trace is independent of any transformation ortho-gonal axes of this is also true for the diagonalized tensors giving the principal stresses. Thus, the principal strains can be written in terms of the deviatory strain, $\varepsilon_{i_D i_D}^{Dev.}$, and the hydrostatic strain, $(\varepsilon_{x_D x_D} + \varepsilon_{y_D y_D} + \varepsilon_{z_D z_D})/3$, as

$$\varepsilon_{i_D i_D} = \varepsilon_{i_D i_D}^{Dev} + \left(\varepsilon_{x_D x_D} + \varepsilon_{y_D y_D} + \varepsilon_{z_D z_D}\right)/3 \quad i = x, y, z \qquad (5.6)$$

Substituting in Equations (5.5a), (5.5b), and (5.5c), the principal stress tensor becomes

$$\sigma_{i_D i_D} = \frac{E}{(1+v)} \varepsilon_{i_D i_D}^{Dev} + \frac{E}{3(1-2v)} \left(\varepsilon_{x_D x_D} + \varepsilon_{y_D y_D} + \varepsilon_{z_D z_D}\right) \quad i = x, y, z \qquad (5.7)$$

where the first term is the deviatory stress and the second the hydrostatic stress. It is worth noting that if only relative stress components are of interest (i.e., only the deviatory stress), the strain-free lattice spacing, d^0, is not required in diffraction measurements. The deviatory stresses are important when considering the onset of plastic flow.

5.1.2   Diffraction-Specific Elastic Constants

As was observed in Figure 4.1 for austenitic steel, the applied stress versus elastic strain response characteristic of each lattice plane family *hkl* is usually different, because in general, the stiffness of a single crystal is not isotropic. At first glance, it is tempting to replace the continuum elastic strain in the equations given in the preceding section with the lattice strain $\underline{\varepsilon}^{hkl}$, measured from the *hkl* reflection. However, the question then arises as to the most appropriate elastic constants to replace the isotropic values E, v. As discussed in Sections 5.3 through 5.6, the single crystal values are not representative of the behavior of grains within a polycrystal because of intergranular strains generated between the differently oriented grains. For the moment, we defer a discussion of the subtleties of intergranular strain and take the pragmatic view that representative elastic constants for a polycrystal can be derived that relate the lattice strains to the macrostress during elastic loading. These are termed the diffraction peak-specific elastic constants (DECs), E_{hkl} and v_{hkl} for texture-free materials. If these are substituted in the generalized Hooke's law equations, then — at least in the elastic loading regime — the strain evaluated for each reflection $\underline{\varepsilon}^{hkl}$, can be converted to a single valued estimate of the macrostress, $\underline{\sigma}^I$, where

$$\sigma_{ij}^I = \frac{E_{hkl}}{(1+v_{hkl})}\left[\varepsilon_{ij}^{hkl} + \frac{v_{hkl}}{(1-2v_{hkl})}\left(\varepsilon_{11}^{hkl} + \varepsilon_{22}^{hkl} + \varepsilon_{33}^{hkl}\right)\right] \qquad (5.8)$$

In fact, all the equations in Section 5.1.1 have *hkl* diffraction peak-specific analogs. The diffraction elastic constants can be measured from calibration experiments in which a polycrystalline sample is subjected to known uniaxial loading. They can also be calculated as discussed in Section 5.5, and those based on the Kröner model are given in Tables 5.8 and 5.9.

5.1.3 X-Ray Measurement of Stress

An alternative formulation commonly used when making x-ray diffraction investigations of stresses and strains is based on a definition of the direction in which the strain is measured through the two angles ϕ and ψ, as shown in Figure 5.1. Henceforth, the explicit labeling of strain, ε, and lattice spacing, d, with the reflection *hkl* will be dropped — it is taken as understood. Using this system, the expression relating the strain, ε'_{33}, along the direction of the scattering vector, $-Q$, parallel to L_3 as a function of strain components, ε_{ij}, in the sample coordinate system, P_1, P_2, P_3, is given by substituting

$$l = \cos\phi.\sin\psi, \ m = \sin\phi.\sin\psi, \text{ and } n = \cos\psi, \qquad (5.9)$$

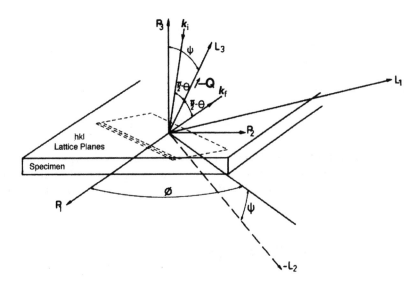

FIGURE 5.1
Definition of sample coordinate system, P_i, and the angles ϕ and ψ. The strain is measured along direction L_3, which is anti-parallel to the scattering vector, Q. (Based on Noyan and Cohen [5].)

in Equation (5.3):

$$\left(\varepsilon_{33}'\right)_{\phi\psi} = \frac{d_{\phi\psi} - d^0}{d^0} = \varepsilon_{11}\cos^2\phi\sin^2\psi + \varepsilon_{22}\sin^2\phi\sin^2\psi + \varepsilon_{33}\cos^2\psi +$$

$$\varepsilon_{12}\sin 2\phi\sin^2\psi + \varepsilon_{23}\sin\phi\sin 2\psi + \varepsilon_{13}\cos\phi\sin 2\psi \qquad (5.10)$$

For materials isotropic at the continuum level, using the generalized form of Hooke's law and the diffraction elastic constants E_{hkl} and v_{hkl}, we can relate the strain component $(\varepsilon'_{33})_{\phi\psi}$ to the components of the stress tensor as

$$\left(\varepsilon_{33}'\right)_{\phi\psi} = \frac{d_{\phi\psi} - d^0}{d^0} = \left(\frac{1+v_{hkl}}{E_{hkl}}\right)\left[\sigma_{11}\cos^2\phi + \sigma_{12}\sin 2\phi + \sigma_{22}\sin^2\phi - \sigma_{33}\right]\sin^2\psi +$$

$$\left(\frac{1+v_{hkl}}{E_{hkl}}\right)\sigma_{33} - \left(\frac{v_{hkl}}{E_{hkl}}\right)\left[\sigma_{11} + \sigma_{22} + \sigma_{33}\right] + \left(\frac{1+v_{hkl}}{E_{hkl}}\right)\left[\sigma_{13}\cos\phi + \sigma_{23}\sin\phi\right]\sin 2\psi$$

$$(5.11)$$

As noted above, in general six independent components of lattice strain must be measured in order to derive the full strain tensor. However, for special stress states, fewer measurements are needed. A free surface is a case in point because the normal stress is zero, $\sigma_{33} = 0$, and there are no σ_{13} and σ_{23} shear components. Here the normal direction, P_3, is a principal axis and Equation (5.11) becomes linear in $\sin^2\psi$:

$$\left(\varepsilon_{33}'\right)_{\phi\psi} = \frac{d_{\phi\psi} - d^0}{d^0} = \left(\frac{1+v_{hkl}}{E_{hkl}}\right)\sigma_\phi\sin^2\psi - \left(\frac{v_{hkl}}{E_{hkl}}\right)\left[\sigma_{11} + \sigma_{22}\right] \qquad (5.12a)$$

where σ_ϕ is given by

$$\sigma_\phi = \sigma_{11}\cos^2\phi + \sigma_{12}\sin 2\phi + \sigma_{22}\sin^2\phi \qquad (5.12b)$$

This expression forms the basis for the so-called $\sin^2\psi$ technique, which has been used extensively for stress measurement using x-ray diffraction since the mid-1920s. Often it is used with the elastic constants expressed in the form

$$S_1 = -\frac{v_{hkl}}{E_{hkl}}, \text{ and } \frac{1}{2}S_2 = \left(\frac{1+v_{hkl}}{E_{hkl}}\right) = \frac{1}{2\mu_{hkl}} \qquad (5.13)$$

S_1 and S_2 are sometimes called x-ray elastic constants (XECs) or diffraction elastic constants (DECs) (see, for example, Hauk and Macherauch [4], Noyan and Cohen [5], and Cullity [6]), although we shall reserve the latter term for the constants E_{hkl}, ν_{hkl}, and μ_{hkl}. S_1 and S_2 are not to be confused with elements of the compliance tensor S_{ijkl}. The attractive nature of Equation (5.12a) is the linearity in $\sin^2\psi$.

$$(\varepsilon'_{33})_{\phi\psi} = \frac{d_{\phi\psi} - d^0}{d^0} = \frac{1}{2}S_2\sigma_\phi \sin^2\psi + S_1\left[\sigma_{11} + \sigma_{22}\right] \qquad (5.14a)$$

Now at $\psi = 0$,

$$(\varepsilon'_{33})_{\phi\psi=0} = \frac{d_{\phi\psi=0} - d^0}{d^0} = S_1\left[\sigma_{11} + \sigma_{22}\right] \qquad (5.14b)$$

Thus,

$$\frac{d_{\phi\psi} - d_{\phi\psi=0}}{d^0} = \frac{1}{2}S_2\sigma_\phi \sin^2\psi \qquad (5.14c)$$

Hence, from a series of measurements of $d_{\phi\psi}$ at a number of different angles ψ, including $\psi = 0$, for a specific angle ϕ, we may generate a linear plot of $[d_{\phi\psi} - d_{\phi\psi=0}]/d^0$ versus $\sin^2\psi$. The slope of this line renders a measure of the in-plane stress component σ_ϕ. Although this procedure requires the knowledge of the reference lattice plane spacing, d^0, since d^0 is simply a multiplying factor of the slope when determining σ_ϕ, it may be approximated by $d_{\phi\psi=0}$, giving rise to errors in the stress of only fractions of a percent. If the principal stress directions in the plane are unknown, they may be determined by making measurements at a number of different angles ϕ and solving Equation (5.12b). This approach has been used in numerous applications throughout the years, and procedures have been proposed for its use in the more general case of bulk measurements, and in the case of textured materials. For further information, see for instance, Noyan and Cohen [5]. Of course, for triaxial stress fields, when $\sigma_{33} \neq 0$, the full expression given in Equation (5.11) must be used, and an accurate value of d^0 is required, as in the case of most neutron diffraction strain measurements.

Although this method cannot be used in neutron diffraction stress measurement so easily because of the depth penetration of the sampled gauge volume, it can be applied to the case of very thin plates when the normal stress can be taken to be zero through the thickness, that is, in the case of plane stress. An example using the transmission technique was discussed in Section 4.6.3. The basic $\sin^2\psi$ principle can also be used when measuring the strain variation between two known principal axes, denoted 1 and 2, at a point within a sample, since from Equation (5.3) there are only two terms for strain measurement perpendicular to the third principal axis:

$$\varepsilon(\psi) = \varepsilon_{11}\cos^2\psi + \varepsilon_{22}\sin^2\psi = \varepsilon_{11} + [\varepsilon_{22} - \varepsilon_{11}]\sin^2\psi \qquad (5.15)$$

Here ψ is the angle between the scattering vector Q and principal axis 1 in the sample. Since it involves a number of measurements at different angles ψ, this method can give very accurate results [7].

5.2   Introduction to Mechanics of Crystallite Deformation

Having selected the atomic lattice planes as an internal gauge for monitoring strains by diffraction techniques, a thorough examination of the mechanics of materials at the appropriate length scale is demanded in order to ensure reliable interpretation of the data. As emphasized in Section 1.4.2, diffraction is a selective technique that monitors an average response over the sampled gauge volume of a selected subset of typically several thousand grains, identified by their identity in *hkl* and similarity in orientation. Because crystals have anisotropic elastic and plastic properties, only very rarely does such a subset faithfully reproduce the macroscopic deformation of the body as a whole.

Consequently, in bridging the gap between probing strains at a scale relevant to the microstructure, and the engineers' vision of macroscopic stresses and strains in structures, it is necessary to acquire both theoretical and experimental insight into the micromechanics of crystalline solids. In the following sections, these aspects are first addressed through a review of the theory of the elastic anisotropy of single crystals. This single crystal anisotropy lies at the root of differences between the strains measured by diffraction and the macroscopic engineering strains in polycrystalline samples, due to the so-called *intergranular strains*. As engineering components are only rarely single crystalline, it is therefore necessary to study the micromechanics of polycrystal deformation to support the interpretation of diffraction data. Hence, models for the deformation of polycrystals will need to be introduced. Such models can guide experimental research, which in turn may provide valuable insight into the validation and refinement of the models.

The importance of the anisotropy in lattice strain response, observed when performing diffraction measurements, has become increasingly evident to those using the neutron diffraction technique to measure strain. Although appreciated in the early days of the technique [8], from the mid-1990's it has drawn increasing attention, both from the point of view of understanding deformation mechanisms and for the correct interpretation and analysis of diffraction data.

Of course, identification of the effects of anisotropy goes back further than the use of the neutron diffraction technique. Glocker [9] addressed the issue of elastic anisotropy in x-ray diffraction experiments as early as 1938, and in several subsequent publications Bollenrath et al. [10–12] and Smith and Wood [13,14] observed anisotropy-related effects following plastic deformation during bending, tension, and compression. The first direct demonstration of the existence of residual *intergranular stresses* following plastic strain caused by interactions between grains of different orientations, came in 1946 by Greenough [15], followed by a number of subsequent publications [16–18]. In fact, much of this work was probably inspired by Heyn [19], who already in 1921, in studying work hardening of metals, arrived at the conclusion that following plastic deformation, some kind of internal self-equilibrating stress–strain state pertained. Despite the fact that Heyn named these stresses and strains 'hidden elastic strains' and 'hidden (latent) stresses,' they were more commonly known as 'Heyn stresses' in Germany. He wrote:

> When fully unloading a metallic material, which has been strained beyond the yield point, a certain amount of elastic strain and therefore stress is still apparent. This is due to a frictional force acting as an obstacle and keeping the system in balance. It is this elastic strain and stress that I call hidden strain and hidden (latent) stress. (Translation from German courtesy of Dr. M. Preuss.)

The issue of intergranular stress, a Type II microstress, has therefore been acknowledged since the early 1920s, but only in recent years has the problem of measuring and calculating its effects in detail been taken up by those using neutron diffraction to measure strain and interpreting the data in terms of stress. This interplay between experiment and modeling has improved our understanding of the mechanics of the deformation of polycrystalline materials, and the field advanced rapidly toward the end of the 20th century.

In general, intergranular stresses generated during loading have two origins:

- Elastic anisotropy due to variation in elastic properties of single crystals with direction
- Plastic anisotropy due to the fact that crystals do not deform homogeneously, but rather only on specific well-defined crystal planes (slip planes) in specific directions (slip directions)

That different Bragg reflections behave differently under plastic strain has been catalogued by those using neutron diffraction for some time, perhaps beginning with Sayers in 1984 [8], who addressed the topic from a theoretical perspective, while realizing its practical implications. An example of some of these early observations is given in the work of Allen et al. [20], which is summarized in Figure 5.2 showing how the diffraction peaks for different *hkl* reflections of a duplex alloy respond differently to an externally applied load.

The duplex alloy is a two-phase system, but even within each phase of the alloy, the anisotropy in lattice strain between different reflections is marked. The observed *hkl* dependency may be presented by plotting the effective stiffness of the various *hkl* reflections as a function the anisotropy parameter

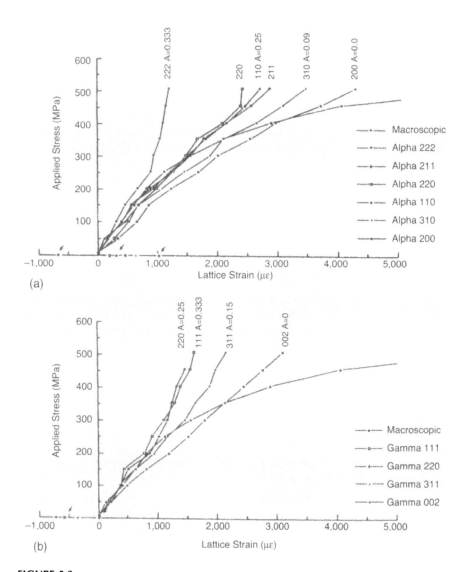

FIGURE 5.2

Lattice strain response from several Bragg reflections measured parallel to an applied uniaxial stress in each of the two phases of a duplex alloy: (a) α-ferritic phase, and (b) γ-austenitic phase. Residual strains upon unloading are marked on the abscissa and the anisotropy factors, A_{hkl}, (Equation (5.24)) are also shown [20].

A_{hkl} defined in Section 5.3.1, Equation (5.24). This is shown in Figure 5.3, from which we observe a large variation in the effective modulus with A_{hkl} for strain responses in both the longitudinal and transverse directions to an applied load, discussed more detail in Section 5.5.

Nonlinearities in the applied stress–lattice strain response occurring as a consequence of the onset of plastic flow have been well documented. An

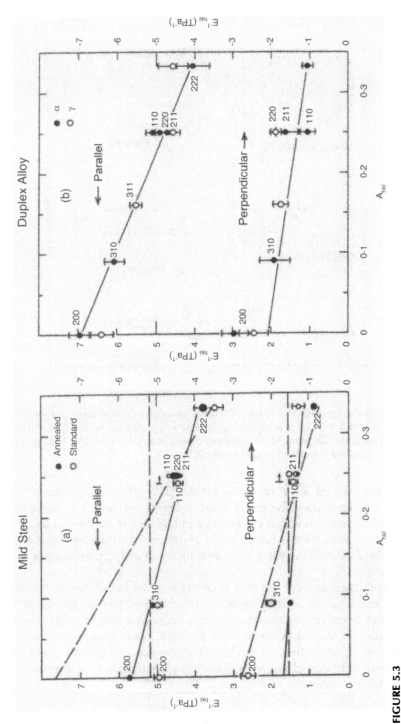

FIGURE 5.3

(a) Anisotropic variation of elastic compliance constants parallel and perpendicular to an applied uniaxial tensile stress for as-received (unannealed), and annealed mild steel. (b) Anisotropic variation of elastic compliance constants for the α and γ phase of duplex steel alloy. Solid lines are drawn through the measured points, and are related by ν = 0.3. Broken lines are calculated from the Voigt and Reuss expressions [20].

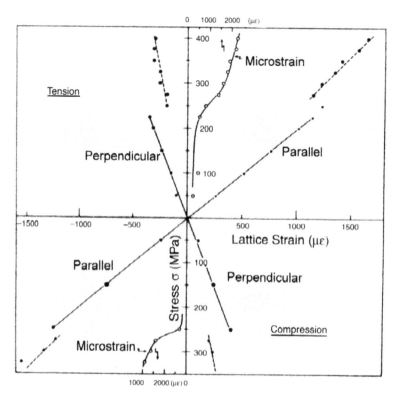

FIGURE 5.4
Lattice strain response from the (310) Bragg reflection in annealed mild steel measured parallel and perpendicular to a uniaxial applied load. Uncertainties are denoted by point sizes. The microstrain (Types II and III) parallel to the stress, evaluated in terms of peak broadening (full width at half of maximum intensity), is also shown [20].

example is the marked 'kink' observed for the (310) reflection for annealed mild steel upon exceeding the elastic limit in Figure 5.4, which is possibly associated with the macroscopic phenomena of Lüders bands. We will return to these aspects of nonlinear lattice strain response when discussing theoretical means of addressing the plastic anisotropy of polycrystalline aggregates in Section 5.6.

Many other observations have now been made of the anisotropy in lattice strain response, leading to discussions of how to interpret measured *hkl*-specific strains in order to obtain macrostrain. Without trying to cover every possible model, in the following sections the basic grain-to-grain micromechanics are introduced through models of increasing complexity, moving from the elastic behavior of single crystals through to the plastic deformation of polycrystalline materials, as outlined in Figure 5.5.

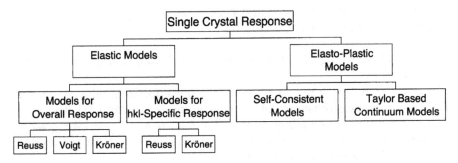

FIGURE 5.5
Established models for describing anisotropies in polycrystals based on single crystal behavior.

5.3 Elastic Anisotropy of Single Crystals

In order to understand the mechanics of elastic deformation of polycrystals from the perspective of strain measurement by diffraction, the most important property is the anisotropy in elastic constants. In an actual diffraction experiment, the *effective* elastic properties of a selected subset of grain orientations is sampled, and identified by a specific *hkl* reflection from a polycrystalline aggregate. Before addressing these *effective* properties, it is helpful to understand the underlying anisotropy in single-crystal elastic properties. The treatment here does not attempt to treat elastic anisotropy of crystal structures in general, which is well covered in texts such as Nye [1]. Instead, we focus on cubic and hexagonal structures that cover the majority of engineering materials. The discussion of the elastic anisotropy of single crystals and polycrystals in this and the next two sections closely follows that given by Noyan and Cohen [5].

The standard way of expressing the elastic properties of such crystals is to relate the strain components, ε_{ij}, to the stress components, σ_{ij}, through the fourth-rank *compliance tensor*, S_{ijkl}, or through the inverse expression giving the relation in terms of the fourth-rank *stiffness tensor*, C_{ijkl}, as given in Equation (5.1). These relations may be written in the form where a summation over repeated indices, here k and l, shown explicitly in Equation (5.1), is implied:

$$\varepsilon_{ij} = S_{ijkl}\,\sigma_{kl} \text{ and } \sigma_{ij} = C_{ijkl}\,\varepsilon_{kl} \tag{5.16}$$

Note that the conventional symbol for these fourth-rank tensors is just the opposite of the first letter in their names! It is common practice to express the fourth-rank tensor quantities $\underline{S}$ and $\underline{C}$, in terms of *matrices*, S_{mn} and C_{mn}, and the second-rank tensors $\underline{\underline{\sigma}}$ and $\underline{\underline{\varepsilon}}$, as σ_m and ε_m where the former and latter pair of suffixes reduce according to

$$11 \Rightarrow 1;\ 22 \Rightarrow 2;\ 33 \Rightarrow 3;\ 23\ \&\ 32 \Rightarrow 4;\ 13\ \&\ 31 \Rightarrow 5;\ 12\ \&\ 21 \Rightarrow 6 \qquad (5.17)$$

In going from tensor to matrix notation, a few additional rules are needed for the compliance constants, but not the stiffness constants [1]. These are that $S_{ijkl} = 1/2\ S_{mn}$ when either m or n is 4, 5, or 6; and $S_{ijkl} = 1/4\ S_{mn}$ when both m and n are 4, 5, or 6. Similarly, the strain components are adjusted so that $\varepsilon_{ij} = 1/2\ \varepsilon_m$ when m is 4, 5, or 6.

It should be noted that when the fourth (second) rank compliance or stiffness (stress and strain) tensors, S_{ijkl} and C_{ijkl} (σ_{ij} and ε_{ij}), are converted into matrix notation, following the rules given above, then the matrices S_{mn} and C_{mn} (σ_m and ε_m) are not tensors. When it is necessary to transform these properties to another reference frame, it is necessary to return to the fourth (second) rank tensor notation, and follow the standard rules for tensor transformation.

5.3.1   Elastic Anisotropy of Cubic Single Crystals

The conventional method for denoting or indexing lattice planes using Miller indices, *hkl*, was described in Sections 2.3.1 and 2.3.3. In the cubic system, whether it is face-centered cubic, body-centered cubic, or primitive cubic, the *hkl* lattice direction, denoted [*hkl*], and the normal to the *hkl* lattice planes, are parallel. To illustrate the Miller indexing notation, some commonly used planes are indexed in Figure 5.6. Due to symmetry in the properties of cubic crystal structures, the number of independent elastic constants is only three. This is described in greater detail in Nye [1], who also provides a graphical matrix representation of the distribution and relation between these constants, as reproduced in Figure 5.7 for some of the simpler crystal structures.

For a specific crystal direction [*hkl*] relative to crystal axes along direction 1 of a new reference coordinate system, the strain component is related to the stress component in that direction by Hooke's Law:

$$\varepsilon_{11} = (1/E)_{hkl}\ \sigma_{11} = S'_{1111}\ \sigma_{11} \qquad (5.18)$$

The reciprocal of the appropriate Young's modulus — the compliance S'_{1111} here indicated by the prime — may be expressed in terms of the compliance tensor referred to the crystal axes, S_{mnop}, following standard transformation rules for rotating the axes of fourth-rank tensors, with summation implied over *m*, *n*, *o*, and *p*:

$$(1/E)_{hkl} = S'_{1111} = l_{1m}l_{1n}l_{1o}l_{1p}\ S_{mnop} \qquad (5.19)$$

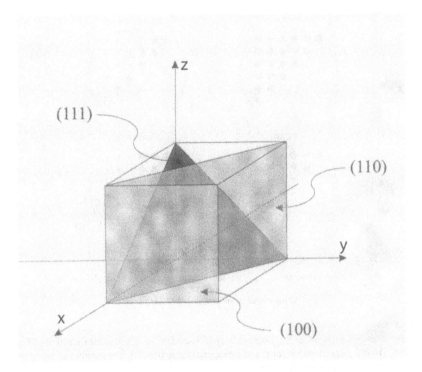

FIGURE 5.6
Selected crystallographic planes of cubic system indexed by Miller indices (*hkl*) notation.

Here l_{1j}, $j = 1, 2,$ or 3, are the direction cosines between the specific direction in question [*hkl*] and the crystal axes. Adopting the tensor contraction rules given above, and expressing the transformation in matrix notation, one finds for cubic crystals that $S_{1111} = S_{11}$, $S_{1122} = S_{12}$, and $S_{1212} = S_{66}/4 = S_{44}/4$. Thus, the reciprocal Young's modulus is given by:

$$(1/E)_{hkl} = S'_{1111} = S_{11} - 2[S_{11} - S_{12} - 1/2\ S_{44}]\ (l_{11}l_{12} + l_{12}l_{13} + l_{13}l_{11}) \quad (5.20)$$

The orientation dependency of the elastic anisotropy with direction [*hkl*], normally denoted A_{hkl}, is governed by the latter term in round brackets, which varies between 0 and 1/3.

$$A_{hkl} = (l_{11}l_{12} + l_{12}l_{13} + l_{13}l_{11}), \quad (5.21)$$

whereas the amount of anisotropy is governed by the term:

$$S_0 = [S_{11} - S_{12} - 1/2\ S_{44}] \quad (5.22)$$

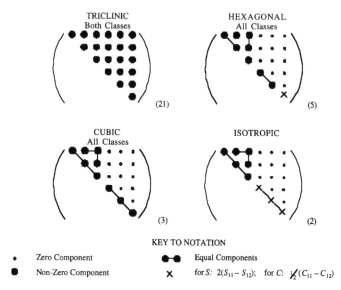

FIGURE 5.7
A "matrix" representation of the number and distribution of independent elastic coefficients for the S_{mn} and C_{mn} matrices for cubic and hexagonal structures [1]. The number of independent elastic constants is indicated in brackets.

The former is dependent on the crystallographic plane, and the latter on the specific material. The degree of anisotropy can be expressed in terms of the ratio

$$2(S_{11} - S_{12})/S_{44} \qquad (5.23)$$

This is equal to 1 for isotropic crystal properties, and deviates increasingly from 1 with increasing degree of anisotropy.

The above expressions are all given in terms of direction cosines of any direction in the crystal. However, with regard to diffraction experiments, only specific crystal directions are of interest, namely those directions corresponding to allowed reflections *hkl* according to the selection rules. Expressing the direction cosines in Equation (5.21) in terms of these Miller indices, the A_{hkl} parameter is given by

$$A_{hkl} = (h^2k^2 + k^2l^2 + l^2h^2)/(h^2 + k^2 + l^2)^2 \qquad (5.24)$$

This quantity is seen to vary from a value of 0 for the [100] crystal direction to 1/3 for the [111] direction. It should be noted that although some reflections, such as the *fcc* (333) and (511), may have the same stress-free lattice spacing, their lattice planes have different elastic anisotropies, since $A_{333} \neq A_{511}$. The most common situation for most metals is that the term in square

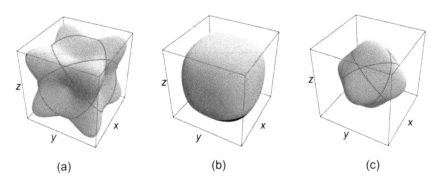

FIGURE 5.8
The radius vector in each figure is proportional to the magnitude of Young's modulus relating uniaxial stress to strain in that direction for various cubic crystal structures: (a) copper, (b) diamond, and (c) rock salt [21]. The axes are defined in Figure 5.6.

brackets, given in Equation (5.22), which precedes the orientation term in Equation (5.20), is positive, in which case the [111] direction is found to be the most stiff and the [100] direction the most compliant. However, for certain combinations of compliance constants, this term will be negative, in which case the stiff and compliant directions are interchanged. The variation in Young's modulus with direction of the strain and stress for some cubic crystals is shown in Figure 5.8.

Typical values of the elastic constants and Young's modulus for selected crystallographic directions [hkl] in *fcc* crystals [22–24], as well as values of the ratio given in Equation (5.23) as a measure of the anisotropy, are summarized in Table 5.1. The materials included in Table 5.1 represent a broad range in terms of the degree of anisotropy, from aluminum, which is nearly isotropic, to copper and austenitic stainless steel, here denoted Fe^γ, which are highly anisotropic. For aluminum, we note that the maximum deviation in stiffness from the typical isotropic macroscopic value of 70 GPa, only amounts to around 8% to 9%, whereas for stainless steel, with a typical bulk polycrystalline Young's modulus of, say, 195 GPa, the deviation in stiffness exceeds 50%. Note also that for all the four materials selected here, the parameter describing the degree of anisotropy is positive, with the consequence that the [111] is the stiffest direction, and the [200] the most compliant one.

In Table 5.2, corresponding data are given for common *bcc* materials. Note that in contrast to the *fcc* materials, not all of *bcc* materials show the [111] direction to be the stiffest. For materials such as molybdenum (Mo) and chromium (Cr), the degree of anisotropy is less than unity so that the [111] direction becomes the most compliant one, and the [200] direction the stiffest. Vanadium, V, and niobium, Nb, also fall into this category of *bcc* elements. For ferritic iron (Fe^α), the maximum deviation from a typical bulk polycrystalline modulus of 210 GPa, amounts to 30% to 40%; hence, ferritic steel

TABLE 5.1

Values of Single-Crystal Elastic Constants and E_{hkl} for Selected hkl Reflections of Common *fcc* Cubic Crystals

Modulus	Units	Al[a]	Cu[a]	Ni[b]	Fe$^{\gamma}$[c]
C_{11}	$\times 10^{10}$ Pa	10.82	16.84	24.40	20.46
C_{12}	$\times 10^{10}$ Pa	6.13	12.14	15.80	13.77
C_{44}	$\times 10^{10}$ Pa	2.85	7.54	10.20	12.62
S_{11}	$\times 10^{-11}$ Pa^{-1}	1.57	1.50	0.83	1.07
S_{12}	$\times 10^{-11}$ Pa^{-1}	−0.57	−0.63	−0.33	−0.43
S_{44}	$\times 10^{-11}$ Pa^{-1}	3.51	1.33	0.98	0.79
$2(S_{11}-S_{12})/S_{44}$		1.22	3.20	2.37	3.80
E_{200}	GPa	63.7	66.7	120.5	93.5
E_{311}	GPa	69.0	96.2	161.4	138.3
E_{420}	GPa	69.1	97.0	162.4	139.6
E_{531}	GPa	71.1	113.6	182.9	165.9
E_{220}	GPa	72.6	130.3	202.0	193.2
E_{331}	GPa	73.6	143.6	216.2	215.5
E_{111}	GPa	76.1	191.1	260.9	300.0

[a] From Hertzberg [22].
[b] From Ledbetter [24].
[c] From Simmons and Wang [23].

TABLE 5.2

Values of Single-Crystal Elastic Constants and E_{hkl} for Selected hkl Reflections of Common *bcc* Cubic Crystals

Modulus	Units	Fe$^{\alpha}$[a]	V[b]	Mo[a]	Cr[b]
C_{11}	$\times 10^{10}$ Pa	23.70	19.60	46.0	35.0
C_{12}	$\times 10^{10}$ Pa	14.10	13.30	17.6	6.78
C_{44}	$\times 10^{10}$ Pa	11.60	6.70	11.0	10.08
S_{11}	$\times 10^{-11}$ Pa^{-1}	0.80	1.13	0.28	0.30
S_{12}	$\times 10^{-11}$ Pa^{-1}	−0.28	−0.46	−0.08	−0.05
S_{44}	$\times 10^{-11}$ Pa^{-1}	0.86	1.49	0.91	0.99
$2(S_{11}-S_{12})/S_{44}$		2.51	2.13	0.79	0.71
E_{200}	GPa	125.0	88.5	357.1	333.3
E_{310}	GPa	146.4	102.3	336.6	306.7
E_{411}	GPa	149.8	104.4	334.1	303.5
E_{321}	GPa	210.5	141.3	305.3	268.5
E_{112}	GPa	210.5	141.3	305.3	268.5
E_{110}	GPa	210.5	141.3	305.3	268.5
E_{222}	GPa	272.7	176.5	291.3	252.1

[a] From Hertzberg [22].
[b] From Simmons and Wang [23].

displays a smaller degree of elastic anisotropy than does austenitic steel. Tungsten, for which the ratio in Equation (5.23) is equal to 1, is the ultimate case having perfectly isotropic properties. Accordingly, a plot similar to those in Figure 5.8 would give a sphere.

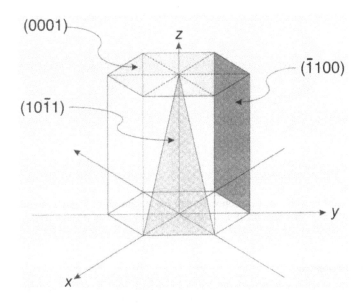

FIGURE 5.9
Selected crystallographic planes in hexagonal system indexed by the Miller–Bravais indexing system.

5.3.2 Elastic Anisotropy of Hexagonal Crystals

The most common way of identifying crystal directions and lattice plane normals in the hexagonal crystal structure is to introduce an extra index, i, described in Section 2.3.4. To illustrate the indexing of lattice planes in the hexagonal crystal structure, some commonly used planes are indexed in Figure 5.9. Symmetry considerations mean that the number of independent elastic constants for the hexagonal system is five, and the matrix representation is shown in Figure 5.7.

Following the approach given above for the cubic system, the strain component due to stress applied along a specific direction of the crystallite given by the unit vector, $\hat{l}$, is given by the elastic compliance [1]:

$$(1/E)_l = (1 - l_3^2)^2\, S_{11} + l_3^4\, S_{33} + l_3^2\, (1 - l_3^2)(2\, S_{13} + S_{44}), \qquad (5.25)$$

where l_3 is the cosine of the angle between the unit vector, $\hat{l}$, and the c-axis of the crystal. It is evident from the above equation that the Young's modulus varies only with angle between the direction of the strain and stress and the c-axis. This is illustrated in Figure 5.10 for selected hexagonal crystals.

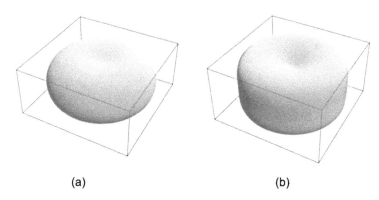

(a)   (b)

FIGURE 5.10
The radius vector in each figure is proportional to the magnitude of Young's modulus relating the stress to strain in that direction for hexagonal (a) cadmium and (b) zinc [21].

5.4   The Bulk Elastic Response of Polycrystals

In Section 4.2.2 and in Figures 5.2, 5.3, and 5.4, it was seen that even under elastic loading, the strains measured using different lattice planes in a poly-crystalline material are different. In fact, it is found that the slope of the lattice strain response to an applied stress usually lies between the bulk value and the single crystal value. While the strains ε_{ij}^{hkl} can be related to a single value of the applied (macro)stress using diffraction elastic constants (Section 5.1.2) calibrated by measuring the lattice strains induced under uni-axial loading of a sample of the material, their variation with *hkl* can tell us much about fundamental aspects of the level of inhomogeneous formation developing within the polycrystal.

In modeling the mechanical response of a polycrystalline aggregate, we have to distinguish between whether the aim is to predict macroscopic properties, such as the *bulk* Young's modulus and shear modulus, or whether the aim is to predict or estimate the *hkl* lattice strain response of selected families of grains having a specific orientation as measured in diffraction experiments. The latter case is relevant even from an engineering perspective, as we are interrogating the mechanical response of a polycrystalline aggregate through the *hkl* reflections, and failure to understand this can lead to misinterpretation. It also provides insight into how microstructural aspects, such as anisotropy in single crystal properties and nonrandom dis-tribution of grain orientations, affect the macroscopic response of the mate-rials. Nevertheless, we shall begin by outlining the theoretical approaches to predicting the overall bulk macroscopic properties. Of the many schemes that have been proposed, the most important are those of Voigt and Reuss, which render theoretical bounds on stiffness, and the more sophisticated approach of Kröner.

5.4.1 Voigt and Reuss Models for Macroscopic Bulk Properties

5.4.1.1 Voigt Model for Bulk Properties

In the Voigt modeling scheme, each grain in the aggregate is assumed to experience the same uniform strain, and the aggregate's elastic properties are calculated by averaging the elastic stiffness, C_{ij}, over all elements of the aggregate [25]. This results in expressions for the macroscopic bulk modulus, K, and shear modulus, μ, written with superscript V to denote Voigt [23]:

$$K^V = \frac{F+2G}{3} \tag{5.26}$$

$$\mu^V = \frac{F-G+3H}{5} \tag{5.27}$$

where the constants F, G, and H are given in terms of the elastic properties of the single crystal by

$$3F = C_{11} + C_{22} + C_{33} \tag{5.28a}$$

$$3G = C_{23} + C_{31} + C_{12} \tag{5.28b}$$

$$3H = C_{44} + C_{55} + C_{66} \tag{5.28c}$$

In this isotropic polycrystalline average case, the bulk and shear moduli are directly related to Young's modulus, E^V, and Poisson's ratio, v^V, through the expressions connecting the moduli of isotropic materials [26]:

$$E^V = 3K^V (1 - 2v^V) = 2(1 + v^V)\mu^V \tag{5.29}$$

$$v^V = \frac{3K^V - 2\mu^V}{2(3K^V + \mu^V)} \tag{5.30}$$

For cubic materials these become

$$K^V = (C_{11} + 2C_{12})/3 \tag{5.31}$$

$$\mu^V = [(C_{11} - C_{12}) + 3C_{44}]/5 \tag{5.32}$$

TABLE 5.3

Bulk, Shear, Young's Moduli, and Poisson's Ratio for Polycrystalline Aggregates Following the Voigt Modeling Scheme

	Units	Al	Cu	Ni	Fe$^\gamma$	Fe$^\alpha$	V	Mo	Cr
K^V	GPa	76.9	137.10	186.7	160.0	173.0	154.0	270.7	161.9
μ^V	GPa	26.5	54.6	78.4	89.1	88.8	52.8	122.8	116.9
E^V	GPa	71.3	144.7	206.3	225.5	227.5	142.2	320.0	282.7
v^V		0.35	0.32	0.32	0.27	0.28	0.35	0.30	0.21

Based on these expressions, and the elastic constants of Table 5.1 and Table 5.2, the overall elastic properties have been calculated for a range of common *fcc* and *bcc* materials, and the results are presented in Table 5.3.

5.4.1.2 Reuss Model for Bulk Properties

In the Reuss modeling scheme, each grain is assumed to experience the average stress applied to the aggregate, and the aggregate properties are calculated by averaging the elastic compliances, S_{ij}, over all elements of the aggregate [27]. However, the model does not allow for continuity of displacement at grain boundaries, as the strains in two adjacent crystallites are different. The bulk and shear modulus are given by expressions of the form [23]:

$$K^R = \frac{1}{(3a + 6b)} \tag{5.33}$$

$$\mu^R = \frac{5}{(4a - 4b + 3c)} \tag{5.34}$$

where the constants a, b, and c are given in terms of the elastic properties of the single crystal by

$$3a = S_{11} + S_{22} + S_{33} \tag{5.35a}$$

$$3b = S_{23} + S_{31} + S_{12} \tag{5.35b}$$

$$3c = S_{44} + S_{55} + S_{66} \tag{5.35c}$$

The expressions for E^R and v^R follow from the isotropic bulk average expressions Equations (5.29) and (5.30). For polycrystalline cubic materials,

$$K^R = (C_{11} + 2C_{12})/3 = K^V \tag{5.36}$$

$$\mu^R = 5/[4(S_{11} - S_{12}) + 3S_{44}] \tag{5.37}$$

TABLE 5.4

Bulk, Shear, Young's Moduli, and Poisson's Ratio for Polycrystalline Aggregates Following the Reuss Modeling Scheme Using Data from Tables 5.1 and 5.2

	Units	Al	Cu	Ni	Fe$^\gamma$	Fe$^\alpha$	V	Mo	Cr
K^R	GPa	76.9	137.1	186.7	160.0	173.0	154.0	270.7	161.9
μ^R	GPa	26.2	40.0	66.0	59.7	72.5	46.2	119.9	114.4
E^R	GPa	70.6	109.4	177.9	159.2	185.2	126.3	314.5	279.3
ν^R		0.35	0.37	0.34	0.33	0.32	0.36	0.31	0.21

The overall elastic properties have been calculated for selected *fcc* and *bcc* materials, and the results are presented in Table 5.4. The Voigt and Reuss models have been shown by Hill [28] to render lower and upper bounds on the aggregate elastic properties, and in fact the average of the two is in quite close correspondence with experimental observations for texture-free aggregates. Hashin and Shtrikman [29,30] proposed more narrow bounds based on a variational principle, but by far the most accepted approach is the one suggested by Kröner in 1958 [31].

5.4.2 Kröner Model for Macroscopic Bulk Properties

In contrast to the idealized approach by Voigt and Reuss, which prescribes either the strain or the stress to be identical in all constituents of the aggregate, the Kröner model allows both stresses and strains to vary from grain to grain. In the self-consistent scheme proposed by Kröner [31], it is merely prescribed that the whole aggregate is exposed to a specific homogeneous average stress field and an associated homogeneous average strain field. In essence, this average stress, $\bar{\sigma}$, for a single-phase material is analogous to the concept of a Type I macrostress, $\underline{\sigma}^I$, such as would arise throughout the gauge section of a tensile test specimen. This average stress does not prevent the stress or the strain of the individual constituent grains, Ω, $\sigma_{ij}(\Omega)$ and $\varepsilon_{ij}(\Omega)$, from varying with grain orientation or shape. Indeed, in the elastic regime a linear relationship exists between this homogeneous average stress, or strain, field, and the stress, or strain, of the individual constituent grains:

$$\sigma_{ij}(\Omega) = p_{ijkl}\varepsilon_{kl}^I \qquad (5.38)$$

$$\varepsilon_{ij}(\Omega) = q_{ijkl}\sigma_{kl}^I \qquad (5.39)$$

Specifically, Kröner chose to express this linear relationship, that is, p_{ijkl} and q_{ijkl}, in terms of the average elastic properties of the aggregate, C_{ijkl} and S_{ijkl}, respectively, modified or perturbed by the addition of grain interaction tensors that take into account the anisotropy and particular shape of the grains. This modification can be interpreted by saying that apart from the

TABLE 5.5

Bulk, Shear, Young's Moduli, and Poisson's Ratio for Polycrystalline Aggregates
Following the Kröner Modeling Scheme Using Data from Tables 5.1 and 5.2

	Units	Al	Cu	Ni	Fe$^\gamma$	Fe$^\alpha$	V	Mo	Cr
K^K	GPa	76.9	137.1	186.7	160.0	173.0	154.0	270.7	161.9
μ^K	GPa	26.4	48.2	72.8	75.6	82.1	49.9	121.9	115.3
E^K	GPa	71.0	129.4	193.3	195.9	212.7	135.1	318.0	279.5
ν^K		0.35	0.34	0.33	0.30	0.30	0.35	0.30	0.21

normal Hooke's Law relationship between $\varepsilon_{ij}(\Omega)$ and σ_{kl}^I through the average elastic compliance tensor of the aggregate, S_{ijkl}, an additional contribution stems from the fact that the elastic constants of the specific grain deviates from the values of the aggregate. Considering the special case of an aniso-tropic spherical inclusion in a homogeneous effective medium, the solution to this interaction problem was proposed by Eshelby in 1957 [32], in terms of the elastic properties of the grain, those of the effective medium, and the grain shape.

Restricting our discussion to cubic crystallites, the bulk modulus is iden-tical to the single crystal bulk modulus:

$$K^K = \frac{(C_{11} + 2C_{12})}{3} = K^V = K^R \tag{5.40}$$

A more involved relation exists for μ^K, which can be solved either graphically or numerically [5,33]. Having determined the overall isotropic bulk and shear modulus, Young's modulus and Poisson's ratio are found through Equations (5.29) and (5.30). Using the single crystal elastic constants of Tables 5.1 and 5.2, these aggregate properties have been calculated for selected materials, and the results are given in Table 5.5.

Note the slight variations in the bulk modulus, K, between the Reuss model and the results of the Voigt and Kröner models, given in the tables. All models ought to render an identical bulk modulus, and the differences stem from the fact that the Reuss model predictions are based on compliance constants, whereas the latter two model predictions are based on stiffness constants. The differences in K indicate that the single crystal data reported in Tables 5.1 and 5.2 are not precisely the inverse of one another, but rather originate from different sources.

On comparing the predictions of the three modeling schemes, we observe that average Reuss and Voigt model results are at no point more than 3% away from the predictions of the more realistic and elaborate Kröner model, which is known to render properties closely resembling experimental obser-vations. As Hill suggested, using the average of the Reuss and Voigt models would be sufficiently accurate for all practical purposes.

Note that the macroscopic bulk aggregate moduli have been calculated in the three cases above on the basis of negligible preferred orientation or texture. It is possible to modify the Kröner model and the other models to take texture into account by performing averaging integrals over the orientation distribution function rather than over all random orientations [34].

5.5 *hkl*-Specific Response in a Polycrystal Undergoing Elastic Deformation

In the previous section, we examined the overall bulk elastic moduli of homogeneous polycrystalline aggregates predicted by three models, which by averaging all crystallite orientations are isotropic for a nontextured material. However, of more direct relevance to the interpretation of lattice strain measurements by diffraction are the model predictions of the elastic response of the specific *hkl* reflections used in converting strain to stress. It is important to remember that in collecting diffracted intensity from a reflection *hkl*, one is measuring the average lattice strain response of all crystallites whose *hkl* plane normals are in the appropriate direction, irrespective of the crystal orientations perpendicular to the *hkl* plane normal. Here we consider the plane-specific Young's modulus E_{hkl} parallel to an applied uniaxial load, and Poisson's ratio v_{hkl} giving the strain measured perpendicular to an applied loading stress in terms of the strain measured parallel to the loading direction.

5.5.1 Voigt and Reuss Models

5.5.1.1 *Voigt Model*

As the Voigt model rests on the assumption that all grains in a polycrystalline aggregate experience the *same* uniform strain, this model does not provide any orientation dependency, and all *hkl* reflections inherently render the same lattice strain response parallel to an external load. Thus, Young's modulus and Poisson's ratio are isotropic and the same for all *hkl*, and are given by Equations (5.29) through (5.32).

5.5.1.2 *Reuss Model*

In applying the Reuss model, all crystallites will experience the same stress. In order to determine the strain response to an applied uniaxial stress experienced by cubic crystallites giving rise to the *hkl* reflection, the contributions from all crystallites with *hkl* planes perpendicular to the [*hkl*] direction are averaged. Two important cases are when these *hkl* planes are perpendicular to the applied stress (i.e., when the lattice strain is measured in the loading stress direction), and when the *hkl* planes contain the loading stress direction (i.e., when this is perpendicular to the [*hkl*] direction).

The polycrystalline average strain measured parallel to the applied uniaxial stress σ^A in the [hkl] direction, $\varepsilon_{//}^{hkl}(poly)$, will be the same as that for a single crystal given by Equations (5.18), (5.20), and (5.24), since the response is the same for all orientations of the contributing crystallites perpendicular to the stress direction [hkl]. From Equation (5.20),

$$\varepsilon_{//}^{hkl}(poly) = \sigma^A\left(S_{11} - 2S_0 A_{hkl}\right) \tag{5.41}$$

and Young's modulus E_{hkl}^R is given by

$$E_{hkl}^R = \frac{1}{\left(S_{11} - 2S_0 A_{hkl}\right)} \tag{5.42}$$

The strain response perpendicular to a loading stress, $\varepsilon_{\perp}^{hkl}(poly)$, is more complicated, and is the equivalent to that obtained by averaging the strain in the [hkl] direction of the crystallites for all load directions in the plane normal to [hkl]. Averaging over all these possible rotations about the hkl lattice plane normal renders the average strain, $\varepsilon_{\perp}^{hkl}(poly)$, of the aggregate as function of the parameter A_{hkl} as

$$\varepsilon_{\perp}^{hkl}(poly) = \sigma^A(S_{12} + S_0 A_{hkl}) \tag{5.43}$$

We can define a Poisson's ratio relating the lattice strains perpendicular and parallel to the loading stress as follows:

$$v_{hkl}^R(poly) = -\frac{\varepsilon_{\perp}^{hkl}(poly)}{\varepsilon_{//}^{hkl}(poly)} = -\frac{(S_{12} + S_0 A_{hkl})}{(S_{11} - 2S_0 A_{hkl})} \tag{5.44}$$

The hkl specific elastic moduli of the Reuss modeling scheme have been calculated for a selected range of *fcc* and *bcc* materials representing a wide range in degrees of elastic anisotropy and given in Tables 5.6 and 5.7. Recalling the conceptual flaws inherent in the Reuss modeling scheme, these numerical estimates provide only a bound on the elastic properties. More refined approaches and algorithms for the calculation of the hkl specific lattice strain response of polycrystalline aggregates have therefore been developed. The Kröner approach is discussed below.

5.5.2   Kröner Model

The derivation of hkl-specific elastic moduli following the Kröner model rests on the basic principles given for the overall bulk response reviewed in Section 5.4.2. An expression for the strain in a specific sample direction, the measurement direction [hkl], is found by averaging this strain over contributing

TABLE 5.6

Tabulated Values of *hkl*-Specific E_{hkl}^R for Selected *hkl* (in GPa) and Poisson's Ratio for Common Polycrystalline *fcc* Engineering Materials Following the Reuss Modeling Scheme

		(200)	(311)	(420)	(531)	(220)	(422)	(331)	(111)
E_{hkl}^R	Al	63.7	69.0	69.1	71.1	72.6	72.6	73.6	76.1
	Cu	66.7	96.2	97.0	113.6	130.3	130.3	143.6	191.1
	Ni	120.5	161.4	162.4	182.9	202.0	202.0	216.2	260.9
	Feγ	93.5	138.3	139.6	165.9	193.2	193.2	215.5	300.0
ν_{hkl}^R	Al	0.36	0.35	0.35	0.35	0.34	0.34	0.34	0.34
	Cu	0.42	0.38	0.38	0.36	0.34	0.34	0.33	0.27
	Ni	0.40	0.36	0.36	0.34	0.33	0.33	0.32	0.28
	Feγ	0.40	0.35	0.35	0.33	0.30	0.30	0.27	0.19

TABLE 5.7

Tabulated Values of *hkl*-Specific E_{hkl}^R for Selected *hkl* (in GPa) and Poisson's Ratio for Common Polycrystalline *bcc* Engineering Materials Following the Reuss Modeling Scheme

		(200)	(310)	(411)	(420)	(321)	(112)	(110)	(222)
E_{hkl}^R	Feα	125.0	146.4	149.8	168.9	210.5	210.5	210.5	272.7
	V	88.5	102.3	104.4	116.3	141.3	141.3	141.3	176.5
	Mo	357.1	336.6	334.1	322.2	305.3	305.3	305.3	291.3
	Cr	333.3	306.7	303.5	288.7	268.5	268.5	268.5	252.1
ν_{hkl}^R	Feα	0.35	0.32	0.32	0.30	0.25	0.25	0.25	0.17
	V	0.41	0.39	0.39	0.38	0.35	0.35	0.35	0.31
	Mo	0.29	0.30	0.30	0.31	0.32	0.32	0.32	0.33
	Cr	0.17	0.19	0.20	0.21	0.23	0.23	0.23	0.25

crystallites with all orientations in the plane perpendicular to this direction. This is related to the average stress in a similar way to the one used for the bulk response. The somewhat complicated calculation procedure has been given explicitly for the case of crystalline aggregates having a cubic structure [35].

The Kröner elastic modulus and Poisson's ratio have been calculated as a function of A_{hkl} for selected materials, and are listed in Tables 5.8 and 5.9. The results are also illustrated graphically in Figure 5.11 for three materials: aluminum (Al), ferritic iron (Feα), and stainless steel (Feγ) representing a range in degree of elastic anisotropy. Observe the monotonic variation in modulus and Poisson's ratio with A_{hkl}, as well as the large differences in magnitude between the different materials. As already mentioned in discussing single-crystal properties, the dependency of Young's modulus and Poisson's ratio on A_{hkl} can be positive or negative, according to the specific values of the elastic constants. Accordingly, the corresponding variations for molybdenum and chromium display the opposite functional dependency on A_{hkl}. The Kröner predictions of the *hkl*-specific lattice strain response find widespread use in the analysis of diffraction data, as will be discussed in further detail in Section 5.6.4.

TABLE 5.8

Tabulated Values of *hkl*-Specific E_{hkl}^K for Selected *hkl* (in GPa) and Poisson's Ratio for Common *fcc* Engineering Materials Following the Kröner Modeling Scheme

		(200)	(311)	(420)	(531)	(220)	(422)	(331)	(111)
E_{hkl}^K	Al	67.6	70.2	70.3	71.2	71.9	71.9	72.3	73.4
	Cu	101.1	122.0	122.5	131.5	139.1	139.1	144.3	159.0
	Ni	160.0	185.0	185.6	195.6	203.9	203.9	209.5	224.6
	Feᵧ	149.1	183.5	184.4	199.5	212.7	212.7	221.8	247.9
v_{hkl}^K	Al	0.35	0.35	0.35	0.35	0.34	0.34	0.34	0.34
	Cu	0.38	0.35	0.35	0.34	0.33	0.33	0.32	0.31
	Ni	0.36	0.33	0.33	0.33	0.33	0.33	0.31	0.30
	Feᵧ	0.34	0.31	0.31	0.29	0.28	0.28	0.27	0.24

TABLE 5.9

Tabulated Values of *hkl*-Specific E_{hkl}^K for Selected *hkl* (in GPa) and Poisson's Ratio for Common *bcc* Engineering Materials Following the Kröner Modeling Scheme

		(200)	(310)	(411)	(420)	(321)	(112)	(110)	(222)
E_{hkl}^K	Feᵅ	173.3	189.1	191.3	203.0	225.5	225.5	225.5	250.6
	V	114.0	122.6	123.9	130.3	141.7	141.7	141.7	154.1
	Mo	336.1	327.7	326.6	321.4	313.7	313.7	313.7	306.9
	Cr	305.8	293.4	291.8	284.4	273.7	273.7	273.7	264.4
v_{hkl}^K	Feᵅ	0.33	0.32	0.32	0.30	0.28	0.28	0.28	0.26
	V	0.38	0.37	0.37	0.36	0.35	0.35	0.35	0.33
	Mo	0.29	0.30	0.30	0.30	0.31	0.31	0.31	0.31
	Cr	0.19	0.20	0.20	0.21	0.22	0.22	0.22	0.23

5.6   *hkl*-Specific Response in a Polycrystal Undergoing Plastic Deformation

As seen in Figures 5.2 and 5.4, with the onset of plastic deformation the lattice responses of different *hkl* reflections to an applied uniaxial load become markedly different. This is due to the creation of intergranular stresses caused by the heterogeneous nature of plastic flow at the grain level. Without a proper understanding of how the various reflections are affected by plastic deformation, there is a high risk of unreliable interpretation of strain data for samples that have been plastically deformed at some stage in their life. In addition to the increasing range of experiments, modeling plays an important role in helping to understand these effects. To date, most of the experimental and modeling work has focused on characterization of plastic anisotropy effects in cubic materials under uniaxial tension. It is expected that the scope of study will be widened as understanding of the phenomena improves.

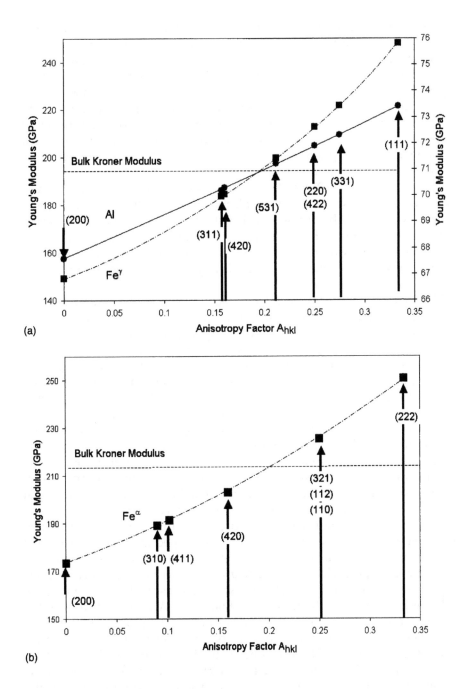

FIGURE 5.11
Young's modulii calculated from the Kröner model, E_{hkl}^{K}, as a function of the orientation parameter A_{hkl} for (a) *fcc* Fe$^{\gamma}$ (referred to left axis) and Al (referred to right axis), and (b) for *bcc* Fe$^{\alpha}$ compared to the Kröner prediction of the overall macroscopic aggregate stiffness, E^{K}.

One of the most fundamental issues in modeling the mechanics of polycrystal deformation is to take proper account of the interaction between a grain and its surroundings. This is a complex task, as it depends on the orientation relationship of the grain with each of its neighbors and the shape of the grain boundaries, as well as their crystallographic orientations. Given this complexity, it is not surprising that the most common approach to modeling polycrystal deformation is through so-called one-site modeling schemes, which do not consider direct grain-to-grain interactions explicitly, but rather take each grain in the aggregate to interact, in some more or less idealized way, with its surroundings.

The early models of Taylor [36] and Sachs [37] constitute simplified representations. In a sense, these are plastic analogs of the Voigt and Reuss models for elastic behavior, since either strain or stress, respectively, is considered uniform throughout the aggregate in these models. Similarly, their description of the way that grains interact with their surroundings is seriously flawed. Hence, like their elastic analogs, they serve as bounds for the real behavior of a polycrystal.

The Taylor–Bishop–Hill [38,39] approach is based on the assumption that every grain in the polycrystal experiences the same strain and is not allowed to deviate from the average strain in the aggregate. This approach neglects the fact that the behavior of individual grains may be affected by the elastic anisotropy in the grain, and is not solely dictated by its surroundings.

An important step toward a more realistic consideration of the interaction between a grain and its surroundings was taken by Kröner in 1961 [35] and Budianski and Wu in 1962 [40]. They considered the surroundings to be an elastic medium, retaining the initial elastic properties of the aggregate with which the individual grain interacted. They allowed stresses and strains in each grain to deviate from the average, taking into account the anisotropy in and orientation of the individual grain. This appears to be a very reasonable assumption in the elastic regime of deformation, but naturally it becomes less realistic as the plastic part of deformation becomes dominant.

The Kröner and Budianski–Wu approach renders a very stiff coupling between a grain and its surroundings, and, especially when considering stresses and strains at a grain size scale, more compliant models are required. More realistic approaches to the modeling of the mechanics of polycrystalline aggregates are the so-called elastoplastic self-consistent (EPSC) modeling schemes. In contrast to the Kröner and Budianski–Wu approach, these EPSC schemes take full account of the changes in material properties of the surroundings brought about as a consequence of progressive plastic deformation of the aggregate. The principles behind them rest on the equivalent inclusion method by Eshelby [32,41], which formulates how an elastically anisotropic inclusion in a homogeneous effective medium accommodates deformation brought about by some external loading of the medium. Among the many reviews of the Eshelby formulation, especially noteworthy are Mura [42], Taya and Arsenault [43], and Clyne and Withers [44].

The EPSC approaches are described in some detail in the following sections as a means of exploring phenomena that depend on the average orientation of the grains giving rise to the diffraction peaks *hkl*. However, the EPSC models neglect nearest-neighbor grain-to-grain interactions. Finite element modeling, on the other hand, enables one to consider the effect on each grain, or subvolume of a grain, of its specific neighbors. Although finite element approaches require consideration of more grains to achieve a statistically representative ensemble, they do enable actual intergranular, grain to grain, interactions to be described. By using constitutive equations based on slip deformation, crystal-plasticity finite element modeling (CPFEM) has been used to reveal the significance of intergranular strain nonhomogeneity on the development of texture through deformation [45–47], as well as to model the development of internal strains [48,49]. This method holds tremendous promise for the future as computing power increases, but it is not referred to further in this section because the computationally simpler EPSC models demonstrate the important phenomena more transparently.

5.6.1 Elastoplastic Self-Consistent Models

In EPSC models, each grain is considered as an inclusion of regular shape (i.e., ellipsoidal or spherical) embedded in a homogeneous effective medium having the instantaneous properties of the polycrystalline aggregate. Neither stresses nor strains are prescribed *a priori* in the individual grains; instead, the boundary conditions on the aggregate are prescribed, as well as the external loading in the form of stress or strain increments. In a stepwise manner, each grain is then considered successively, with the underlying requirement that the weighted average of the stress increments and strain increments of all grains corresponds to the macroscopic stress and strain increments imposed on the aggregate.

EPSC models are essentially all based on the same formalism given by Hill in 1965–1967 [50–53], which was addressed comprehensively by Hutchinson in 1970 [54]. The original focus in the earlier modeling approach was on predicting macroscopic stress–strain curves, texture, and yield loci [55–58]. Not until the late 1980s were modeling schemes proposed with specific emphasis on lattice strains and stresses in polycrystalline aggregates [59,60]. Two implementations that specifically addressed the lattice strain evolution under elastoplastic loading of polycrystalline aggregates have found widespread use in recent years with regard to neutron diffraction strain–stress measurement. Both of these address the interpretation of *hkl*-specific lattice strains as monitored by diffraction techniques. They are based on essentially the same foundations as those provided by Hill and Hutchinson. Lebensohn and Tomé presented a modeling scheme in 1993 [57]; a detailed description of its implementation was provided by Turner and Tomé in 1994 [61], and subsequently by Turner et al. in 1995 [62]. The second modeling scheme was given by Clausen and Lorentzen in 1997 [63,64]. In the following sections, the mechanics of plastic deformation are discussed using the latter formulation. While much can be gained from a consideration of the model, those interested only in its predictive capabilities could move straight to Section 5.6.3.

5.6.2   Implementation of EPSC Model

The basic principles and physics behind the EPSC modeling scheme involve consideration of the crystallographic slip mechanisms by which the grains of the polycrystalline material, with lattice planes, *hkl*, having a specific orientation to an applied uniaxial load, accommodate plastic deformation. The elastic part of deformation is modeled by standard continuum mechanics principles, whereas the modeling of plastic deformation considers crystallographic slip only along specific slip directions on specific slip planes. Provided that these slip systems are known, a material with any crystal structure can be modeled. However, in some materials the available slip mechanisms are more easily characterized experimentally than in others. It is for this reason that the initial implementations of the method have focused on materials with crystal structures having well-defined slip mechanisms such as in *hexagonal* and *fcc* materials. For purpose of clarity, we will limit the discussion here to *fcc* systems, where the primary mode of crystallographic slip is on the four {111} planes, each having three close-packed <110> directions.

Having defined the permissible modes of crystallographic slip, each grain in the aggregate is considered as a spherical inclusion having the elastic anisotropy of a single crystal. Each is embedded in a homogeneous effective medium, with the effective isotropic properties derived as an average of all grains in the aggregate. This situation is exactly the one pictured by Eshelby through his equivalent inclusion theory, which is shown schematically in Figure 5.12.

Eshelby realized that when an inclusion misfitting by $\underline{\varepsilon}^T$ (top left in Figure 5.12) is placed into an ellipsoidal hole, the strain field inside it is the same throughout; that is, an ellipsoid when constrained in a matrix takes up an ellipsoidal shape, $\underline{\varepsilon}^C$ (top right). This means that for any constrained uniform ellipsoidal inclusion, it is always possible to identify a 'ghost' or 'equivalent' inclusion (bottom left) *having the same elastic properties as that of the matrix* such that when it is deformed to the same shape as the constrained inclusion it has the same stress field (bottom right). Thus, the two inclusions could be interchanged without disturbing the stress in the matrix. The equivalent elastically homogeneous problem is relatively easy to solve analytically. The trick is to find the elastically homogeneous inclusion that is equivalent to a given situation. The relationship between the equivalent stress-free misfit and the constrained shape is given by the Eshelby tensor, $\underline{\underline{S}}^E$, where $\underline{\varepsilon}^C = \underline{\underline{S}}^E \underline{\varepsilon}^T$. Eshelby's continuum approach is scale independent, so that there is information about grain shape but not size. The resulting stress in the inclusion

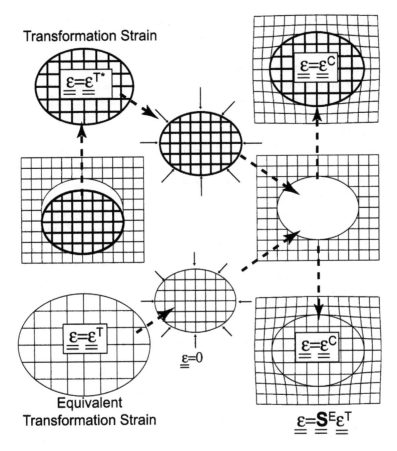

FIGURE 5.12
Illustration of Eshelby's cutting and welding exercise for an idealized stress-free misfit, $\underline{\varepsilon}^{T*}$, between an elastically different grain (above) and the effective medium representing all other grains. The equivalent elastically homogeneous inclusion (below) having the same elastic properties as the effective medium has a stress-free misfit strain $\underline{\varepsilon}^{T}$. This misfit is chosen so as to give the same constrained stress field as the elastically different grain. Eshelby related the stress-free shape misfit of the equivalent inclusion, $\underline{\varepsilon}^{T}$, to the final elastically constrained inclusion strain, $\underline{\varepsilon}^{C}$. This relationship is described by the Eshelby tensor, $\underline{\underline{S}}^{E}$ [44], and lies at the core of elastoplastic self-consistent models.

caused by this constraint is then $\underline{\underline{C}}_M(\underline{\varepsilon}^C - \underline{\varepsilon}^T)$, where $\underline{\underline{C}}_M$ is the stiffness tensor representative of the aggregate. This simple situation is only appropriate for a single grain embedded in a matrix. In practice, for a polycrystalline material a solution must be found iteratively by modifying the properties of the matrix, sometimes termed the effective medium, until it is consistent with properties of all the grains in the polycrystal, as discussed in Section 5.6.1.

Having identified the available modes of crystallographic slip, plasticity is considered in a stepwise fashion. The increment in the plastic strain tensor of each constituent grain, $\Delta\underline{\varepsilon}\,_c^{\,P}$, is determined as the sum of all increments in crystallographic slip, $\Delta\underline{\gamma}^i$, where i represents all the available slip systems.

The total increment in strain of the grain is then found by adding the plastic part and the elastic part, where the latter is defined by the stress increment, $\Delta \underline{\underline{\sigma}}_c$. The increments $\Delta \underline{\underline{\varepsilon}}_c^{\,P}$ and $\Delta \underline{\underline{\sigma}}_c$ represent the increments during a step in the calculation procedure. (An alternative notation in which these increments are represented by a dot above the variable is sometimes used, and the term "rate" is used rather than "increment." However, this should not be confused with a time derivative, as the modeling scheme in this formulation is time independent).

As in the models of Taylor and Sachs, the initiation of plastic flow is governed by the specification of a critical resolved shear stress, τ^i, for each slip system i. In the current incremental form, this is through a definition of the current increment in this quantity, $\Delta \tau^i$. It appears natural to relate the incremental growth of $\Delta \tau^i$ to the deformation history analogous to conventional work hardening. This has been expressed in terms of summing the contribution of slip on all slip systems, i, in the grain [52]. At this point, it is important to ask whether hardening on one slip system affects hardening on the other slip systems, so-called *latent hardening*, or whether slip on each slip system should be completely independent of the history of slip on other slip systems, so-called *self-hardening*.

Deformation history is introduced through the definition of h_γ, a parameter describing the hardening of slip on each system. One choice is to make the hardening of system i depend linearly on the accumulated crystallographic slip on all slip systems. This has proven to be very accurate in reproducing both the elastic-plastic transition and the plastic regime [65]. The exact form of h_γ is a matter of choice [64]. It should be remembered that while the calculation procedure is based on a physically meaningful definition of crystallographic slip, the definition of the hardening behavior is essentially empirical in nature. The relationship between the increment in critical resolved shear stress, $\Delta \tau^i$, and the accumulated slip is an empirical expression of the fact that the critical resolved shear stress ought to increase with plastic deformation much like the description of work hardening in a continuum mechanics sense. The various formulations are not founded upon the physical mechanisms of hardening by crystallographic slip and the associated pile-up of dislocations, they merely express an empirical rule for the evolution of the critical resolved shear stress. Nor do the various hardening parameters involved have any physical meaning, and as will become apparent in the subsequent description of the practical calculation procedure, they are considered as free-fitting parameters.

Having defined the relationships between the increments in shear and the plastic strain increment, the hardening behavior and dependency on deformation history, we can now turn to the calculation of the stress and strain increments of the constituents, which is done via the so-called concentration tensor proposed by Hill in 1965 [50,51]. The procedure incorporates the Eshelby formulation for the equivalent inclusion problem, which expresses the stress arising from a given misfit between the grain and its neighborhood.

Once the stress and strain increments in the individual constituent grains have been determined, the polycrystal stress and strain increments are found as the weighted averages of the stress and strain increments of all grains. In the actual procedure of calculating the aggregate response, the loading is simulated by prescribing an increment, $\Delta \underline{\varepsilon}^I$, in the overall strain $\underline{\varepsilon}^I$. At a certain point of loading, the current stress states in all constituent grains are known, and the potential active slip systems can be specified. From this the increment in stress and strain within the grain can be determined, $\Delta \underline{\sigma}_c$ and $\Delta \underline{\varepsilon}_c^F$, as well as the instantaneous modulus, L_c. This makes it possible to calculate the overall aggregate stress increment, $\Delta \underline{\sigma}^I$, associated with the prescribed strain increments, $\Delta \underline{\varepsilon}^I$, as well as to determine the current aggregate stiffness tensor, $\underline{\underline{L}}$. More detailed descriptions of this procedure can be found in [64], and flow diagrams of the entire procedure are given in Clausen and Lorentzen [63].

Having established an operational method, a realistic representation of the orientation of the grains in the polycrystalline aggregate is required. The ODF described in Section 2.6.2. represents the preferred orientation within the material, and forms the basis for defining a discrete set of grain orientations in the model that represent the true texture of the material. For a random homogeneous isotropic assembly of grains, this function is unity.

With the orientations of the grains within the polycrystalline aggregate defined, the calculations can begin, based in this case on the three single crystal stiffness constants, C_{ij}, an initial value of the critical resolved shear stress, τ_0, a quantitative expression determining the extent of latent and self-hardening, and a constitutive equation for hardening response. Rather than selecting these parameters at random, they are chosen so as to simulate the macroscopic stress-strain curve. Based on a reduced set of grains, such calculations are rather quick, and through an iterative approach the parameters are easily established. This approach can be justified as such a model must be able to capture the macroscopic deformation characteristics before it can be used to make predictions at a microstructural level about stresses and strains in individual grains or families of grains.

5.6.3 Predicting Crystallographic Slip

While the main aim is to obtain model predictions for the interpretation of lattice strain evolution during polycrystal deformation, it is useful to pay attention to some of the crystallographic features associated with the complex anisotropic evolution of lattice strains. Since plastic deformation occurs by crystallographic slip, it is useful to know for each grain the number of slip systems that have become active at any given stage, and in particular the orientation dependency of the slip activity necessary to accommodate the imposed overall deformation (Figure 5.13). The most noticeable feature is the large percentage of grains with three or four slip systems activated. At about 0.5% plastic strain, approximately 90% of the grains accommodate the deformation using less than the active slip systems.

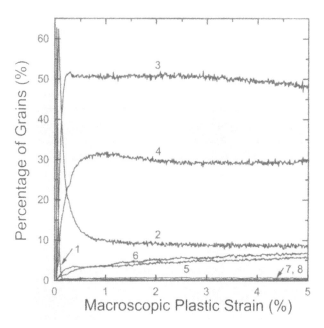

FIGURE 5.13

Percentage of grains having different numbers of active slip systems shown as a result of macroscopic uniaxial plastic strain for a typical *fcc* material [66].

An example of the orientation dependency of slip activity is given in Figure 5.14 for aluminum and copper. From this early stage of plastic deformation, it can be seen that for aluminum (Figure 5.14a) slip commences in crystallites with <100> and <110> directions along the uniaxial loading axis, which have identical Schmid factors describing the resolved shear stress on the slip plane. A large Schmid factor indicates an orientation such that the applied stress exerts a large shear stress component in a slip direction on a slip plane [67]. On the other hand, slip activity is minimal near the <111> corresponding to a lower Schmid factor. The picture is quite different for copper because it has a much higher degree of elastic anisotropy (Figure 5.14b). At this early stage of plasticity, activity is concentrated around the <111> orientation, with hardly any grains oriented with <100> parallel to the tensile axis slipping despite the much larger Schmid factor. Although grains are less favorably oriented for slip near the <111> orientation than near the <100> orientation, slip starts here, in contradiction to the concept behind the Taylor model. This difference is related to the higher elastic anisotropy of copper (see Table 5.1), which means that this direction is much stiffer so that it attains a higher stress despite being less favorably oriented for slip.

Another common measure of the slip activity is the so-called Taylor factor, or *m* factor. It is given by

$$m = \sum_i \Delta\gamma^i / \Delta\varepsilon^P \qquad (5.45)$$

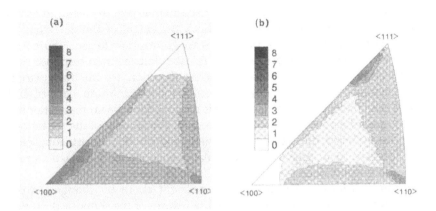

FIGURE 5.14
Number of active slip systems plotted according to the direction in the crystallite lying parallel
to the tensile axis for (a) aluminum and (b) copper, at a plastic strain of 0.011% [66].

which is the ratio of the sum of the incremental slips to the incremental
plastic strain. It can be interpreted as a reciprocal of the efficiency to accommodate deformation by crystallographic slip. A large m factor shows that a
high degree of slip is required to accommodate a specific increment in plastic
strain in a specific grain. In other words, the orientation is rather inefficient
in accommodating the strain increment by slip. The cubic directions <110>
and <111> have high m factors, and <100> a low m factor. This correlates
nicely with the self-consistent calculations of the slip activity. The results for
copper are contrary to this at the very early stage of plastic straining, where
elastic anisotropy plays a role. These features, attributed to the differences
in elastic anisotropy, are only significant at the very early stage of plasticity.
In fact, even before 0.5% strain, results for aluminum and copper become
similar. Thus, the plastic deformation characteristics, as measured by the
number of active slip systems and their orientation dependency, appear to
be a characteristic of the *fcc* crystal structure rather than the degree of elastic
anisotropy.

5.6.4 EPSC Predictions of Lattice Strains under Plastic Straining

The EPSC models described above provide a description of lattice strain
evolution of direct relevance to the interpretation of lattice strains measured
by neutron diffraction. Indeed, as noted above, the development of the
neutron diffraction technique for measuring lattice strains has stimulated
interest in these modeling schemes from an entirely new perspective.

As the principles behind EPSC schemes are based on the Eshelby–Kröner
model, the predictions in the elastic regime are the same as from that model.
As discussed in Section 4.2.2, an *hkl* reflection can be used for strain measurement in the linear regime of the stress–lattice strain relation provided

that the appropriate effective stiffness is used to interpret the result in terms of stress. Much more important is to check the extent of the deformation regime over which these elastic properties are appropriate to use. Figure 5.15 shows self-consistent calculations of the (elastic) lattice strain response parallel and perpendicular to the loading stress direction, for the deformation of three materials of varying degree of anisotropy: aluminum, copper, and stainless steel [64,68]. In each case, there is a marked nonlinear repartitioning of stress upon the onset of plastic deformation. While none of the reflections are absolutely linear throughout both elastic and plastic deformation, some are sufficiently so to enable one to make a wise choice for elastic strain determination in different materials, as listed in Table 4.1.

The nonlinearities in Figure 5.15 are brought about by changes in the intergranular stresses. The most noticeable feature of the numerical predictions is the rather strong nonlinearity of the 111 reflection in aluminum, and of the 200 reflection in copper and stainless steel. At the onset of plasticity (i.e., the onset of slip), grains with these orientations increase their rate of lattice strain accumulation. This nonlinearity can be understood in terms of Figure 5.14, which indicates that grains close to these orientations are the last to slip, and as a result the load is repartitioned toward them away from the slipping grains. This increases the proportion of the external load that they carry. The nonlinearity thus does not reflect plasticity in these specific *hkl* grain orientations themselves, but rather the contrary, that is, an ability to continue to strain elastically with little accumulated plasticity and associated hardening, as grains of other orientations begin to slip. With further plastic straining, all *hkl* reflections appear to enter another regime of essentially linear behavior as grains of all orientations slip, as indicated in Figure 5.15.

Since the onset of nonlinear behavior closely correlates with the onset of crystallographic slip at the local scale (i.e., the first slip activity in any grain), the evolution of nonlinear lattice strain response is a sensitive indicator of the onset of plasticity. In fact, the nonlinear behavior commences well before reaching the macroscopic engineering definition of yielding at 0.2% permanent offset. This is exemplified by the stainless steel response shown in Figure 5.15e and f, where the 0.2% macroscopic offset, or proof, yield stress, $\sigma_{0.2}$, is 265 MPa, while the onset of nonlinearity is predicted at an applied stress of only 160 MPa.

As an illustration of how this nonlinear effect can mislead interpretation of a measured lattice strain in terms of a Type I macrostress, we consider the lattice strains parallel to the applied stress in a uniaxially loaded stainless steel component, where the predicted strains are shown in Figure 5.15e. Suppose that a lattice strain of 2000 µε was measured using the 200 reflection. Using an estimate of the effective elastic modulus based the Kröner model of 149 GPa, this would suggest a stress level of 298 MPa. However, from Figure 5.15e it can be seen that 2000 µε is associated with an applied stress of only 239 MPa. As a result, the effective elastic modulus gives an error of 25%. Considering that these nonlinearities occur well in advance of the $\sigma_{0.2}$

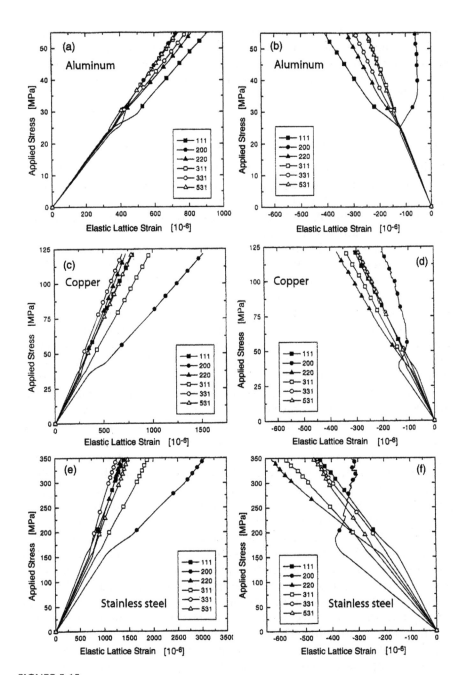

FIGURE 5.15
The evolution of lattice strain component parallel to the load axis predicted by elastoplastic self-consistent model, as a function of applied uniaxial tensile stress, for texture-free polycrystalline aggregates of (a) aluminum, (c) copper, and (e) stainless steel. The corresponding lattice strain components perpendicular to the load axis are shown in (b), (d), and (f) respectively. In each case, the deviations from linearity occur at the onset of plastic deformation. The "points" simply label each curve [64,66,68].

limit, this severely limits the window for safe conversion of lattice strain to stress through the use of diffraction elastic constants for those reflections that strongly show nonlinear behavior through the plastic regime. This is the basis for discouraging the use of certain reflections for stress measurement in Table 4.1.

5.6.5   EPSC Predictions of Residual Intergranular Strains

The illustrative 'thought experiment' of the previous section indicates that unloading a stainless steel sample from an applied load of 239 MPa would lead to the existence of residual lattice strains. This is because the nonlinearities in Figure 5.15 during straining are caused by intergranular shape misfits generated by differential slipping during deformation. Unless there is reverse slip, these will be entirely retained upon unloading. In other words, if there is no reverse plasticity the unloading curve will be linear, with a slope equal to that of the elastic response for each particular reflection. Hence, the deviation from linearity during forward loading corresponds exactly to the level of residual intergranular lattice strains retained upon unloading. From Figure 5.15e upon unloading from 239 MPa, the 200 reflection from stainless steel would thus record axial tensile strains of around +400 µε. Taken at face value, such a measured strain in this sample would be indicative of an applied elastic stress of 59 MPa, whereas in fact no applied stress is acting. We term such a strain due to plasticity and the resulting intergranular stresses a *"pseudo strain,"* and the inferred stress as a *"pseudo stress."*

By measuring the deviation of the predicted lattice response from the extrapolated elastic response for each reflection in Figures 5.15, one may determine how the residual intergranular *"pseudo"* lattice strains accumulate as a function of a prior forward tensile-loading lattice strain. These deviations from linearity, which represent the residual lattice strains on unloading, are plotted as a function of calculated macroscopic plastic strain, that is, the deviation from linearity of the bulk macroscopic strain or equivalently, the residual bulk macroscopic strain on unloading in Figure 5.16 a through f. The most noticeable feature of these predictions is the partitioning of residual intergranular strains into tensile or compressive strains, and a group of lattice planes showing essentially no residual intergranular strains at all. In terms of the residual strains along the loading axis, shown in Figures 5.16a, c, and e, copper appears to behave much like stainless steel, whereas the grouping is different for aluminum. For aluminum, the 111 reflection develops the highest tensile residual intergranular strain, while the 200 reflection develops the most compressive, and the 331 reflection appears to develop essentially no residual strains. In other words, the 331 reflection is largely insensitive to plastic strain, having no memory of the plastic strain that the polycrystalline aggregate has experienced. However, these observations of the partitioning of residual intergranular strains parallel to the loading axis cannot be generalized to any state of deformation. This is illustrated by considering

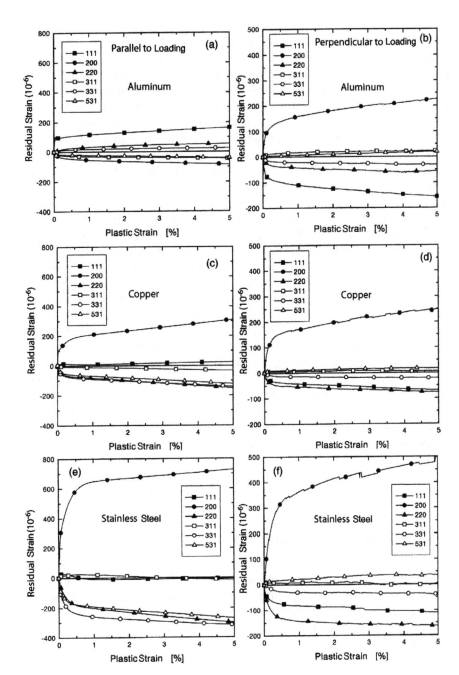

FIGURE 5.16
Evolution of residual intergranular lattice strains parallel to the tensile stress axis predicted by the elastoplastic self-consistent model, assuming elastic unloading for six reflections, as a function of the macroscopic plastic strain experienced during forward loading for (a) aluminum, (c) copper, and (e) stainless steel. Corresponding strain components perpendicular to the load axis are shown in (b), (d), and (f) respectively. The "points" are merely labels [64,66,68].

the strain response perpendicular to the loading axis, shown in Figure 5.16b, d, and f, where the grouping of *hkl* reflections into those with tensile and compressive residual intergranular strains is different. It is worthwhile noting that Dawson et al. [49] obtain the same trends for the 111, 200, 220, and 311 reflections in the axial and transverse directions for aluminum 5182 alloy using a crystal plasticity finite element model. However, reasonable agreement with neutron diffraction data was only found when a single-crystal anisotropy higher than that reported for pure aluminum single crystals was used.

As illustrated in our thought example for a stainless steel above, residual intergranular 'pseudo strains' may be present at zero load if the sample has been previously plastically deformed, and these will superimpose on those introduced by subsequent applied loading. They may be present in 'stress-free' reference samples cut from a previously plastically deformed sample, giving rise to a false reference. However, as mentioned in Section 4.6.3, if such a reference sample is taken in the form of a comb cut to relax the macrostresses from an identical sample to that being measured, the same pseudo strains will be present in both and will cancel out. Furthermore, they will vary spatially with any spatial variation of accumulated plastic strain throughout the sample. Clearly, when unaccounted for, they can lead to large errors when converting the measured lattice strain to macrostress if one simply uses the *hkl*-dependent, elastic effective stiffness of the polycrystal. If the sole purpose of an experiment is to determine the macrostress, then a reflection showing minimal nonlinearity in both the axial and transverse response should be selected. On the other hand, reflections that show large nonlinearities retain information about the plastic strain history. By comparing measurements of strain using different reflections, it may be possible *a priori* to deduce the extent of plastic strain that a sample has undergone, as well as its macrostress distribution [69]. In this respect, time-of-flight data are ideal because many peaks are measured simultaneously.

5.6.6 Analysis of Experimental Data

In this section, examples are given of the neutron diffraction measurement of lattice strains in samples under controlled loading in both the elastic and plastic regimes. Further examples are given in Section 6.4. These examples serve to highlight the important role of modeling polycrystalline deformation in interpreting lattice strain data from both macroscopic and microscopic perspectives.

5.6.6.1 Uniaxial Tensile Loading of Aluminum

In many ways, aluminum is an ideal material for studying the effects of plastic anisotropy. Most importantly, it has a very low degree of elastic anisotropy (Table 5.1) so that plastic effects can be clearly discerned in simple

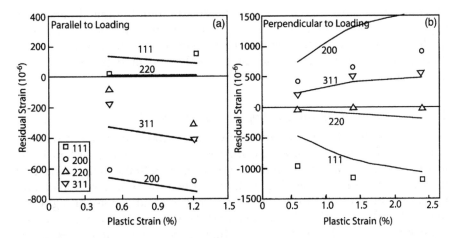

FIGURE 5.17

Residual lattice strains measured at zero load after plastic straining to various extents (a) parallel and (b) perpendicular to the straining direction at the Chalk River facility on an Al7050 sample. The predictions of an elastoplastic self-consistent model are shown as continuous lines [70].

uniaxial loading experiments monitored *in situ* by neutron diffraction. As discussed in the previous section, one way in which the intergranular strains may be isolated is to subtract from the measured lattice strain the linear, fully elastic response to the applied stress (σ^A). In other words, $E_{hkl}^{-1}\sigma^A$ is subtracted from the axial response, and $v_{hkl}E_{hkl}^{-1}\sigma^A$ is added to the transverse response. An experiment was undertaken on aluminum 7050 by Pang et al. [70], in which a rod sample, cut from a plate in the normal direction, was subjected to progressive loading and unloading cycles. It was found that, as one would expect, during solely elastic loading no plastic intergranular strain was generated with increased loading. When the applied stress exceeded 300–400 MPa, intergranular strains were found to develop. The intergranular lattice strain component measured at the Chalk River facility parallel and perpendicular to the loading direction for four reflections (111, 002, 113, and 220) as a function of bulk macroscopic plastic strain are shown in Figure 5.17. Note that the intergranular strains are much larger than in Figure 5.16a and b, presumably because 7050 Al alloy is a high-strength alloy and so can maintain large intergranular strains.

The predicted intergranular lattice strains were also calculated using an EPSC model constructed along the lines described in Section 5.6.2. The model appears to be in qualitative agreement with experiments and captures the main features, but for specific reflections large discrepancies were observed. This is most noticeable for the 220 reflection, and is probably attributable to uncertainties in selecting an appropriate crystallographic hardening behavior, as discussed in Section 5.6.2, and the difficulty in modeling the initial strain state of a sample taken from a rolled plate. Discrepancies of a similar magnitude have also been observed between neutron diffraction data and multisite models [71], as well as crystal-plasticity finite element models, both of which take into account nearest-neighbor grain-to-grain interactions explicitly [49]. At the present time, neutron diffraction strain data of sufficient quality have not been acquired to determine incontrovertibly whether these discrepancies are due to shortcomings in the slip models on which the models are based.

5.6.6.2   *Uniaxial Tensile Loading of Stainless Steel*

For stainless steel, the elastic anisotropy is much more pronounced than for aluminum. Figure 5.18 shows a comparison between self-consistent EPSC model predictions of lattice strain components in both the longitudinal and the transverse direction arising from an applied uniaxial tensile load, and the results of neutron diffraction lattice strain measurements for eight different *hkl* reflections [72]. The measurements were made using a dedicated stress rig on the NPD powder time-of-flight diffractometer at LANSCE, Los Alamos National Laboratory (Los Alamos, New Mexico).

Comparing the lattice strain data with the predicted curves for the direction parallel to the tensile loading axis, shown in Figure 5.18a and b, the most striking feature is the capacity of the EPSC model to capture the highly nonlinear behavior of the 200 reflection. Also, results for the 311, 420, 220, and 531 reflections are in excellent agreement, although minor discrepancies are seen to develop at the higher loads. Note, however, that the model fails to capture the experimental observations for the 331 reflection. For the transverse response shown in Figure 5.18c and d, the discrepancies are more striking. In particular, large deviations from the experimental data are seen to develop for the 331, 420, and 531 reflections.

It is possible to extend such studies to include preferred texture in a sample [73,74]. Other studies of intergranular strains in *fcc* materials include those on Monel and Inconel [75–77].

5.6.6.3   *Hexagonal Materials*

The measurement of residual stress in hexagonal systems has been studied for a range of materials, including titanium [78,79], beryllium [80], and zirconium [81–84]. The latter has attracted particular interest because the presence of residual stress affects creep, fracture toughness, corrosion, and dimensional stability. This is of especial concern in the reactor industry where the use of zirconium and its alloys is widespread. For hexagonal materials such as Zr, Be, and Zn, anisotropic thermal expansion can generate large intergranular stresses even in the absence of plasticity. Hence, whereas annealing is normally used to relieve residual stresses, for these materials it is the cause of residual stresses, and may lead to unacceptable changes in dimension. Another consequence is progressive distortion under thermal cycling. Furthermore, because hexagonal materials have just one close-packed slip plane, other mechanisms of slip are required to achieve three-dimensional deformation of a polycrystal. The unavailability of low-energy

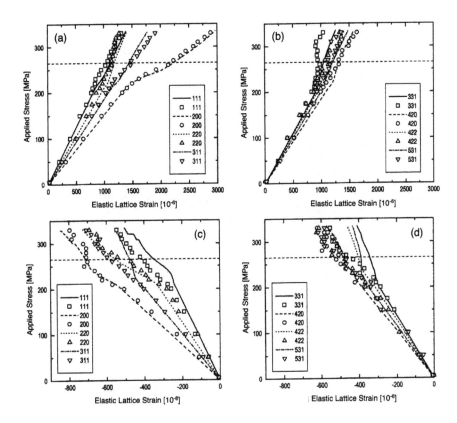

FIGURE 5.18
Lattice strain response to an applied uniaxial tensile stress for eight reflections from a texture-free polycrystalline stainless steel sample. The lattice strains were measured parallel (a and b), and perpendicular (c and d), to the tensile loading axis. The symbols denote measurements and the lines denote elastoplastic self-consistent model predictions. The horizontal dashed line at ~265 MPa represents the 0.2% proof macroscopic yield stress [64,72].

slip mechanisms leads to highly anisotropic and grain-orientation-dependent plastic behavior, which further affects the development of residual intergranular stresses.

The measurements of residual intergranular strain as a function of plastic deformation on Zircaloy were among the first measurements of intergranular effects [81–84]. The measured residual strains perpendicular to the uniaxial loading axis along a rod sample proved to be highly orientation dependent (Figure 5.19). Note how grains oriented with their <0002>, <10$\bar{1}$5>, <10$\bar{1}$4>, or <10$\bar{1}$3> directions perpendicular to the rod axis develop compressive residual intergranular strains monotonically, while counterbalancing tensile strains develop in grains with their <10$\bar{1}$2>, <10$\bar{1}$1>, or <10$\bar{1}$0> direction perpendicular to the rod axis. Comparison of Figure 5.19 with corresponding curves for *fcc* materials in Figure 5.16 emphasizes the magnitude of intergranular stresses. Indeed, residual strains exceeding 2000 × 10⁻⁶ are seen for

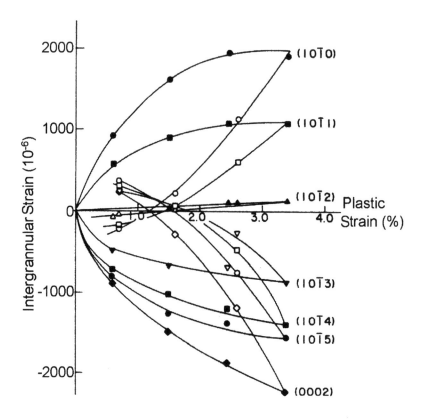

FIGURE 5.19
Evolution of intergranular residual strains due to various macroscopic *plastic* deformations measured in annealed Zircaloy-2. The strain component transverse to the loading direction is shown for various *hkil* planes in grains belonging to the <11 $\overline{2}$0> fibre component. The solid symbols correspond to measurements made on specimens loaded in tension to 1%, 2%, 3%, and 4% *total* macroscopic strain, and then unloaded. The open symbols correspond to measurements carried out on specimens first deformed to 4% strain in tension, and then compressed by 1%, 2%, and 3% before unloading [83,84].

certain reflections following a macroscopic plastic deformation of just 3.5%. Such strains correspond to stresses in excess of 200 MPa compared with 0.2% yield stress of 190 MPa in tension and 325 MPa in compression.

5.6.6.4 Cyclical Loading

To date, most investigations of lattice strain development under plastic deformation have been limited to monotonic uniaxial loading in tension, and subsequent unloading. Little work has focused on cyclic loading. One example of cycling between tension and compression is provided by the tensile/compressive straining of a stainless steel sample [85]. The sample was cycled between fixed total strain limits of +0.4% and −0.5% for eight successive cycles. The corresponding plastic strain envelope extended to +0.3 and −0.4%. The sample was loaded using a dedicated Instron hydraulic load frame on the ENGIN station of the PEARL beam line at the ISIS facility of the Rutherford Appleton Laboratory. In total, about eight complete tension/compression cycles were applied. The cycling was very slow due to the need to halt the deformation for 40 minutes in order to acquire diffraction data at each load level during the deformation cycle. During the initial three cycles, diffraction patterns were monitored at approximately 24 load levels during each cycle. Thereafter, diffraction patterns were collected only at the extreme tensile and compression levels, and when passing through zero load. Single peak fitting of six to eight statistically well-defined reflections *hkl* was made for each load step. Typical experimental results for the measured lattice strain component in the loading direction are shown in Figure 5.20a and b, while the corresponding transverse results are shown in Figure 5.20c and d. The experimental data show very noticeable differences in the lattice strain loops. The loop corresponding to the longitudinal 200 reflection (Figure 5.20a) displays an open loop with clear hysteresis-like behavior, as to a lesser extent do those for 331 and 220 (not shown). However, the strain measured from the 111, 311, and 420 reflections shows an essentially linear, closed-loop behavior. For strain components perpendicular to the loading axis, these features are still found, but to a lesser extent. These differing hysteresis curves arise because some grain families such as 111 for stainless steel generate very little intergranular strain when strained to 0.4%, whereas others such as 200 generate a more significant plastic misfit, as seen in Figure 5.16e. As discussed in Section 5.6.4, this is due to the fact that some families of grain yield at low applied loads, while others are deforming only elastically, causing load repartitioning and nonlinearities in the applied stress–lattice strain curves.

One of the most striking features is that the lattice strain loops measured over eight consecutive cycles essentially stabilize during the first cycle, and do not evolve significantly with the number of cycles. It would appear that work hardening does not cause the loops to narrow or the peak loads to rise with increasing numbers of cycles. These experimental results have been compared to self-consistent model predictions, where they have guided the specification of cyclic hardening rules [85].

5.6.6.5 Rietveld Analysis of Diffraction Patterns

Although there has been good progress in understanding the effects of intergranular stresses on lattice strain, they are not yet so fully understood that the relationship between intergranular stresses and deformation history can be incorporated quantitatively and accurately into the analysis of diffraction results from samples that have been plastically strained. In the standard Rietveld refinement of powder diffraction patterns, as described in Sections 3.2.2 and 4.5, the measured intensity distribution is assumed to resemble an ideal crystal structure, such as *fcc*, *bcc*, or *hcp*, and the best fit to

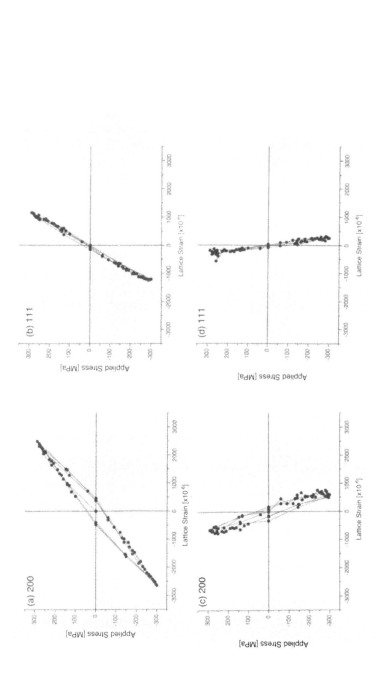

FIGURE 5.20
The measured lattice strain component parallel and perpendicular to the uniaxial loading direction of a stainless steel sample during plastic cycling. The response is shown for reflections (a) 200 strain parallel to applied load, (b) 111 parallel strain to applied load, (c) 200 strain perpendicular to applied load, and (d) 111 strain perpendicular to applied load [85].

match this structure is determined by adjusting the lattice parameters and positions of atoms in the structure. As discussed in Section 4.5, this represents a weighted average of the strain taken over all the peaks in the profile. However, as a consequence of deformation of a sample, whether it is elastic or plastic, the basic physical relationship between the lattice spacing of various *hkl* diffraction peaks and the strain changes. One method of analysis, mentioned in Section 4.5, is, in the case of cubic systems, for example, to simply fit an overall lattice parameter, *a*, and to obtain an overall strain from this. Indeed, it is found that by using this overall strain and the macroscopic bulk elastic moduli, a good value of the stress is obtained even into the plastic regime, the anisotropy having been averaged out.

Another approach is to realize that each reflection will behave differently as a consequence of the elastically and plastically generated intergranular misfits, and to incorporate these effects into the refinement. As a first step, attempts have been made to incorporate the inherent elastic anisotropy, which is relatively well understood [69]. Again, the results of the lattice strain behavior of samples stressed *in situ* by uniaxial loading has proved useful. Because the elastic anisotropy would be expected to account for the deviation from a stress-free refinement correctly only for solely elastic deformation, it is possible to fit the diffraction pattern by a model for the lattice spacings exhibiting the expected elastic anisotropy. From the quality of the fit obtained, one may deduce whether previous deformation has exceeded the elastic limit [69]. For example, for a uniaxial elastic deformation, the diffraction peaks shift according to their specific stiffness, and this may be described by the Kröner relations for the elastic moduli that involve the A_{hkl} parameter (Equation (5.24)).

Alternatively, the elastic anisotropy may be introduced into the Rietveld refinement by incorporating a new fitting parameter, γ, called the *"anisotropy strain"* parameter, into the refinement. The way in which specific diffraction peaks shift according to their elastic stiffness is governed by the term γA_{hkl}. Rather than refining the diffraction pattern toward an average ideal crystal structure, it is now modified so that the lattice spacings track the strain in the <*h*00> direction. The other reflections are assumed to deviate from this ε^{h00} term according to the relation

$$\varepsilon^{hkl} = \varepsilon^{h00} - (\gamma A_{hkl}/C_0) \qquad (5.46)$$

where C_0 is an instrument-related constant.

In the elastic regime, this anisotropy strain parameter, (γ/C_0), should develop linearly and correlate solely as a multiplicative factor, with the elastic anisotropy factor A_{hkl}. As soon as deformation exceeds a limit where elastic anisotropy given by A_{hkl} can no longer explain all the elasticity related deviations from the ideal, unstrained crystal structure, the fitted anisotropy strain is expected to start deviating from this linear behavior. This has been tested on data measured from *in situ* uniaxial loading of a stainless steel

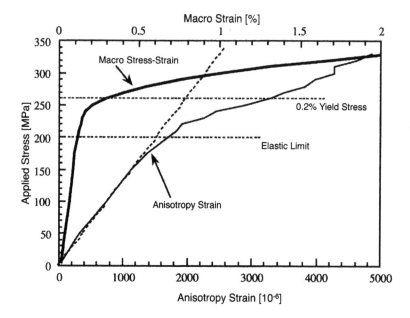

FIGURE 5.21
Evolution of the anisotropy strain parameter (γ/C_0) (lower abscissa) with applied uniaxial tensile stress on a stainless steel sample. Also plotted is the macrostrain obtained from a strain gauge (upper abscissa) [69].

sample; results are shown in Figure 5.21. The expected deviation from the initial linear behavior (broken line) correlates directly with the commencement of a noticeable deviation from linearity in the *hkl*-specific lattice strain response as seen in Figure 5.18.

To incorporate plastic strain, it has therefore been suggested that the fitting parameter, γ, could be split into an elastic, γ^{el}, and a plastic, γ^{pl}, component,

$$\gamma = \gamma^{el} + \gamma^{pl} \tag{5.47}$$

The result is illustrated in Figure 5.22, where the anisotropy strain is plotted against the elastic strain derived from an overall traditional Rietveld refinement performed without taking account of the elastic anisotropy. This empirical factor, γ^{pl}, clearly captures the transition from elastic to plastic deformation, providing a metric for the extent of the deviation from an elastic response. This suggests that one might be able to determine the extent of plastic deformation from a single diffraction profile. To date, only semiquantitative success has been achieved.

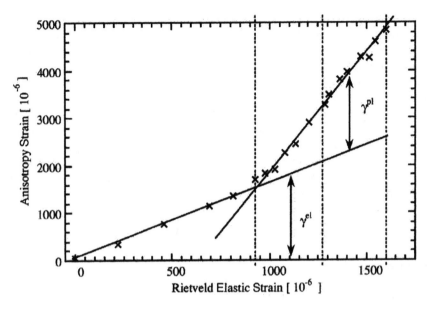

FIGURE 5.22
The variation of the anisotropy strain parameter (γ/C_0) in the elastic and plastic regimes of a stainless steel sample under applied uniaxial tensile stress with the internal phase strain obtained from an overall Rietveld fit without including elastic anisotropy. The vertical broken lines indicate the Rietveld elastic strains corresponding to macroscopic plastic strains of 0%, 0.2%, and 2% [69].

5.6.7 Summary

Elastic anisotropy strongly affects the magnitude of measured lattice strains, and when materials have been exposed to plastic deformation, this leads to the development of residual intergranular stresses even in the case of homogeneous plastic deformation of the sample. The interaction between neutron diffraction lattice strain measurement and modeling has provided valuable insights into the micromechanics of polycrystal deformation, and has shed light on some of the uncertainties in the implementation of polycrystal modeling schemes. Whereas existing analytical theories of the anisotropy of polycrystals under elastic strain account well for experimental lattice strain data, self-consistent modeling schemes are currently being developed to explore the main effects of plastic anisotropy. In many cases, these can provide numerical results in good quantitative agreement with experimental observations. They may be used as a guide in the interpretation of diffraction data on residual stresses and strains in engineering components, but they are not, at the present stage, refined to a state where they can be relied on quantitatively in the general case. When wishing to determine the macrostress in an engineering sample, one is not aware *a priori* of its previous strain history, or whether the elastic limit has been exceeded in all or parts of the component. It is therefore recommended that the Bragg reflections chosen to measure the lattice strain should exhibit, experimentally or theoretically, minimal intergranular stresses under uniaxial loading. These are listed for some common materials in Table 4.1.

Deformation is a highly complicated process. Much remains to be explored in terms of the micromechanical mechanisms of accommodating slip, the influence of texture and grain size, and other aspects influencing the heterogeneity of deformation for polycrystalline aggregates. Future model refinements are expected to come about through a symbiotic interplay between diffraction-based experiments and modeling work, including self consistent and finite element models of crystal plasticity, which are able to capture individual grain to grain interactions as well as the behavior of assembly in polycrystals.

5.7 Analysis of Bragg Peak Broadening

Most of the discussion thus far has been concerned with the measurement and interpretation of small shifts in Bragg diffraction peak angles, or flight times, from a sample under examination compared with those from a stress-free reference. However, the diffraction peak profile itself may contain important information for both the materials scientist and the engineer, relating to the microstructure and the inhomogeneous state of stress within the gauge volume. Even if the main object is to extract a measurement of macrostrain, it is important to understand how changes in peak shape may affect reliable analysis of the data. A change in profile might be interpreted by a curve-fitting routine as a shift in peak position. An appreciation of the reasons why particular diffraction peaks or peaks measured under specific conditions may appear to change shape is therefore important.

Relatively poorer instrument resolution causes neutron diffraction peaks typically to be broader than those recorded by x-rays, as discussed in Chapter 3. For example, on a continuous source instrument, the full width at half the maximum intensity (FWHM), is typically at best ~0.2° in scattering angle ϕ^s at $\phi^s = 90°$. Although intrinsic broadening can be observed at this resolution, if peak profile changes are of primary interest one should consider using laboratory x-ray, or better still synchrotron x-ray, studies. In the latter case, peak widths as narrow as FWHM ~0.001° in ϕ^s at $\phi^s = 10°$ have been recorded from engineering materials [86].

Intrinsic peak shape broadening can have several origins, and the actual profile may be a convolution of more than one of these with the instrumental broadening. Their individual separation is not always possible. Major contributions include:

- Particle size broadening
- Dislocations
- Stacking faults
- Microstrain broadening
- Steep strain gradients.

The last of these, steep macrostrain gradients within the sampled gauge volume, is the most straightforward contribution to peak broadening. It has been known since the early part of the 20th century that peak widths are affected by the size of the coherently diffracting volumes, which can give rise to so-called particle size broadening [87]. Particle size broadening, as measured by the increase in FWHM over the instrument value, may be taken to be inversely proportional to the mean size of the crystallites making up the polycrystalline aggregate. As discussed in Section 2.3.2 and expressed in Equation (2.17), the width of the peak in scattering angle, ϕ^s, given by FWHM = $\lambda/(t_c \cos\theta^B)$, is inversely proportional to the size of the crystal t_c. Consequently, for small crystallite sizes, the Bragg peaks become sharper with increasing crystallite size, until the peak width becomes limited by the instrument resolution. In fact, it is not normally the size of the crystallites that is important, but rather the mosaic spread within which the crystallite or grain can be considered to act as a single diffracting entity. Any increase in dislocations or stacking faults will reduce the size of this entity. For plastically deformed materials, this can be substantially smaller than the grain size, and consequently, gives a contribution to peak broadening which increases with plastic deformation. The angular width due to particle size broadening alone is independent of the scattering angle, or equivalently, the scattering vector.

Strain fields that vary over a scale much smaller than the gauge volume also give rise to broadening, as mentioned in Section 1.4.2, and discussed in Section 6.3.8. An example of this is the grain interaction stresses caused by anisotropy in elastic, plastic, and thermal properties at the grain size scale. These interaction stresses are brought about by mismatches in stiffness, plasticity, coefficient of thermal expansion, and so on among individual grains. For example, peak broadening is observed to depend on plastic deformation, as illustrated in Figures 5.4 and 5.23, and this can be due to a combination of smaller diffraction entities and microstress variation within them. In the simplest case, strain broadening alone will give an angular width that is proportional to the scattering vector **Q** for different reflections.

Strain broadening is often categorized by the root mean square (rms) strain $\sqrt{\Delta\varepsilon_i^2}$, and termed the microstrain. Here $\Delta\varepsilon_i$ is the deviation from the long-range macrostrain at points i within the grains. As the characteristic length over which this average is carried out decreases, the root mean square strain increases. In fact, because a dislocation causes a very severe short-range elastic strain, it is possible to infer the dislocation density from measurements of the rms strain [88]. However, this is not valid if the dislocations are

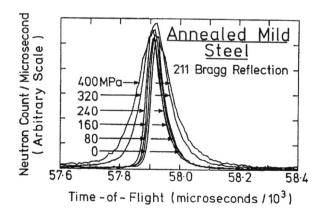

FIGURE 5.23

The 211 Bragg reflection profile from an annealed mild steel sample subjected to successive stress levels, observed by time-of-flight diffraction on the High-Resolution Powder Diffractometer instrument at the ISIS facility. The strain component perpendicular to the applied stress is measured, and the peak is seen to broaden rapidly above the yield stress of 240 MPa [97].

clustered into cells or boundaries. Stacking faults upset the natural sequence of the atomic lattice planes causing broadening. Because stacking faults only occur on specific lattice planes, their effect is *hkl* related.

Peak broadening due to plastic deformation can also be highly grain-orientation specific, or equivalently, *hkl* specific. This is illustrated in Figure 5.24, which shows experimental data for the uniaxial deformation of annealed polycrystalline hexagonal Zircaloy-2. As in Figure 5.19, data are given for both forward and reverse loading, and show a monotonic increase in width with plastic straining. The rate of increase in FWHM with strain is observed to be highly orientation specific. Depending on the strain level, several different features can be observed. Most noticeable is the strong increase in the rate of basal plane peak broadening (0002) in the interval between 0.5% and 1.5%, and the effect of load reversal, which initially tends to reduce the peak width; however, upon further straining into compression, it broadens the peak again.

One of the fundamental analytical schemes for separating several of the microstructure-related effects that alter the peak shape, including particle size broadening, strain broadening, dislocation density effects, and stacking faults, is a conventional Fourier expansion of the observed profile developed by Warren and Averbach [89–93]. Greater detail on this issue is beyond the scope of this text; the reader is referred to the original references for further information. A more detailed historical overview of the field of Bragg peak profile analysis is provided by Genzel [94]. Klimanek [95,96] has proposed one of the most recent methods for peak profile analysis based on so-called diffraction multiplets, which gives further insight into the phenomena.

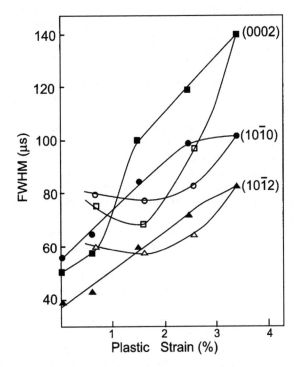

FIGURE 5.24
Effect of uniaxial plastic deformation on the measured FWHM in time of flight of the (0002), (10$\bar{1}$0), and (10$\bar{1}$2) diffraction peaks of annealed hexagonal Zircaloy-2 measured perpendicular to the applied stress. Solid symbols correspond to measurements made on specimens loaded in tension to *total* strains of 1%, 2%, 3%, and 4%, and then unloaded. The open symbols correspond to measurements carried out on specimens first deformed to strains of 4% in tension, followed by compressive strains of 1%, 2%, and 3%, and then unloaded [84].

References

1. J.F. Nye, *Physical Properties of Crystals — Their Representation by Tensors and Matrices*, Clarendon, Oxford, 1985.
2. S.P. Timoshenko and J.N. Goodier, *Theory of Elasticity*, McGraw-Hill, New York, 1982.
3. T. Lorentzen and T. Leffers, Strain tensor measurements by neutron diffraction, in *Measurement of Residual and Applied Stress Using Neutron Diffraction*, M.T. Hutchings and A.D. Krawitz, Eds., NATO Series E: Applied Sciences, vol. 216, Kluwer, Dordrecht, 1992, 253–261.
4. V.M. Hauk and E. Macherauch, A useful guide for x-ray stress evaluation (XSE), *Adv. X-ray Anal.*, 27, 81–99, 1983.

5. I.C. Noyan and J.B. Cohen, *Residual Stress — Measurement by Diffraction and Interpretation*, in *Materials Research and Engineering*, B. Ilschner and N.J. Grant, Eds., Springer-Verlag, New York, 1987.

6. B.D. Cullity, *Elements of X-ray Diffraction*, 2nd ed., Addison-Wesley, Reading, MA, 1978.

7. C. Genzel, W. Reimers, R. Malek, and K. Pohlandt, Neutron and x-ray residual stress analysis of steel parts produced by cold forward extrusion and tube drawing, *Mater. Sci. Eng.*, A205, 79–90, 1996.

8. C.M. Sayers, The strain distribution in anisotropic polycrystalline aggregates subjected to an external stress field, *Phil. Mag. A*, 49, 243–262, 1984.

9. R. Glocker, Einflüs einer elastischen anisotropie auf die röntgenographische messung von spannungen, *Z. F. Techn. Phys.*, 19, 289–293, 1938.

10. F. Bollenrath and E. Schiedt, Röntgenographische spannungsmessungen bei uberschreiten der fliessgrenzen an biegestaben aus flüsstah, *VDI Z.*, 82, 1094–1098, 1938.

11. F. Bollenrath and E. Osswald, Über den beitrag einzelner kristallite eines vielkristallinen körpers zur spannungsmesung mit röntgenstrahlen, *Z. Metallkde*, 31, 151–159, 1939.

12. F. Bollenrath, V. Hauk, and E. Osswald, Röntgenographische spannungsmessungen bei überschreiten der fliëssgrenzen an zugstäben aus unlegiertem stahl, *VDI Z.*, 83, 129–132, 1939.

13. S.L. Smith and W.A. Wood, A stress-strain curve for the atomic lattice of iron, *Proc. R. Soc.*, A178, 93–106, 1941.

14. S.L. Smith and W.A. Wood, Internal stresses created by plastic flow in mild steel, and stress-strain curves for the atomic lattice of higher carbon steels, *Proc. R. Soc.*, A182, 404–414, 1944.

15. G.B. Greenough, Residual lattice strains in in plastically deformed metals, *Nature*, 160, 258–260, 1947.

16. G.B. Greenough, Residual lattice strains in plastically deformed polycrystalline metal aggregates, *Proc. R. Soc.*, A197, 556–567, 1949.

17. G.B. Greenough, Internal stresses in worked metals, lattice strains measured by x-ray techniques in plastically deformed metals, *Metallurgical Treatment A: Drop Forging*, 16, 58–64, 1949.

18. G.B. Greenough, Quantitative x-ray diffraction observations on strained metal aggregates, *Proc. Metal. Phys.*, 3, 176–219, 1952.

19. E. Heyn, Eine theorie der vervestigung von metallischen stoffen infolge kaltreckens, in *Festschrift der Kaiser Wilhelm Gesellschaft zur Forderung der Wissenschaften, zu Ihrem Zehnjahrigen Jubilaum*, Verlag von Julius Springer, Berlin, 1921, 121–131.

20. A.J. Allen, M. Bourke, W.I.F. David, S. Dawes, M.T. Hutchings, A.D. Krawitz, and C.G. Windsor, Effect of elastic anisotropy on the lattice strains in polycristalline metals and composites measured by neutron diffraction, in Proceedings, 2nd International Conference on Residual Stresses, C. Beck, S. Denis, and A. Simon, Eds., Elsevier Applied Science, London, 1989, 78–83.

21. J.F. Nye, Elastic behaviour of single crystals: anisotropy, in *Encyclopedia of Materials: Science & Technology*, vol. 4, K.H.J. Buschow, et al., Eds., Elsevier, Oxford, 2001, 2415–23.

22. R.W. Hertzberg, *Deformation and Fracture Mechanics of Engineering Materials*, Wiley, New York, 1976.

23. G. Simmons and H. Wang, *Single Crystal Elastic Constants and Calculated Aggregate Properties: a Handbook*, M.I.T. Press, Boston, 1971.
24. H.M. Ledbetter, Monocrystals: polycrystal elastic constants of a stainless steel, *Phys. Stat. Solidi*, A85, 89–96, 1984.
25. W. Voigt, *Lehrbuch der Krystallphysik*, B.G. Teubner, 1928.
26. J.W. Dally and W.F. Riley, *Experimental Stress Analysis*, McGraw-Hill, New York, 1978.
27. A. Reuss, Berechnung der fliessgrenze von mischkristallen auf grund der plastizitötsbedingung fur einkristalle, *Z. Angew. Math. Mech.*, 9, 49–58, 1929.
28. R. Hill, The elastic behaviour of a crystalline aggregate, *Proc. Phys. Soc. London*, A65, 349, 1952.
29. Z. Hashin and S. Shtrikman, On some variational principles in anisotropic and nonhomogeneous elasticity, *J. Mech. Phys. Solids*, 10, 335–343, 1962.
30. Z. Hashin and S. Shtrikman, A variational approach to the theory of the elastic behaviour of multiphase materials, *J. Mech. Phys. Solids*, 11, 127–140, 1963.
31. E. Kröner, Berechnung der elastischen konstanten des vielkristalls aus den konstanten des einkristalls, *Z. Physik*, 151, 404–418, 1958.
32. J.D. Eshelby, The determination of the elastic field of an ellipsoidal inclusion and related problems. *Proc. R. Soc. London*, Ser. A, 241, 376–396, 1957.
33. G. Kneer, Über die berechnung der elastizitätsmoduln vielkristalliner aggregate mit textur. *Phys. Stat. Solidi*, 9, 825, 1965.
34. C. Tomé, Elastic behaviour of polycrystals, in *Encyclopedia of Materials: Science and Technology*, K.H.J. Buschow, et al., Eds., Elsevier, Oxford, 2001, 2409–2415.
35. E. Kröner, Zur plastichen verformung des vielkristalles, *Acta Metall.*, 9, 155–161, 1961.
36. G.I. Taylor, Plastic strain in metals, *J. Inst. Met.*, 62, 307–324, 1938.
37. G. Sachs, Zur ableitung einer fliessbedinung, *Z. VDI*, 72, 734–736, 1928.
38. J.F.W. Bishop and R. Hill, A theory of the plastic distortion of a polycrystalline aggregate under combined stresses. *Phil. Mag.*, 42, 414–427, 1951.
39. J.F.W. Bishop and R. Hill, A theoretical derivation of the plastic properties of a polycrystalline face-centered metal. *Phil. Mag.*, 42, 1298–1307, 1951.
40. B. Budiansky and T.T. Wu, Theoretical prediction of plastic strains of polycrystals, in Proceedings of the 4th U.S. National Congress of Applied Mechanics, University of California, Berkeley, vol. 1, American Society of Mechanical Engineers, 1962, 1175–1185.
41. J.D. Eshelby, Elastic inclusions and inhomogeneities, in *Progress in Solid Mechanics*, vol. 2, I.N. Sneddon and R. Hill, Eds., 1961, 89–140.
42. T. Mura, *Micromechanics of Defects in Solids*, The Hague, Nijhoff, 1987.
43. M. Taya and R.J. Arsenault, *Metal Matrix Composites: Thermomechanical Behaviour*, Pergamon, Oxford, 1989.
44. T.W. Clyne and P.J. Withers, *An Introduction to Metal Matrix Composites*, Cambridge Solid State Series, Cambridge University Press, Cambridge, 1993.
45. P. Bate, Modelling deformation microstructure with the crystal plasticity finite element method, *Phil. Trans. Roy. Soc.* (London), A357, 1589–1601, 1999.
46. A.J. Beaudoin, K.K. Mathur, P.R. Dawson, and G.C. Johnson, 3-Dimensional deformation process simulation with explicit use of polycrystal plasticity models, *Int. J. Plasticity*, 9, 833, 1993.
47. R. Becker, Analysis of texture evolution in channel die compression, 1. Effects of grain interaction, *Acta Metall. Mater.*, 39, 1211–1230, 1991.

48. N. Barton, P. Dawson, and M. Miller, Yield strength asymmetry predictions from polycrystal elastoplasticity, *J. Eng. Mat. Techol. Trans.*, 121, 230–239, 1999.

49. P. Dawson, D. Boyce, S. MacEwen, and R. Rogge, On the influence of crystal elastic moduli on computed lattice strains in AA–5182 following plastic straining, *Mat. Sci. Eng.*, A313, 123–144, 2001.

50. R. Hill, Continuum micromechanics of elastoplastic polycrystals, *J. Mech. Phys. Solids*, 13, 89–101, 1965.

51. R. Hill, A self-consistent mechanics of composite materials, *J. Mech. Phys. Solids*, 13, 213–222, 1965.

52. R. Hill, Generalized constitutive relations for incremental deformation of metal crystals by multislip, *J. Mech. Phys. Solids*, 14, 95, 1966.

53. R. Hill, Essential structure of constitutive laws for metal composites and polycrystals, *J. Mech. Phys. Solids*, 15, 79–85, 1967.

54. J.W. Hutchinson, Elastic-plastic behaviour of polycrystalline metals and composites, *Proc. R. Soc.*, A319, 247–272, 1970.

55. S.V. Harren, The finite deformation of rate-dependent polycrystals, I: A self-consistent framework, *J. Mech. Phys. Solids*, 39, 345–360, 1991.

56. S.V. Harren, The finite deformation of rate-dependent polycrystals, II: A comparison af the self-consistent and Taylor methods, *J. Mech. Phys. Solids*, 39, 361–383, 1991.

57. R.A. Lebensohn and C.N. Tomé, A self-consistent anisotropic approach for the simulation of plastic deformation and texture development of polycrystals: application to zirconium alloys, *Acta Metall. Mater.*, 41, 2611–2624, 1993.

58. S. Nemet-Nasser and M. Obata, Rate-dependent, finite elastoplastic deformation of polycrystals, *Proc. R. Soc.*, A407, 343–375, 1986.

59. F. Corvasce, P. Lipinski, and M. Berveiller, Intergranular residual stresses in plastically deformed polycrystals, in Proceedings, 2nd International Conference on Residual Stresses, Elsevier Applied Science, London, 1989, 535–541.

60. F. Corvasce, P. Lipinski, and M. Berveiller. Second order residual and internal stresses in inhomogeneous granular materials under thermomechanical loading, in Proceedings, 2nd International Conference on Residual Stresses, ICRS–2, Elsevier Applied Science, London, 1989, 399–404.

61. P.A. Turner and C.N. Tomé, A study of residual stresses in Zircaloy–2 with rod texture, *Acta Metall.*, 42, 4143–4153, 1994.

62. P.A. Turner, N. Christodoulou, and C.N. Tomé, Modeling the mechanical response of rolled Zircaloy–2, *Int. J. Plasticity*, 11, 251–265, 1995.

63. B. Clausen and T. Lorentzen, A self-consistent model for polycrystal deformation: Description and implementation, Risø National Laboratory, Report R–970(EN), Roskilde, Denmark, 1997.

64. B. Clausen, *Characterisation of Polycrystal Deformation by Numerical Modeling and Neutron Diffraction Measurements*, Ph.D. thesis, Technical University of Denmark, Risø National Laboratory, 1997.

65. B. Clausen and T. Lorentzen, Polycrystal deformation; experimental evaluation by neutron diffraction, *Metall. Mater. Trans.*, 28A, 2537–2541, 1997.

66. B. Clausen, T. Lorentzen, and T. Leffers, Self-consistent modelling of the plastic deformation of fcc polycrystals, and its implications for diffraction measurements of internal stresses, *Acta Mater.*, 46, 3087–3098, 1998.

67. R.W.K. Honeycombe, *The Plastic Deformation of Metals*, Edward Arnold, London, 1968.

68. T. Lorentzen and B. Clausen, Self-consistent modeling of lattice strain response and development of intergranular residual strains, in Proceedings, 5th International Conference on Residual Stresses, Institute of Technology, Linkoping University, 1997, 454–459.

69. M.R. Daymond, M.A.M. Bourke, R.B. Von Dreele, B. Clausen, and T. Lorentzen, Use of Rietveld refinement for elastic macrostrain determination and for the evaluation of plastic strain history from different spectra, *J. Appl. Phys.*, 82, 1554–1562, 1997.

70. J.W.L. Pang, T.M. Holden, and T.E. Mason, In situ generation of intergranular strains in an Al7050 alloy, *Acta Mater.*, 46, 1503–1518, 1998.

71. L. Delannay, R.E. Loge, Y. Chastel, and P.V. Van Houtte, Prediction of intergranular strains in cubic metals using a multisite elastic-plastic model, *Acta Mater.*, 50, 5127–5138, 2002.

72. B. Clausen, T. Lorentzen, M.A.M. Bourke, and M.R. Daymond, Lattice strain evolution during tensile loading of stainless steel, *Mat. Sci. Eng.*, A259, 17–24, 1999.

73. M.R. Daymond, C.N. Tomé, and M.A.M. Bourke, Measured and predicted intergranular strains in textured austenitic steel, *Acta Metall.*, 48, 553–564, 2000.

74. M.R. Daymond, M.A.M. Bourke, B. Clausen, and C.N. Tomé, Effect of texture on hkl dependent intergranular strains measured in-situ at a pulsed neutron source, in Proceedings, 5th International Conference on Residual Stresses, Institute of Technology, Linköping University, 1997, 577–585.

75. C.N. Tomé, T.M. Holden, P.A. Turner, and R. Lebensohn, Interpretation of intergranular stress measurements in Monel–400 using polycrystal models, in Proceedings, 5th International Conference on Residual Stresses, Institute of Technology, Linköping University, 1997, 40–45.

76. T.M. Holden, A.P. Clarke, and R.A. Holt. The impact of intergranular strains on interpreting stress fields in Inconel–600 steam generator tubes, in Proceedings, 5th International Conference on Residual Stresses, Institute of Technology, Linköping University, 1997, 1066–1070.

77. T.M. Holden, A.P. Clarke, and R.A. Holt, Neutron diffraction measurements of intergranular strains in Monel 400, *Metall. Mater. Trans.*, 28A, 2565–2576, 1997.

78. D. Dye, H.J. Stone, and R.C. Reed, A two-phase elastic-plastic self-consistent model for the accumulation of microstrains in Waspaloy, *Acta Mater.*, 49, 1271–1283, 2001.

79. M.R. Daymond and N.W. Bonner, Lattice strain evolution in IMI 834 under applied stress, *Mater. Sci. Eng.*, 340A, 272–280, 2003.

80. D.W. Brown, M.A.M. Bourke, B. Clausen, T.M. Holden, C.N. Tomé, and R. Varma, A neutron diffraction and modeling study of uniaxial deformation in polycrystalline beryllium, *Metall. Mater. Trans.*, 34A, 1439–1449, 2003.

81. S.R. McEwen, J. Faber Jr., and A.P.L. Turner, The use of time-of-flight neutron diffraction to study grain interaction stresses, *Acta Metall.*, 31, 657–676, 1983.

82. S.R. McEwen and C.N. Tomé, Residual stress in textured Zirconium alloys, in Proceedings, 7th International Conference on Zirconium in the Nuclear Industry, American Society for Testing and Materials, 1987, 631–652.

83. S.R. McEwen, N. Christodoulou, C. Tomé, J. Jackman, T.M. Holden, J. Faber Jr., and R.L. Hitterman, The evolution of texture and residual stress in Zircaloy-2, in Proceedings, 8th International Conference on Texture of Materials (ICOTOM–8), Warrendale, PA, 1988, 825–836.

84. S.R. McEwen, N. Christdoulou, and A. Salinas-Rodriguez, Residual grain-interaction stresses in zirconium alloys, *Metall. Mater. Trans.*, 21A, 1083–1095, 1990.

85. T. Lorentzen, M.R. Daymond, B. Clausen, and C.N. Tomé, Lattice strain evolution during cyclic loading of stainless steel, *Acta Mater.*, 50, 1627–1638, 2002.

86. P.J. Withers, Use of synchrotron x-ray radiation for stress measurement, in *Analysis of Residual Stress by Diffraction using Neutron and Synchrotron Radiation*, M.E. Fitzpatrick and A. Lodini, Eds., Taylor & Francis, London, 2003, 170–189.

87. P. Scherrer, *Nachr. Ges. Wiss. Gottingen*, 26, 98–100, 1918.

88. D.E. Mikkola and J.B. Cohen, in *Local Atomic Arrangements Studied by X-ray Diffraction*, J.B. Cohen and J.E. Hilliard, Eds., Gordon and Breach, New York, 1966, 289–348.

89. B.E. Warren, A generalized treatment of cold work in powder patterns, *Acta Crystallogr.*, 8, 483–486, 1955.

90. B.E. Warren and B.L. Averbach, The separation of cold-work distortion and particle size broadening in x-ray patterns, *J. Appl. Phys.*, 23, 497, 1952.

91. B.E. Warren and B.L. Averbach, The effect of cold work distortion on x-ray patterns, *J. Appl. Phys.*, 21, 595–599, 1950.

92. B.E. Warren, *X-ray Diffraction*, Addison-Wesley, Reading, MA, 1969.

93. B.L. Averbach and B.E. Warren, Interpretation of x-ray patterns of cold-worked metals, *J. Appl. Phys.*, 20, 885–886, 1949.

94. C. Genzel, Line broadening by non-oriented micro RS, in *Structure and Residual Stress Analysis by Nondestructive Methods*, V. Hauk, Ed., Elsevier, Amsterdam, pp. 435–460, 1997.

95. P. Klimanek, X-ray diffraction line profiles due to real polycrystals, *Mater. Sci. Forum*, 1, 73–84, 1991.

96. P. Klimanek, Problems in diffraction analysis of real polycrystals, in *X-ray and Neutron Structure Analysis in Materials Science*, J. Hasek, Ed., Plenum Press, New York, 1989, 125–138.

97 M.T. Hutchings, Neutron diffraction measurement of residual stress fields: Overview and points for discussion, in *Measurement of Residual and Applied Stress Using Neutron Diffraction*, M.T. Hutchings and A.D. Krawitz, Eds., NATO ASI Series E: Applied Sciences vol. 216, Kluwer, Dordrecht, 1992, 3–18.

6

Applications to Problems in Materials Science and Engineering

In this chapter, a range of applications for strain measurement by neutron diffraction is described. These examples have been selected across numerous fields in order to demonstrate the scope of problems that can be tackled, and to illustrate practical applications of the methods introduced in earlier chapters. Even so it is not possible to cover all possible applications of neutron strain measurement. In each section, the important aspects that must be considered to give reliable strain measurement are emphasized, and potential avenues for further exploitation outlined. These include many future exciting areas of application that will emerge from the new dedicated facilities now becoming available.

6.1 Introduction

The importance of both long- and short-range residual stresses on component performance means that considerable effort is now being devoted to the development of methods by which a knowledge of the residual stress field can be incorporated into design, especially in aerospace, nuclear and other safety critical engineering industries. As seen in earlier chapters, neutron diffraction is one of very few techniques that enables the engineer or materials scientist to obtain information about the state of residual stress relatively deep within the material, absolutely and nondestructively. In this chapter, we will review the application of neutron diffraction techniques to real engineering problems, with emphasis on the challenges that arise from measurement and interpretation, and how to meet such challenges.

There has been a marked increase in the use of neutron diffraction for engineering stress measurements in recent years as demonstrated in Figure 6.1, which summarizes publications by year. Recently, the number of synchrotron x-ray diffraction measurements of strain, which have much in

TABLE 6.1

Percent of Publications[a] on Strain Measurements by Neutron and Synchrotron X-Ray Techniques, with Areas of Application, 1982–2002[b]

	Welds and Components	Multiphase	Intergranular	Surfaces and Coatings	General
Neutron techniques	24	30	22	5	19
Synchrotron x-ray techniques	9	21	20	32	18

[a] Publications pertaining to neutron and synchrotron techniques totalled 413 and 92, respectively.

[b] As registered by Web of Science online citation service.

common with neutron diffraction methods, has also increased sharply, boosted by the arrival of third-generation sources, such as the European Synchrotron Radiation Facility in Grenoble. The trends in the use of neutron and synchrotron x-ray strain mapping usage in recent years are summarized in Table 6.1, which highlights the differences in their capabilities. Through their phase- and grain-orientation selectivity, both are well suited to the measurement of interphase stresses in composites, and of intergranular stresses. With respect to the first of these, attention has traditionally been focused on metal matrix composites, but functional composites such as shape memory and ferroelectric materials are currently receiving increased attention. Neutron diffraction is currently the method of choice for measuring long-range stresses in large components, because of the greater penetration depth and higher diffraction angles, which allow deeper measurements in reflection geometry than is possible with high-energy x-rays. Synchrotron x-ray methods, on the other hand, are perhaps better suited to measurements on thin surface coatings or treatments because of the smaller gauge volumes possible and faster data acquisition times. The neutron diffraction work listed in the general classification in Figure 6.1 includes papers developing new techniques or reviewing existing capabilities.

Each of the generic areas listed in Table 6.1 is examined in turn in the following sections. The examples chosen are not exhaustive, but rather illustrative of the opportunities and challenges posed in applying the neutron diffraction technique to stress measurement. In each section, the origin and nature of the different stress fields are examined, and through practical examples the measurement options and challenges of each are discussed before reviewing the current state of the art and potential future directions. Some basic ideas and principles given in earlier chapters are repeated and enlarged as when appropriate.

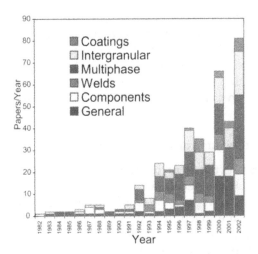

FIGURE 6.1
Number of papers that describe engineering residual stress measurements by neutron diffraction published per year in major journals. The papers are categorized as follows: general papers, measurements on engineering components, welds, multiphase materials (composites), intergranular stresses, and coatings and near surface regions.

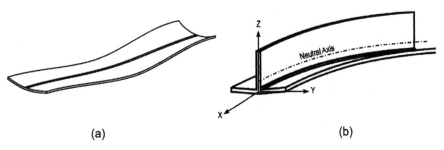

(a)   (b)

FIGURE 6.2
Examples of various types of distortion caused by weld stresses, for (a) butt, and (b) fillet welds [1].

6.2 Welded Structures

Welds and other types of joints are often locations of stress concentration. Add to this the fact that the localized heating required to make a weld can lead both to high levels of stress and softening of the parent metal within the heat-affected zone (HAZ), and one has all the ingredients necessary for unexpected failure. Sometimes this occurs by sudden fracture, but is much more likely to occur by fatigue, stress corrosion cracking, or even creep cavitation. Not only can residual stress give rise to failure but it can also lead to weld distortion, and thus nonconformance causing costly rejection (Figures 6.2 [1] and 6.13). It is therefore not surprising that studies of stress in weldments and similar components represent a quarter of all publications on neutron diffraction in Table 6.1. The task of making accurate measurements of residual stress in weldments presents particular challenges.

6.2.1   Origin and Nature of Weld Stresses

In order to understand the origin of residual stress in a weldment, it is helpful to consider the changes in temperature that occur for a weld bead laid down on a plate, either at a given point as a function of time, or alternatively, at different locations at a given instant in time, as shown in Figure 6.3. Well ahead of the welding arc (A–A) in Figure 6.3, the section is uniformly at ambient temperature and no residual stress is present. Across the line of the weld pool (B–B), the temperature distribution is necessarily very steep. In and near the molten zone the stresses are essentially zero because of the zero or very-low-yield stress. As a result, misfits near the weld are accommodated by local plastic deformation. Compressive stresses develop in the HAZ because of the expansion in the vicinity of the weld pool, while stress balance requires that this region be constrained by tensile forces farther from the weld. Behind the weld zone (C–C), the temperature profile is less severe and tensile stresses begin to develop near the weld line. This occurs because as the weld cools and the weld metal and HAZ begin to shrink, tensile stresses are introduced locally due to the misfit between the plastically deformed region created when the near-weld region was hot and the remainder of the plate. Once the weld torch has passed and the weldment has returned to room temperature (D–D), the final distribution of the longitudinal stress distribution is obtained, being tensile in the weld and HAZ, and compressive in the far field.

It is important to remember that residual stresses arise from misfits (Section 1.4). Consequently, while localized heating does create transient residual stresses, these would disappear on cooling back to room temperature were it not for the permanent misfit between the near-weld region and rest of the plate created by these transient stresses. In other words, if the spatial gradient in temperature were insufficient to cause plastic deformation, then there would be no final residual stresses. This is the basis of various strategies developed to limit weld residual stress. Clearly, the thermal and mechanical behavior of the weld is complex, depending on the thermomechanical properties of the parent plate, plate size and thickness, heating characteristics of the torch, jigging of the plates, metal flow within the molten region, and so on. This makes the construction of finite element models to predict weld stresses in weldments very difficult without information from direct measurements. Neutron diffraction can provide just the right level of two- or three-dimensional information to compare with the predicted strain or stress

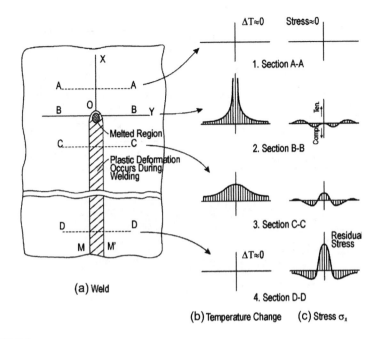

FIGURE 6.3
(a) Various positions at a given instant of time along a welding line caused by forming a typical bead on plate weld progressing along OX. The locations A-A and so on are discussed in the text. (b) Distribution of temperature, T, and (c) longitudinal residual stress component, σ_x [1].

fields, leading to more robust finite element models. Indeed, finite element models are now able to capture the essence of both the thermal profile and the resultant stress distribution [2,3]. An example is the prediction of temperature and stress profiles for a TIG bead on plate weldment in aluminum shown in Figure 6.4. In this case, the thermocouple readings have been used to refine the heat sink model of the copper-backing plate. It can be seen in Figure 6.4b that the backing plate gets progressively warmer as the TIG weld proceeds. This causes a reduction in the final near-weld stresses along the plate.

In reality, a number of complicating factors can affect the reliability of weld residual-stress predictions, most notably:

- Transformations can occur within the weld zone as the metal cools.
- Softening due to annealing, recrystallization, grain growth, and, for heat-treatable alloys, dissolution of the strengthening precipitates.
- Addition of weld filler material, which is usually of a different composition than the parent metal.

Unfortunately, as well as complicating predictive modeling, each of these factors also complicates the accurate measurement of residual stress by

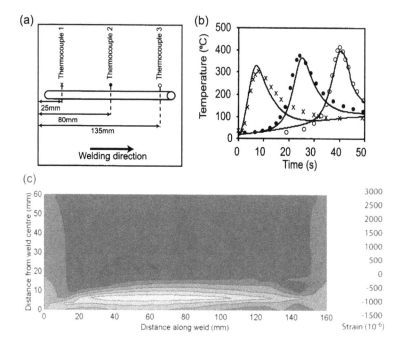

FIGURE 6.4

(a) Schematic of an approximately 8 mm wide autogenous bead on plate TIG weld in 3.2 mm Al plate from 10 to 150 mm showing thermocouple positions and representative symbols. (b) Three thermocouple readings plotted against time. The finite element calculations that include gradual warming of the copper baseplate are shown by the continuous lines. (c) The longitudinal elastic strain field component, in units of 10^{-6}, predicted from finite element analysis for one-half of plate [4].

neutron diffraction by influencing the reference lattice spacing as discussed in Section 4.6.

6.2.2 Comparison of Various Stress Measurement Techniques

Traditionally, many measurement techniques have been used to characterize the state of stress in welded structures. A brief outline of these is useful to put the neutron diffraction technique into context. Masubuchi [5] classified these techniques into three types: (a) stress relaxation [6]; (b) diffraction, which we expand here to include other nondestructive methods; and (c) cracking-related methods. The techniques are listed in Tables 6.2, 6.3, and 6.4, respectively. The first are essentially destructive methods, relying on measuring the distortion brought about when cutting disturbs the stress field [7], while the last group involves the deduction of aspects of the stress field from the observation of cracks on the surface of the sample. These cracks may occur by stress corrosion or be hydrogen induced [8]. However, this method can only provide qualitative information. One method receiving

TABLE 6.2

Destructive Stress Determination Techniques Using Stress Relaxation Applicable to Stress in Weldments

Technique	Principle	Application	Advantages	Disadvantages
Sectioning using strain gauging	Strain gauges mounted and location cut out completely	Plates	Reliable Simple principle High measurement accuracy	Machining expensive and time consuming Gives average over area removed
Mathar–Soete hole drilling	Small hole is drilled and strain gauge rosette records relaxation. Stresses back calculated [7]	Laboratory and field tool for plates	Simple principle Cheap Causes little damage Accepted by engineers	Drilling causes plastic strains at hole periphery, displacing results Provides information to a depth equal to diameter of hole Provides spot measurement Only in-plane stresses Holes cannot be close together, giving poor spatial resolution Method must be used with extreme care
Contour method	Section is cut by electro-discharge machining, and deviations in flatness from surface interpreted in terms of out-of-plane stresses	Laboratory	Simple apparatus Provides stress field normal to cut	Sample movement during cutting must be prevented Best applied normal to principal axis, as in, for example, measuring longitudinal weld stress for a long weld
Gunnert drilling	Four holes located in circle drilled through plate and diametrical distance between holes measured, and then circular groove trepanned around holes	Laboratory and field tool for plates and three-dimensional solids	Robust and simple apparatus Semidestructive Damage to object can be repaired	Relatively large margin of error in perpendicular direction Underside of weldment must be accessible to apply fixture
Rosenthal–Norton sectioning	Two narrow blocks, one parallel to weld and one transverse are cut, stresses calculated from strain changes while cutting [6]	Laboratory measurements in thick weldments and three-dimensional solids	Fairly accurate	Troublesome, time consuming Completely destructive

Adapted from Masubuchi [5].

TABLE 6.3

Nondestructive Methods of Stress Determination Applied to Welds

Technique	Principle	Application	Advantages	Disadvantages
Magnetic	Magnetic properties and elastic strain are coupled through magnetostriction	Laboratory and field	Cheap Online monitoring Some magnetic techniques — depth sensitive	Sensitive to microstructure and chemistry changes Complex relationship between magnetic parameters and stress No universally agreed method of data interpretation
X-ray diffraction	Often uses sin²ψ technique, which assumes that because of low penetration, the normal stress is zero	Laboratory technique that can also be used in the field	Well-accepted, standard method Nondestructive for surface stresses Does not require strain-free reference Can provide surface strain fields	Provides only near-surface information Sensitive to surface preparation
High-energy synchrotron x-ray diffraction	Determine strain with reference to strain-free zero lattice spacing	Specialized large-scale facility technique	Provide two-dimensional stress fields Quick High resolution (better than 1 mm)	Best applied to plates since low diffracting angles mean transmission geometry only is generally preferable Measuring triaxial strains is difficult Difficult access
Neutron diffraction	Determine strain from measured lattice spacing with reference to strain-free lattice spacing	Specialized large-scale facility technique	Provide three-dimensional stress fields Probes depth of many centimeters Standard definition is underway	Relatively expensive Access difficult

Adapted from Masubuchi [5].

TABLE 6.4

Methods of Stress Determination Based on Cracking Applied to Welds

Technique	Principle	Application	Advantages	Disadvantages
Hydrogen-induced cracks	Assess distribution and level of residual stress from distribution and number of cracks [8]	Laboratory and field	Provides two-dimensional 'map' indicating largest stresses Well suited to environmental fatigue	Not quantitative Expensive, destructive method
Stress corrosion–induced cracks	Assess distribution and level of residual stress from distribution and number of cracks [8]	Laboratory and field	Provides two-dimensional 'map' indicating largest stresses Well suited to environmental fatigue	Not quantitative Expensive, destructive method

Adapted from Masubuchi [5].

considerable attention recently is the *contour method*, which can give a measure of the residual stress field in a plane before making a cut across that plane. It is based on inferring the prior residual stresses from the cut's deviations from flatness. Neutron diffraction methods have been used to validate this method for evaluating longitudinal weld stresses, as shown in Figure 6.5. One of its main advantages over most destructive techniques is that it provides a two-dimensional stress map with a single cut. Of all these methods the most widely used in industry is *hole drilling*. This method can provide subsurface information, but is normally restricted to a depth approximately equal to the diameter of the hole, usually less than 1 mm. The separation of the strain gauges forming the rosette around the hole defines the spatial resolution, typically around 5 mm. Conventional *x-ray diffraction* can also provide near-surface information if combined with layer removal. Portable x-ray units that enable stress measurements to be made *in situ* at the plant are commercially available.

The only truly nondestructive techniques for measurement at positions within a sample are *neutron diffraction* and *high-energy synchrotron x-ray diffraction*. However, magnetic techniques are being successfully developed to measure biaxial stress in magnetic materials to depths of 6 mm or more [10]. Synchrotron x-ray diffraction can provide very good spatial resolution combined with fast measurement times. For example, in Figure 6.6 the longitudinal strain field around the end of an Al bead on a plate TIG weld was mapped over 40×20 mm on the BM16 beam line at the ESRF with a 1 mm beam area size, and involved 800 measurements in 8 hours, exploiting the small probe size and fast data acquisition [11]. While well suited to the study of Al alloys, synchrotron x-ray radiation is rather limited in applicability to other metals because of the increase in attenuation with increasing atomic number, as seen in Table 2.2. At the high energies needed for deep penetration, the corresponding wavelengths are in the 0.3 to 0.08Å range, so that the scattering angles are usually low, from ~4 to 30°. As a result, long path lengths are often unavoidable for strain measurements in certain directions, and the instrumental gauge volume is elongated into a thin diamond shape in the horizontal scattering plane. Consequently, although measurements of in-plane strain components can be made relatively easily in plate samples, because of the long path lengths associated with what are essentially grazing-incidence measurements of out-of-plane strain components cannot be made at significant depths (>1 mm).

Neutron diffraction on the other hand can provide line, two-dimensional (2D) or three-dimensional (3D) strain or stress maps, and is ideal for direct comparison with finite element predictions. A good example of this is provided by measurement of the strain components as a function of distance from the centerline of a bead on a plate electron beam weld (EBW) in a Ni superalloy (waspaloy) plate, following a superficial cosmetic weld pass used to improve the surface finish [12]. The resulting variation of the longitudinal residual stress component is shown in Figure 6.7. The measurements were made at a depth of 0.5 mm beneath the surface by neutron diffraction from

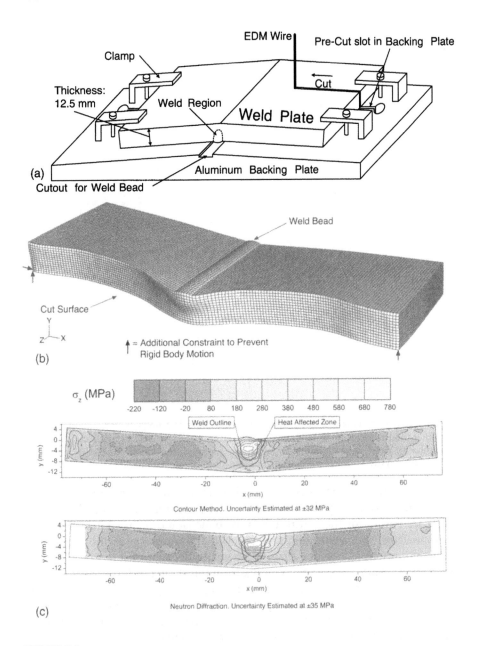

FIGURE 6.5
Contour method for longitudinal weld stress measurement on a 12.5 mm thick ferritic plate containing a 12-pass TIG weld made by the TWI, Cambridge, UK, showing (a) how to electro-discharge machine the weld, (b) the relaxation caused by the cut, magnified 40 times, and (c) the inferred residual stress before making the cut, as compared with results of neutron diffraction measurements [9].

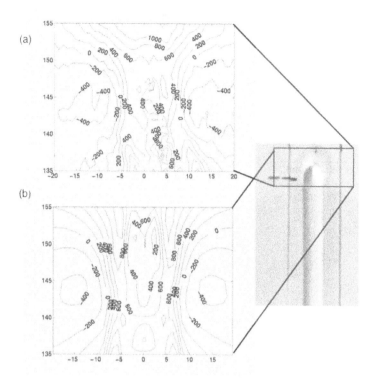

FIGURE 6.6
(a) Contour map of longitudinal residual strain component in units of 10^{-6} around the end, located at (0 mm, 145 mm) of an 8 mm wide bead on plate TIG weld in 3 mm thick Al plate (inset photo). These were measured by synchrotron x-ray diffraction on the BM16 instrument at the ESRF. (b) Comparison with results of finite element calculations [11].

the 111 planes, at the surface by laboratory x-ray diffraction from the 311 planes, and also by the conventional hole-drilling technique. Strain-free reference values for the neutron data were measured using matchstick samples cut from the weldment, and showed the variation of the strain-free lattice parameter across the plate to be small. Because of intergranular and interphase strains, only after changing from the 111 to the 311 reflection using Fe target radiation did the x-ray results come into agreement with the neutron data. These measurements were made in order to provide benchmark results for comparison with finite element models designed to calculate welding stresses and distortions in waspaloy welds [3].

6.2.3 Challenges to Accurate Neutron Diffraction Measurement of Weldment Stresses

The main challenge for accurate measurement of stress fields in weldments using neutron diffraction lies in the accurate determination of strain-free

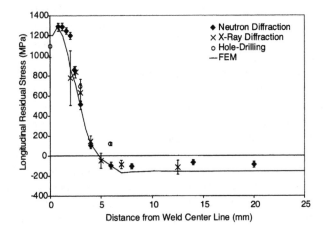

FIGURE 6.7
Variation of longitudinal residual stress component in an electron beam weldment of waspaloy 4.3 mm thick plate as a function of distance lateral from the weld centerline, measured near the top surface using three techniques. The majority of measurements were made outside the ~2 mm wide melted part of the weld. The continuous line refers to a finite element model calculation [13].

lattice spacing d^0. When examining welds, it is not usually advisable to use the balance of stress and bending moments described in Section 4.6.3 to determine a *global* strain-free reference value. This is because the different regions experience very different thermal histories, as illustrated in Figure 6.3b. In addition, variations in composition and microstructure in the near-weld region can occur in cases where filler material has been added. In fact, the strain-free lattice parameter may vary from point to point due to at least three of the mechanisms discussed in Section 4.6.2.

Changes in precipitate fraction and solute composition as a result of heat treatment are particularly important for heat-treatable alloys such as Cu containing Al 2XXX series alloys, since they have been specifically designed so that the precipitate microstructure and the fraction of alloying elements in solution can evolve according to heat treatment. In fact, many Al alloys can age even at room temperature. The changes in matrix composition can have significant effects on the strain-free reference lattice spacings, which if not taken into account, can lead to significant error in the resulting residual stress. The lattice parameter variation for aluminum with copper content, measured by Mohr and Priesmeyer [14], is shown in Figure 4.18. Note that if interpreted directly as an apparent macrostrain, without any correction to account for the Cu in solution, a change of 1% Cu solute content is equivalent to about 1500×10^{-6} error. While it is unlikely that heat treatments would bring about such large changes in the amount of Cu in solution through changes in precipitation, care should always be exercised when examining heat-treatable aluminum alloys. Another system prone to similar errors is that of Fe-C. Small changes in dissolved carbon can again cause large changes in the lattice parameter. This problem can also extend to welds which have been post-weld heat treated (PWHT) to reduce the weld stresses, in that the heat treatment may have altered the relative fractions of elements in solid solution and thus altered the strain-free lattice spacing.

Filler material is often added to improve the quality of the weld. Clearly, filler material can have a drastic effect on composition in the weld zone. Indeed, in many cases certain elements are added to the filler specifically to control weld microstructure. In all cases, this must be accounted for if meaningful strains are to be recorded in this region.

Phase transformation can occur. In certain alloys, especially in steels, the dramatic thermal excursion experienced by the weld and HAZ can cause diffusionless phase transformations to occur. Diffusionless transformations occur by coordinated displacement of the crystalline lattice. Such transformations can have a marked effect on the residual macrostresses, as illustrated by the experiments of Jones and Alberry [15,16] on martensitic (high Cr), bainitic (low Cr), and austenitic (AISI 316) steels whose results are summarized in Figure 6.8a. During cooling of a uniaxially constrained specimen from the austenite phase, tensile stresses develop due to thermal contraction. Between 600 and 300°C, martensitic transformation takes place. Note also that the high-temperature part of the austenite curve follows the yield stress. On the other hand, the ferrite that forms has a much higher yield stress. This means that the slope of the ferrite part of the stress–temperature graph is steeper consistent with the larger thermal contractions strains (thermal expansivity of ferrite $\sim 12 \times 10^{-6}/\text{K}$). In the region of the stress–temperature plot where the transformation occurs, the interpretation is difficult; a schematic interpretation is given in Figure 6.8b. This diffusionless transformation has a shear component that is much larger than the dilatational term, giving rise to significant intergranular stresses and dominating plastic flow. Figure 6.8 shows that that if the transformation is completed at a high temperature, then the ultimate level of stress at room temperature will be large, since the fully ferritic specimen contracts over a large temperature range. Conversely, as work in Japan has demonstrated, when using welding consumables with very low transformation temperatures it is possible not only to reduce the residual tensile stresses, but also even to introduce residual compression into the weld [18].

Besides influencing the macrostresses, the transformation to form a second phase will set up Type II stresses where previously only macrostresses were being generated. Consequently, both the transformed phase and any residual untransformed phase should be considered, and the separation of the two types of stress should be included in the analysis. Finally, not only does the extent of such a transformation affect the magnitude and nature of the stress field, it may also affect the strain-free lattice spacings of the constituent phases, possibly leading to a misleading interpretation. In order to dispel concerns regarding the possible point-to-point variation in the strain-free lattice parameter d^0 in the weld and HAZ, it is therefore highly advisable to

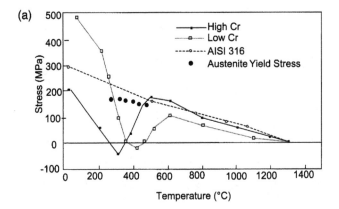

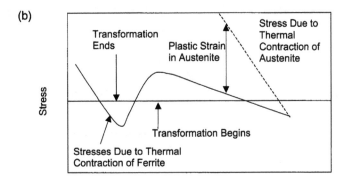

FIGURE 6.8
(a) Axial component of microstress that develops in uniaxially constrained specimens during cooling of martensitic (high Cr), bainitic (low Cr), and austenitic (AISI 316) steels. Also shown are experimental data for the yield stress of austenite in low-alloy steel [15,16]. (b) A schematic interpretation of Jones and Alberry's experiments [17].

measure d^0 as a function of position in the actual weld or a comparison weld. Various methods of doing this were discussed in Section 4.6.3.

An added complication may occur if the weld region contains very large columnar grains that can form during cooling of the weld pool. The extreme texture may cause certain reflections to be unobservable, or the effects of large grains may give rise to anomalous effects as discussed in Section 3.6. In addition, the diffraction elastic constants appropriate to the extreme texture must be used in the conversion of strain to stress. Welds in austenitic steel are particularly prone to these effects. However, if the region of the parent plate outside the HAZ is of primary interest, as is often the case, the strain-free lattice spacing is unlikely to vary appreciably, and so mapping its variation may not be necessary.

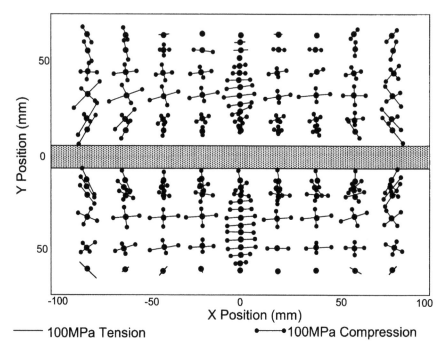

100MPa Tension   **100MPa Compression**

FIGURE 6.9
Magnetic measurements showing the principal in-plane stress vectors for positions in a butt-welded steel plate. Only near the symmetry lines are the principal axes of stress aligned with the weld line and transverse directions of the plate. (Courtesy of D.J. Buttle.)

If the primary interest is the determination of residual stresses rather than strains, an important issue is the number of strain directions in which it is necessary to make measurements in order to accurately convert the strain to stress, as discussed in Section 5.1. In practice, often the strain is measured in only three orthogonal directions that are assumed to be the principal axes, usually the symmetry axes longitudinal and transverse to the weld path, and through thickness. However, these may be the principal axes only down the symmetry line of the weld, although they may also be close to the principal axes in the far-field region. It is certainly not true throughout the near-weld zone, especially for narrow welds in thick plate, or toward the edges of the plate, as seen from the results of magnetic measurements shown in Figure 6.9. For general positions, measurements of strain in at least three, and preferably more, additional directions must be made.

6.2.4   Examples of Stress Measurement in Weldments

6.2.4.1   Alumino-Thermic Welds

Railway rails are now welded together in preference to the traditional method of bolting, giving rise to a smoother ride for the passenger. This process involves the placement of a ceramic mold around the gap between

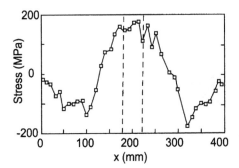

FIGURE 6.10
Variation with position along the rail direction (x) of the longitudinal residual stress component at a vertical height of 76 mm from the base, approximately the mid-height of the rail, after alumino-thermic welding. The weld zone represented by the vertical dashed lines is 40 to 50 mm wide [19].

two rails, and a mixture of iron alloy, iron oxide, and aluminum-based powder being poured into a small hopper above it. The mixture is ignited and the molten steel flows downward to make the weld, as if by casting. This is a process that is found to set up stresses within and near the weld region. Figure 6.10 shows typical results from measurements made on the L3 instrument at the NRU facility in Chalk River, Canada, for the variation along the rail direction of the longitudinal residual stress component at approximately the mid-height on the rail after welding [19]. The weld zone is shown, and the adjacent HAZ is 15 to 25 mm on either side. The mold assembly is about 100 mm wide. Although tensile at mid-height, it was found that the stress component longitudinal to the rail was compressive at the top and bottom of the rail. A strain-free reference sample was taken from each rail and used outside the weld region. By balancing forces and moments and averaging all results for each area, as described in Section 4.6.3, a global strain-free reference was chosen for the weld region. While this may not be exactly correct, it is much more accurate than assuming a global strain-free value over the entire weldment, and does ensure stress equilibrium. It is interesting to note that steep change in stress occurs not at the weld boundaries nor at the heat-affected zone, but at the extremity of the mold assembly. Near the rail at mid-height as shown, it even extends beyond the mold assembly, and is at the location of the steepest longitudinal temperature gradient. For welded joints, the region of steepest thermal gradient is commonly important in determining the extent of the tensile stress region.

6.2.4.2 Manual Metal Arc Repair Welds

In power plant and other high-performance applications, it is not uncommon for welded regions to crack under a combination of the initial residual stresses and applied service stresses. The remedy is not always straightforward. If a repair is undertaken, the residual stresses tend to be regenerated,

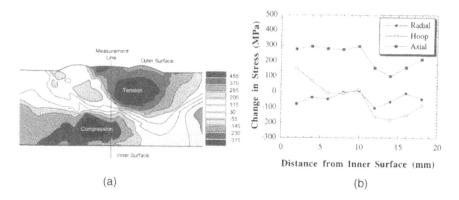

(a)   (b)

FIGURE 6.11

(a) Finite element calculations of the distribution of hoop stress component (in MPa) in a multipass girth weld on a pipe before repair. (b) Measured *change* in residual stress distribution brought about by the repair, as a function of radial distance from inner surface of the tube along line marked in (a) [20].

increasing the magnitude and triaxial nature of the residual stress field in the weld that had, to a large extent, relaxed through the generation of the crack! This may markedly increase the susceptibility of the region to reheat cracking. The following questions must then be answered: "If a crack is found, is it better to leave it and monitor its progress by nondestructive testing, confident that much of the original weld stress has died away, leaving little driving force for growth? Alternatively, should it be repaired, with the risk that the new residual stress field caused by the repair cycle will generate further stresses and foreshorten life?"

Finite element models of the three-dimensional stress state in the vicinity of repair welds have progressed significantly in recent years. Nevertheless, without accurate measurements of stress deep within the welds, it is not possible to validate such models and thus make reliable decisions as to the best strategy. Neutron diffraction is ideal in such cases because it can reveal the state of stress even centimeters beneath the surface. A good example is provided by the study of a 60 mm long, 50% depth, circumferential eight-pass repair weld deposited in an excavated region of a double V-shaped girth weld in a 316L stainless steel pipe of wall thickness 20 mm and 170 mm diameter [20]. Finite element calculations of the hoop stress before repair are shown in Figure 6.11a. The stress field was measured before and after repair along a line 10.5 mm from the weld line. The axial stress component was found to be increased significantly, especially over the inner half of the pipe thickness, that is, deeper than the repair depth. Results for the changes in stress components are shown in Figure 6.11b. The residual stresses formed are also more hydrostatic, leading to creep ductility that may be as little as 20% of that found from uniaxial load testing. This indicates that in this particular case, reparing the weld may not be advisable.

6.2.4.3   Electron Beam Welds

Electron beam welding is an important joining process for the assembly of Ni-base superalloy aeroengine disks. Furthermore, because disk assemblies are safety-critical components, the effect of weld stresses on structural integrity is a key issue. Neutron-diffraction strain measurements have played an important role in the evaluation of residual stresses, and in the development of finite element models of the stress developed during the welding process. In one example, 2 mm wide through-thickness electron beam beads-on-plate welds introduced along the length of 50 mm wide by 200 mm long, 4.3 mm thick, test plates of nickel superalloy were examined by neutron diffraction. The results are compared with those from hole drilling and conventional x-ray diffraction measurements in Figure 6.7. Ni superalloys present some challenges for the measurement of Type I macrostresses because they are essentially two-phase composites, comprising coherent γ' precipitates in a matrix of γ phase. As is discussed in Section 6.3.3, mean phase Type II stresses can arise hindering unambiguous interpretation of peak shifts in terms of the macrostress. However, it has been shown by *in situ* loading experiments [21,22] of the type discussed in Section 5.6.6, that when suitable reflections are used — such as 111 or 311 for neutron diffraction, and 311 for laboratory x-ray diffraction using an iron target tube — good agreement between the techniques can be achieved. This is largely because both the γ and ordered γ' phases contribute to these peaks.

One of the main advantages of using a nondestructive technique such as neutron diffraction is that the evolution of residual stress can be tracked from the initial fully penetrating weld pass, through a cosmetic nonpenetrating pass used to improve top surface finish, to the effect of the final stress relief heat treatment. Results of the three sets of measurements are shown in Figure 6.12. From this figure it is clear that the cosmetic pass increases the tensile residual stress near the top surface in the weld region. In addition, it is also a cause of distortion. The subsequent heat treatment reduces the stresses by about 60%. In relatively thin plates such as these, the misfit strains generated in the weld region are partially accommodated by distortion rather than stress. The level of distortion of the fully penetrating weld pass is shown in Figure 6.13a. This type of distortion is sometimes termed a saddle distortion. Figure 6.13b shows the level of this distortion as a function of the defocus of the electron beam.

6.2.4.4   Friction-Based Welds

A number of friction-based welding methods, including linear, inertial, and friction stir, are being developed at the present time. The joint is formed by friction heating–induced plasticity, which creates clean metallic surfaces capable of metallic bonding to one another without the need for the weld material to reach melting temperature. This has the advantage that weld cracking, often caused by incipient melting or by poor melt pool characteristics, can be avoided. In fact, it has been demonstrated that many alloys which are

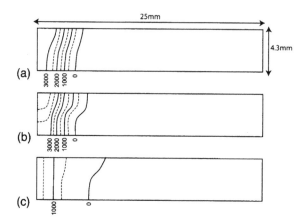

FIGURE 6.12

A 50 mm wide by 200 mm long, 4.3-mm thick, nickel superalloy plate containing a 2 mm wide through-thickness electron beam bead-on-plate weld along its length. (a) The longitudinal residual strain component, in units of 10^{-6}, measured over half the 50 × 4.3 mm cross-section at mid-length (weld at extreme left). The same strain component, (b) after a subsequent partially penetrating cosmetic pass, and (c) after a post-weld heat treatment. (Courtesy H. Stone.)

difficult to fusion weld can be satisfactorily joined by friction-based welding techniques. In addition, it is relatively simple to join dissimilar alloys to one another. Linear friction welds involve the linear rubbing under pressure of two metal surfaces, axisymmetric inertia welds can be formed by bringing into contact a rotating and a stationary part, while for friction stir welding (FSW), a rotating tool piece is inserted into the join line and by a combination of rotational and linear motion "stirs" and extrudes the weld zone. In FSW, a butt weld can be formed by the passage of the rotating tool with a pin that "stirs" the two materials together, as shown in Figure 6.14a. In actual fact, the shoulder in contact with the top surface generates much of the heat, and the plasticized metal "extrudes" past the pin confined to the plane of the plate by a backing plate below (not shown) and the shoulder of the tool above.

A good example of the challenges presented by the accurate measurement of the strain field associated with friction welding is provided by a study of friction-stir butt welding of 13 mm thick 7010 plate [24]. This was undertaken using a rotational speed of 270 rpm, traverse speed 81 mm/min, and vertical (downward) force of approximately 36 kN. The tool had a 18 mm diameter shoulder and an M5-threaded pin 5 mm wide and 0.8 mm pitch. The weld direction was parallel to the rolling direction of the plates. FSW is a candidate for the welding of aerospace alloys such as 7010. However, typical of high-strength aerospace aluminum alloys, 7010 (5.7% to 6.7% Zn, 2.1% to 2.6% Mg, 1.5% to 2.0% Cu, 0% to 0.12% Si, 0% to 0.15% Fe, 0% to 0.10% Mg, 0% to 0.05% Cr, 0.10% to 0.16% Zr) is a heat-treatable alloy. The T7651 thermo-mechanical treatment comprises a solution heat treatment around 465 to 485°C followed by water quenching, controlled stretching (1.5% to 3%), and a two-stage artificial aging procedure (10 h at 120°C + 8 h at 170°C). This

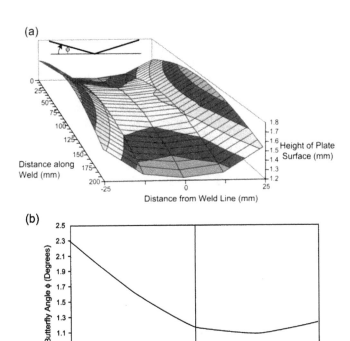

FIGURE 6.13

(a) Topography of a 50 mm wide by 200 mm long, 4.3 mm thick, nickel superalloy plate containing a 2 mm wide through-thickness electron beam bead on plate weld along its length. The longitudinal residual stress and strain components are shown in Figures 6.7 and 6.12, respectively. The definition of the butterfly angle, ϕ is shown. (b) Variation in butterfly angle ϕ across the mid-length, 100 mm position, with electron beam defocus [23].

treatment results in the formation of fine coherent Guinier–Preston zones after the first aging stage, followed by strengthening precipitates of homogeneous η' Mg(Zn,Cu,Al)$_2$ and grain–boundary precipitation of η MgZn$_2$. Now, the unstrained lattice parameter of aluminum varies markedly with the solution of other elements (Figure 4.18). An increase by 1% of the amount of Mg in solution changes the aluminum matrix lattice parameter by approximately 0.00375 Å, equivalent to a spurious strain of 927 $\mu\varepsilon$. Similarly, a 1% increase in the level of Zn in solution is estimated to lead to a decrease of the lattice parameter by 0.00066 Å or 165 $\mu\varepsilon$. Thus, the complete precipitation or dissolution of 2.1% to 2.6%wt of Mg to η MgZn$_2$ would lead to a net change in the lattice parameter in the HAZ or in the thermomechanically affected zone (TMAZ), which corresponded to a spurious strain of between 1250 and 1550 $\mu\varepsilon$. This change is, of course, only an extreme boundary, since these elements are never present fully in solution.

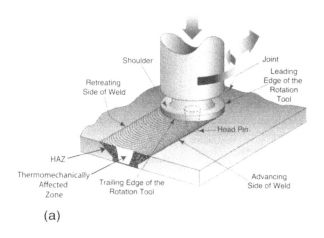

(a)

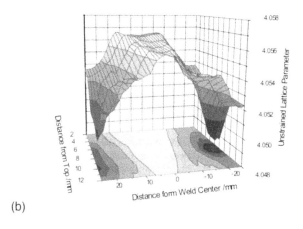

(b)

FIGURE 6.14

(a) Schematic of friction stir welding process. (b) The variation of the unstrained lattice parameter calculated from the measured 311 lattice plane spacings measured by synchrotron X-rays using Equation (4.19) for a thin cross-sectional slice of an AE7010 friction stir weld. (c) The variation of the longitudinal residual strain component 2 mm from the top (2 mm) and bottom (11 mm) surfaces of the plate calculated from the measured lattice spacings on the basis of a global stress-free lattice parameter (left), and taking into account the spatial variation in stress-free lattice parameter shown in (b) (right).

Many of the methods outlined in Section 4.6.3 have been used to determine the variation in the stress-free lattice parameter d^0 across FSWs, including the use of fabricated combs. In the case of the 7010 plate weld described above, a 1 mm thick cross-sectional slice was cut midway along the FSW in order to use the transmission $\sin^2\psi$ method described in Section 4.6.3 using high-energy synchrotron x-rays, and the results for d^0 are shown in Figure 6.14b. It is evident that the changes in stress-free lattice parameter are

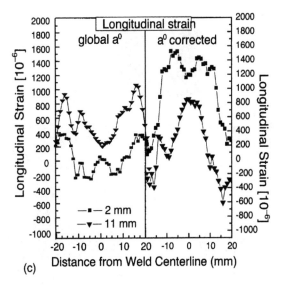

(c)

FIGURE 6.14 (CONTINUED)

significant. If a global value of d^0 were used as a reference, such a variation in stress-free lattice spacing would give a spurious strain of up to 2000×10^{-6}, corresponding to a spurious stress of 140 MPa. The longitudinal component of residual strain across the weld line, determined from neutron diffraction strain measurements, is shown in Figure 6.14c. It is clear that if the variation in stress-free lattice parameter were not taken into account, the tensile strains near the weld centerline would be significantly underestimated. Note that the variation in unstrained lattice parameter has the same shape as the Vickers hardness map shown in Figure 6.15a. Both reflect changes in precipitate microstructure in the TMAZ and the HAZ of the weld shown in Figure 6.15b, c, d, and e. In the scanning electron micrographs, the precipitates in the parent plate are too fine to be resolved, but can be seen in the weld zone. It is also noteworthy that both the microstructure and the residual stress variations show the dimensions of the shoulder of the tool to be more important than the pin width. This is because both are determined primarily by the thermal history rather than by the mechanical deformation. As was the case for the alumino-thermic weldments, the major cause of the residual stress is the steep thermal gradient, in this case between the region directly under the 18 mm diameter shoulder and that outside it. The smaller TMAZ cannot be discerned from the measured stress field, although it was found to have a clear effect on the diffraction peak widths.

A recent study of the stresses generated when Ni superalloy is linear-friction welded has been the subject of a round robin study to examine the reproducibility and reliability of neutron diffraction measurements made at facilities around the world under the auspices of the VAMAS TWA 20 working group. One of the challenges of reliable interpretation of the data was

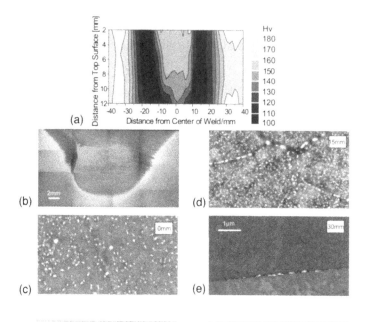

FIGURE 6.15

(a) Variation in Vickers hardness in near-weld zone of friction stir–welded 7010 plate. (b) Low-resolution optical micrograph showing the macrostructure of the weld zone. Field emission gun–scanning electron micrographs of the microstructure taken at mid-thickness: (c) on the weld centerline, (d) 15 mm from the center line, and (e) in the parent plate.

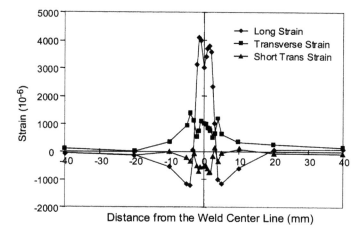

FIGURE 6.16

Variation of three components of residual strain, longitudinal, transverse, and through thickness, across a sample in which forged and cast Ni superalloy blocks were joined by linear friction welding. Different strain free–lattice parameters were used on each side of the weld. [25].

the fact that two different materials, one forged and one cast Ni superalloy, were joined together. In this case, different strain-free zero values were used for the two materials. The results are shown in Figure 6.16, where it can be seen that the residual strain levels are very high, despite the fact that melting did not occur [25].

6.2.5 Future Directions for Neutron Diffraction Studies of Weldments

As seen above, neutron diffraction is proving to be a useful tool for the development of new joining processes, as well as for the optimization of more traditional methods. Further examples of its use include:

- Validation of finite element models used for weld process optimization
- *In situ* studies of postweld heat treatments
- Studies of the effects of phase transformations during welding
- Improved strategies for prolonging component and plant lifetime.

6.2.5.1   *Validation of Finite Element Models*

Commercial pressures mean that new joining processes must be introduced quickly and safely. However, there are often many parameters that can be varied in the attempt to minimize weld residual stress and distortion. Finite element models are key to this optimization procedure. A good example of this is thermal tensioning for low stress and low distortion. This is a technique by which the near-weld zone is either heated or cooled to thermally "tension" the weld during welding, thereby optimizing the final weld stress state. This can be carried out in a number of ways, such as by heating using a laser, and cooling using jets of chilled air or nitrogen. In addition, to the normal process parameters, such as welding speed, jigging, heat extraction, weld speed, and weld power, this introduces new parameters, such as the degree, area, and lateral and anterior location of heating/cooling. Finite element modeling is an important tool for calculating the stress response in such a multiparametric space, and thus of accelerating process optimization. A good example is provided by the modeling of low-stress no-distortion welding of aluminum plates in which two lasers have been run alongside the TIG torch to give a temperature field shown in Figure 6.17a in order to prestress the weld [26]. As is clear from the results of neutron diffraction measurements on a conventionally TIG welded plate shown in Figure 6.17b, without thermal tensioning large stresses are generated. From the calculations, the thermal tensioning is predicted to reduce the stress in the weld, in this case by a factor of 4. Neutron diffraction is proving to be a valuable means of testing and improving the reliability of such models.

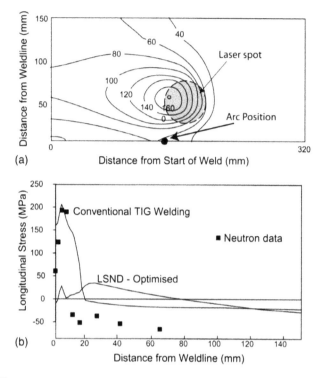

FIGURE 6.17

(a) Finite element predictions of the thermal temperature field in degrees Centigrade for a low-stress, no-distortion (LSND) aluminum bead on plate weldment, 320 mm by 300 mm by 3.2 mm thick, using laser heating. An absorbed laser power of 400 W was assumed to be uniformly distributed over two 60 mm diameter spots, 30 mm ahead of the arc and 50 mm either side of the weld centerline at the bottom of the figure. (b) The variation of the longitudinal stress component laterally from the weld line at a position 160 mm along the weld in a conventional tungsten-inert-gas weld measured by neutron diffraction. The calculations (solid lines) predict that thermal tensioning reduces the tensile peak stresses in the weld by ~150 MPa [26].

6.2.5.2 Postweld Heat Treatment

Residual stress levels resulting from welding in a component are often far too high for practical use, and have to be reduced by effective heat treatment procedures. With the use of neutron diffraction, it is now possible to study the effect of postweld heat treatments in real time. This may offer a means of optimizing heat treatments so as to maximize stress relief and minimize microstructural changes occurring during the process [27].

6.2.5.3 Phase Transformations

Relatively little work has been carried out specifically on welds in which significant phase transformations occur in the weld zone. Here the ability to use the relative peak intensities to measure the volume fraction of the transformed products, in addition to measuring the stresses in each phase, makes neutron diffraction a very important tool. In tandem, finite element modeling packages now exist that can take into account changes brought about by phase transformations.

6.2.5.4   Strategies for Improved Lifetime

Another area ripe for exploitation is the definition and testing of better lifing models based on residual stress maps provided by a combination of finite element modeling and neutron diffraction measurements. Until recently, it has not been possible to obtain such detailed information about the three-dimensional state of stress in the vicinity of welds, and so at the present time lifing models tend to be relatively unsophisticated and thus unnecessarily conservative.

6.3   Composites and Other Multiphase Materials

The presence of residual (internal) stresses must be taken into account when considering the mechanical properties of almost any system that is mechanically nonhomogeneous. This is especially true of composite materials used in engineering applications. Indeed, the term "reinforcement" via inclusions in a composite implies an unequal sharing of any applied load toward the stronger or stiffer phase. For the majority of composites, the second phase is added with a view to bringing to the composite attributes not present in the matrix alone, such as stiffness, strength, low thermal expansivity, and so on. In each case, the degree of success to which this is achieved is, at least partly, dependent on the generation of internal stresses between the phases. Similarly, many of the performance-limiting characteristics are also related to the generation of internal stresses — for example, poor resistance to thermal cycling, premature microyielding, microcracking, and so on. If the behavior of a composite is to be optimized for a particular application, it is therefore important to be able to monitor the levels of internal stress and thus to learn how to tailor them to advantage. This might occur by changing the morphology, diameter, or aspect ratio of the reinforcing phase, by modification of the matrix by alloy or heat treatment, or by modification of the interface (e.g., to encourage crack bridging by pulled-out fibers, or to increase the efficacy of load transfer). In this section, we examine the level of information that neutron diffraction can provide on composite behavior, and the challenges of accurate strain and stress measurement through the reliable interpretation of diffraction data.

6.3.1   Nature and Origins

In the solid state, composite materials are almost never stress-free. In fact, there is a host of ways in which stresses can be introduced into composite materials; an example is shown in Figure 1.3. Most commonly, one must consider

- Elastic mismatch stresses
- Thermal misfits
- Plastic misfits
- Transformations.

Because there are so many different opportunities for generating residual stresses in multiphase materials, the examples of studies in this section are grouped together into subsections according to the above broad categories. Furthermore, these stresses normally arise across the full range of length scales because the heterogeneous nature of a composite material means that uniform stresses almost never arise. According to the scheme presented in Section 1.4.1, it is useful to identify the characteristic length scale l_0 over which the various types of stress vary.

Long-range macrostress (Type I) is the stress that would occur in a *single-phase* homogeneous body. In other words, it is the stress that would be predicted using a continuum finite element model based on the overall homogeneous, but not necessarily isotropic, composite properties. The length scale l_0^I over which the Type I stress field varies is comparable with the scale of the body, and is, in most cases, much coarser than the scale of the reinforcement.

Short-range microstress (Types II and III) comprises all the stress types inherently connected to the inhomogeneity that is introduced by the incorporation of the reinforcing phase. Normally, for single-phase materials, Type II stresses are taken to vary over length scales corresponding to the grain scale. For composites, however, the major microstructural scale corresponds to the scale of the reinforcement. In such cases, it is more sensible to define the Type II stresses as corresponding to the scale of the fibers rather than the grains, since this is the dominant level of heterogeneity. Under this definition, matrix Type II stresses will normally fluctuate over the interparticle spacing, as sketched in Figure 1.6. For single crystal-reinforcing particles, the reinforcement scale and grain scales are the same. In some cases, the reinforcement scale is much greater than the grain scale, as in long fibers such as SCS6 SiC monofilaments that have a nanocrystalline structure. In this case, if the fibers are all aligned along direction 1, the characteristic lengths perpendicular to the fiber direction, $l_{0,2}^{II} = l_{0,3}^{II}$, are approximately equal to the interfiber spacing, but in the fiber direction, the characteristic length, $l_{0,1}^{II}$, is approximately equal to the fiber length and can even be comparable to the scale of the Type I stresses.

6.3.1.1 Mean Phase-Specific Microstresses

While the stress field normally varies sharply from point to point within either phase, it is often helpful to consider the volume averaged stresses in each phase because many mechanical properties such as stiffness and the coefficient of thermal expansion are dependent on the volume averaged stresses. For example, if a metal matrix containing spherical ceramic particles of smaller thermal expansivity is cooled, on average the matrix will be in net hydrostatic tension, with the particles in hydrostatic compression as illustrated in Figure 1.6. In order to balance Type II stresses, the following equation must hold for the mean phase i microstress, denoted by $\langle \underline{\sigma} \rangle_i^{II}$, provided the volume considered is larger than $(L_0^{II})^3$:

$$(1-f).\langle \underline{\sigma} \rangle_M^{II} + f.\langle \underline{\sigma} \rangle_R^{II} = 0 \qquad (6.1)$$

where f is the volume fraction of the reinforcement, and $\langle \underline{\sigma} \rangle_M^{II}$ and $\langle \underline{\sigma} \rangle_R^{II}$ the volume mean phase microstress of the matrix and reinforcement, respectively [28].

6.3.2 Methods for Measurement of Stress in Composites

Only a few direct methods are available for nondestructive measurement of internal phase-specific strains or stresses in composites.

6.3.2.1 Raman Spectroscopy

Raman spectroscopy has been developed to measure the internal strain in organic fiber systems [29–31], and has been extended to study SiC reinforcements. The technique exploits shifts in the Raman spectrum, that is shifts in the characteristic frequency response of inelastically scattered light, under the application of elastic strain. Spatial resolution is excellent at around 1 μm, as is seen from the example of the strain along a carbon fiber in a resin composite under load before and after the fiber is fractured, shown in Figure 6.18. However, the approach is limited to reinforcing species that produce Raman or fluorescence spectra, such as Al_2O_3, and, except for transparent matrices, to the observation of reinforcements near the surface.

6.3.2.2 Photoelasticity

Photoelasticity is an invaluable tool for the measurement of stress in optically transparent systems, and while this precludes its application to most inorganic composites, it has been used to develop and validate predictive models for polymer composites [33].

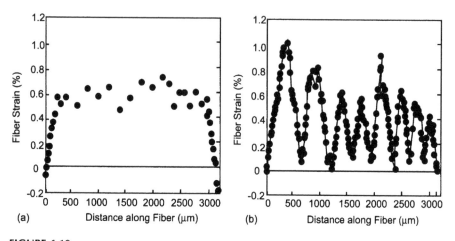

FIGURE 6.18
Distribution of elastic strain component along a reinforcement fiber determined by Raman spectral shifts from a single 3.25 mm long polyacrylonitrile (T50) grade carbon fiber in an epoxy resin under uniaxial loading. (a) Just before first fracture at an applied composite strain of 0.6%. (b) After multiple fracture at 1% applied composite strain [32].

6.3.2.3 Diffraction Techniques

Diffraction techniques have played a major role in increasing our understanding of internal stress partitioning in composites due to their advantage of phase and grain orientation selectivity at depth. They are, of course, restricted to crystalline materials, although it is possible to study the behavior of polymer composites if fugitive crystalline powders are incorporated as strain markers. As discussed in Section 1.4.2, despite the fact that the sampled gauge volume is usually larger than the characteristic volume over which the interphase Type II residual stresses self-equilibrate, $V_v > V_0^{II}$, the fact that each diffraction peak is phase specific means that diffraction can be used to measure the average stresses in each phase of a multiphase material. This makes diffraction an especially useful tool for the study of composite materials as indicated by Figure 6.1.

Electron diffraction is generally of little use due to its poor penetration, of the order of only hundreds of nanometers (Table 1.1) relative to the scale of most composite microstructures. Nevertheless, the method provides a way of measuring Type II and Type III stresses, such as the misfit strains between γ/γ' phases in nickel superalloys [34–37].

X-ray diffraction has found considerable use, and much information has been obtained using x-rays on a wide range of composite systems [38–42]. However, penetration is severely limited, especially with higher-atomic-weight matrix materials such as titanium and nickel, where it is less than 10 μm, which restricts its use with these materials. In all cases, great care must be taken in surface preparation to minimize disruption to the near-surface stress state that is being measured. It is interesting to consider the use of x-

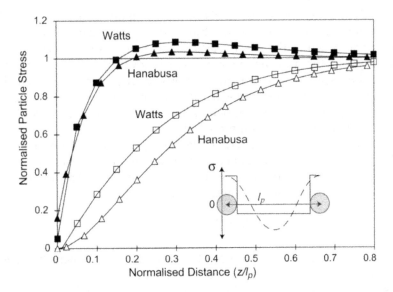

FIGURE 6.19
Predicted profiles of thermally induced particle stress, normalized by that for a particle in the bulk, for SiC particles in an Al matrix with normalized depth, z/l_p, from the surface. The calculations are for the in-plane (filled symbols) and normal (open symbols) components of stress for a three-dimensional arrangement of particles having spacing l_p. They are based on either a sine fluctuation of stress (Hanabusa [43]), or a step function (Watts [44]) as depicted in the inset. A normalized distance of 0.8 corresponds to 50 μm for 20 vol% of 20 μm particles.

ray diffraction in some detail in order to see how it complements the use of neutron diffraction to study composites.

The assumption made when applying the traditional $\sin^2\psi$ approach — that the normal component of the macrostress is zero over the penetration depth of x-rays — is usually valid for single-phase materials, as described in Section 5.1.3. However, it has been realized for some time that it may not be so for composites and other microstructurally nonhomogeneous materials. This is because the mean phase Type II stresses are expected to decay toward the surface of a composite over a distance comparable to the interparticle spacing, l_p, as illustrated in Figure 6.19. Hanabusa et al. [43] were among the first to examine quantitatively the interpretation of x-ray data for composite systems as a function of penetration depth and the scale of the composite microstructure. They modeled the fluctuation in stress from particle center to particle center as a sine wave, while more recent analytical and computational work by Watts and Withers [44] has identified some errors in the original analysis, and has extended the approach to represent more accurately the decrease in normal stress as a free surface is approached; however, the conclusions on the basic trends remain unchanged. Figure 6.19 shows that the assumption that the stress normal to the surface is zero is only acceptable over a distance equal to about a tenth of the interparticle spacing. Even if the x-ray penetration is sufficiently low for this condition to be satisfied, since the in-plane stresses are significantly relieved at this

shallow depth, the measured matrix or particle thermal stresses are unfortunately not representative of the bulk. At depths greater than half the
interparticle spacing, the mean particle and matrix stresses are approximately hydrostatic and the $\sin^2\psi$ method is invalid. Furthermore, partly
because the depth varies as a function of ψ angle, in many cases a $\sin^2\psi$
curve will be found to be nonlinear, making any analysis invalid, and even
if linear the graph may give apparent internal stresses of the *opposite* sign
to those actually existing within the bulk.

In view of the fact that the interparticle spacing of many composite systems
is of the order of the x-ray penetration, it is often difficult to know what
analysis method is appropriate. The fraction, G_z, of the total diffracted intensity that comes from a surface layer of depth z is [45]

$$G_z = (1 - e^{-2\mu_a z / \sin\theta})   \qquad (6.2)$$

For example, for Cu x-ray radiation 50% of the diffracted intensity for Al at
$\theta \approx 90°$ comes from a surface layer of only 25 μm.

In summary, there are three important regimes, depending on the interparticle spacing l_p:

If x-ray penetration depth z < 0.1 l_p: The thermal stress state is almost fully
relaxed over the examined region so that little stress is determined
by either the $\sin^2\psi$ or by a triaxial analysis.

If x-ray penetration depth z lies between 0.1 l_p and 0.5 l_p: Both the $\sin^2\psi$ and
triaxial analyses are invalid because at each different ψ tilt different
depths are sampled, giving nonlinear $\sin^2\psi$ plots. Large errors result
if the measured curves are fitted by a "best" straight line.

If x-ray penetration depth z > 0.5 l_p: The bulk triaxial mean phase microstress state is sampled. The $\sin^2\psi$ method is not valid, since $\sigma_{33} \neq 0$.
If, nevertheless, it were applied, it would indicate zero in-plane
stress. Only a full triaxial method of analysis gives the correct results
[46].

The same problems occur for aligned fiber composites when attempting
to measure thermally induced matrix and fiber stress components transverse
to the fiber direction. However, the interpretation of data is a little more
straightforward for measurement of the axial stress component along the
fiber. In this case, the $\sin^2\psi$ method yields σ_{axial} when the penetration depth
is small, and $(\sigma_{axial} - \sigma_{transverse})$ when it is large compared to the fiber spacing.

Fortunately, for long-range macrostresses rather than thermal stresses, the
errors introduced by the biaxial $\sin^2\psi$ are not so large [44]. As a result, much
has been achieved with conventional x-ray techniques, even for relatively
high atomic number systems such as Ti/SiC. Indeed, x-ray studies have been
providing important insights into the behavior of continuous fiber metal
matrix composites since the 1960s [39].

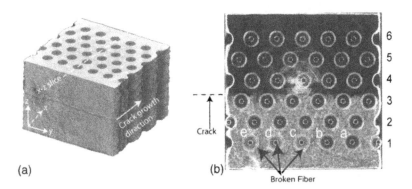

FIGURE 6.20

Reconstructed synchrotron x-ray tomographic images of a region of six-ply, Ti/140 μm diameter, SiC unidirectional fiber composite containing a fatigue crack. (a) Three-dimensional view. (b) The same volume aligned so as to look along the fiber direction, rendered with the matrix and fibers transparent, but with interfaces in sharp relief so that the progress of the matrix crack can be clearly visualized. The three broken fibers in the first ply, caused by fatigue cycling, are indicated [47].

As summarized in Table 1.1, the extremely intense beams of *high-energy synchrotron x-ray diffraction* (very hard x-rays) enable relatively large penetration depths of many millimeters. A good example of the capability of the technique is a study of a Ti/SiC composite that combined three-dimensional tomography and strain measurement at extremely high spatial resolution to analyze fiber bridging of a matrix fatigue crack [47]. Fiber bridging is very important under fatigue because it shields the matrix crack from the full effect of the applied fatigue loading cycle. As the crack grows longer, it becomes bridged by more and more fibers, provided that they do not break, increasing the extent of crack-tip shielding. As a result, the rate of growth of the matrix fatigue crack slows until crack arrest occurs [48]. In this study, a matrix fatigue crack was grown perpendicular to the fiber direction under a loading amplitude just sufficient to cause fiber breakage in the first ply. The fatigue crack was grown until it had advanced past three rows (plies) of fibers, as shown in Figure 6.20b. The tomograph in Figure 6.20 was taken on instrument ID19 at the ESRF with the composite under the maximum crack-opening load encountered during the fatigue cycle. It is clear that three fibers out of five in the first ply are broken.

In order to define a small enough gauge volume to map strains in individual fibers over reasonable time scales, a 5 μm focused incident beam along with narrow (30 μm) receiving slits were used on the ID11 beam line at the ESRF with an x-ray wavelength of 0.24 Å. The sample was aligned with Oz, the fiber axis and loading direction along the scattering vector. By setting the receiving slit height to 30 μm, at a scattering angle of 9.8° about a horizontal axis, a small diamond-shaped gauge volume, 340 μm in the long dimension, along Oy is defined by the intersection of the allowed incident and diffracted beam paths. By ensuring that only one 140 μm diameter fiber

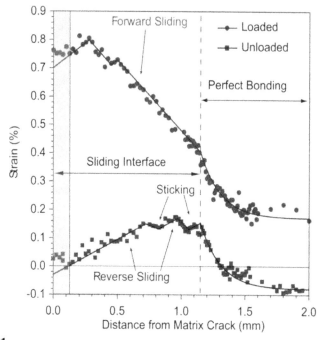

FIGURE 6.21

Axial component elastic fiber strain data (points) and a modified partial debonding model (solid lines) showing the strain distribution in bridging fiber ply 1 labeled as 1b in Figure 6.20, in the loaded and unloaded states. Sliding positions were derived directly from the data. Note that the matrix crack lies at z = 0 [49].

lies within the gauge, data can be collected from one fiber at a time. Each data point took 5 min to collect and by scanning the sample it was possible to measure along each fiber in turn in 10 to 25 µm increments. The measured strain distribution in the intact ply 1 fiber, labeled 1b in Figure 6.20, is shown in Figure 6.21 both at nominal K_{max} (1400 MPa) and minimum tensile load K_{min} (0 MPa). Note the constant gradient in the forward sliding region. This is characteristic of frictional sliding between the matrix and fiber equivalent to a shear sliding stress of around 70 MPa. Upon unloading, reverse sliding has occurred. The sawtooth features are suggestive of a number of sticking points [49]. Figure 6.22a shows the individual fiber axial strain components at the maximum applied load for the intact ply 1 fiber 1b and those immediately behind it in plies 2, 3, 4, 5, and 6, compared in Figure 6.22b with those for the broken fiber 1d and those immediately behind it. The peak strain (~0.8%) in the unbroken ply 1 fibers corresponds to a stress of 3.6 GPa, close to the fiber breaking strength. Unsurprisingly, the stress in the broken fiber is small near the matrix crack. The matrix strain builds up from essentially zero at the point of the fiber break over a distance of around ±1 mm. Were the intact fibers to break, fatigue failure would soon occur. However, for the stress-intensity range studied, they carry significant load, which reduces the opening of the crack and, along with the unbroken fibers in plies 2 and 3,

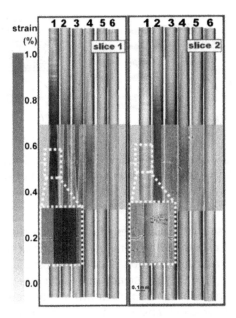

FIGURE 6.22
Tomographic slices through the Ti/SiC fiber composite shown in Figure 6.20 with the fiber strains at K_{max} superimposed (shades) on the locations of each of the fibers. In slice (a), including fiber 1b, all the fibers are intact, whereas in slice (b), the ply 1 fiber, fiber 1d, was broken [49].

shield the matrix crack from the full opening load. This has a significant effect in retarding crack growth. Note also that the ply 2 fiber immediately behind the broken one in Figure 6.22b is more strained than the corresponding fiber behind the intact fiber in Figure 6.22a.

These results illustrate the very high spatial resolution and fast data acquisition times available using synchrotron x-ray diffraction. However, the major disadvantage for mapping the strain in engineering components is that as the diffraction angles can be very small, typically $\phi^s = 4°$ to $20°$, strain measurement in certain directions is not feasible due to the long path lengths in the material.

Neutron diffraction has several major advantages over the use of synchrotron x-rays for the study of crystalline composite materials, and the two techniques complement each other well. The main strengths of neutron diffraction in this context follow:

- Penetration, typically of the order of many centimeters, allowing the nondestructive measurement of phase-specific strains representative of the bulk average
- The capability to monitor strains under in-service conditions, or under load in simple rigs and furnaces
- The opportunity of measuring strains in more than one phase simultaneously, often using time-of-flight methods.

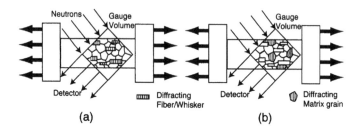

FIGURE 6.23

Schematic illustrating how neutron diffraction can measure volume-averaged (a) fiber, and (b) matrix, phase-longitudinal strains over a large gauge volume for a composite under uniaxial load.

In all but the coarsest of composites, the sampled gauge volumes, V_v, of the order of several cubed millimeters characteristic of neutron diffraction are too large to provide information about the point-to-point variation of microstress, since $V_v >> (l_0^{II})^3$. Instead, in most cases the gauge volume provides information about the strain variation averaged over a significant number of particles, or an extensive region of the matrix with the correct grain orientation, as shown schematically for a whisker reinforced composite under uniaxial load in Figure 6.23. The measured strains are the volume *average* phase-specific strains in the matrix $\bar{\varepsilon}_M$ and reinforcement $\bar{\varepsilon}_R$. Often position-dependent information is not required, and a large sampled gauge volume is used so that the measured internal strains are averaged over many reinforcement particles and matrix grains with the correct grain orientation.

Time-of-flight instruments enable diffraction peaks from both phases to be acquired simultaneously, as illustrated in Figure 6.24, and so in many cases are better suited to studies of residual strain in composites. However, if only one specific peak from each phase is sufficient, an instrument on a continuous source may be just as efficient at acquiring the required data. In both cases, it is important that anomalous shifts due to sample positioning are avoided by careful centering of the sample.

One of the main challenges when evaluating the mean phase strains from the measured diffraction peak lattice spacings is the determination of an appropriate strain-free reference lattice spacing. As explained in Section 4.6.3, it is common practice to use annealed powders in order to avoid macro-stresses, but strain-free reference values are notoriously difficult to obtain, and this situation is worsened when considering metal matrix composites because of the sensitivity of the strain-free lattice parameter to the addition of trace elements in solution. Even if the elemental composition is the same, the proportion of the alloying elements actually in solution is difficult to quantify, especially in cases such as for many aluminum alloys where the ageing response is accelerated in the composite relative to the unreinforced alloy. For ceramics, powders provide somewhat more reliable strain-free

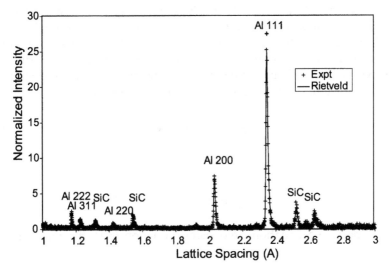

FIGURE 6.24
The time-of-flight diffraction pattern for an Al/SiC particulate composite acquired on ENGIN at the ISIS pulsed neutron source, illustrating the simultaneous measurement of many peaks from both the matrix and reinforcement. The Rietveld fit to the data is shown.

values, both because the chemistry is less variant and because they tend to be simpler microstructurally.

6.3.3 Separation of Different Stress Contributions in Composites

In this section, we consider the various origins of stress in composites. As mentioned in Section 6.3.1, since composites are often inhomogeneous in a number of respects, including elastically, plastically, and in terms of expansivity, there are many opportunities for internal stresses to develop at all levels of scale. Both macrostresses and average phase microstresses can cause the shifting of a diffraction peak, but provided that the average strains are measured in both phases, it is possible to separate them. This is because the *mean* phase microstresses, $\langle \underline{\sigma} \rangle_M^{II}$ and $\langle \underline{\sigma} \rangle_R^{II}$ in the matrix, M, or reinforcing inclusions, R, must average to zero over the sampling volume, as described by Equation (6.1), while the macrostress, $\underline{\sigma}^{Macro}$, must be the same in both phases.

Diffraction measurements yield the *average*, or *total*, lattice strain components for a given *hkl* for each phase, in the three principal strain directions, j, $\bar{\varepsilon}_{Mj}^{hkl}$ and $\bar{\varepsilon}_{Rj}^{hkl}$ (in contracted notation, see Section 5.3), over the sampled gauge volume. The average lattice strains are converted to average phase stress components, $\bar{\sigma}_{Mi}$ and $\bar{\sigma}_{Ri}$, using the relations written generally in the form:

$$\bar{\sigma}_{Mj} = \sum_k C_{Mjk}^{hkl} \cdot \bar{\varepsilon}_{Mk}^{hkl} \text{ , and } \bar{\sigma}_{Rj} = \sum_k C_{Rjk}^{hkl} \cdot \bar{\varepsilon}_{Rk}^{hkl} \tag{6.3}$$

where C_{Mijk}^{hkl} and C_{Rjk}^{hkl} are the relevant hkl (superscript) stiffness matrix components for the individual free matrix and inclusions, respectively, in contracted notation (Section 5.3). Equations (5.2), and (5.5) through (5.7) give these general relations more specifically in terms of the elastic moduli.

The average or total phase stresses, $\bar{\underline{\sigma}}_i$, in the matrix, $i = M$, and reinforcement or inclusion, $i = R$, are given by the tensor relations

$$\bar{\underline{\sigma}}_M = \underline{\sigma}^{Macro} + \langle \underline{\sigma} \rangle_M^{II} \tag{6.4}$$

and

$$\bar{\underline{\sigma}}_R = \underline{\sigma}^{Macro} + \langle \underline{\sigma} \rangle_R^{II} \tag{6.5}$$

Adding these two equations and using Equation (6.1), the macrostress is given by

$$\underline{\sigma}^{Macro} = (1-f)\ \bar{\underline{\sigma}}_M + f\ \bar{\underline{\sigma}}_R. \tag{6.6}$$

The average phase stress tensor components, $\bar{\underline{\sigma}}_i$, are determined from the measured average phase strain tensor components, $\bar{\underline{\varepsilon}}_i$, which are in turn determined from the hkl reflections by using the relations in Equation (6.3), with the corresponding *free* material elastic constants E_{hkl} and v_{hkl}.

In practice, it may be possible to take these elastic constants as isotropic to a good approximation. Thus, if the volume fraction of the reinforcing inclusions, f, is known, the macroscopic stress may be readily determined from the measured strains. Once having converted from strain to stress, it is best to continue the analysis in terms of stress, as the individual different contributions to stress discussed below are not individually related to those of strain. It is more difficult to isolate the individual contributions to the microstress, and these are now considered.

The mean phase microstresses $\langle \underline{\sigma} \rangle_M^{II}$ and $\langle \underline{\sigma} \rangle_R^{II}$ may have a number of possible contributions:

- Elastic mismatch stress ($\langle \underline{\sigma} \rangle_M^{II,El}$) arises from the mismatch in elastic constants and the macrostress.
- Plastic misfit stress ($\langle \underline{\sigma} \rangle_M^{II,P}$) arises from different plastic behaviors of the matrix and reinforcement.
- Transformation misfit stresses ($\langle \underline{\sigma} \rangle_M^{II,Tr}$) arise from misfits generated by phase transformation of one or more of the phases.
- Thermal misfit stresses ($\langle \underline{\sigma} \rangle_M^{II,Th}$) arise after changes in temperature from misfits generated by different expansion coefficients between matrix and reinforcement.

Thus,

$$\langle \underline{\underline{\sigma}} \rangle_M^{II} = \langle \underline{\underline{\sigma}} \rangle_M^{II,El} + \langle \underline{\underline{\sigma}} \rangle_M^{II,P} + \langle \underline{\underline{\sigma}} \rangle_M^{II,Tr} + \langle \underline{\underline{\sigma}} \rangle_M^{II,Th} \tag{6.7a}$$

and

$$\bar{\underline{\underline{\sigma}}}_M = \underline{\underline{\sigma}}^{Macro} + \langle \underline{\underline{\sigma}} \rangle_M^{II,El} + \langle \underline{\underline{\sigma}} \rangle_M^{II,P} + \langle \underline{\underline{\sigma}} \rangle_M^{II,Tr} + \langle \underline{\underline{\sigma}} \rangle_M^{II,Th} \tag{6.7b}$$

with corresponding equations for $\langle \underline{\underline{\sigma}} \rangle_R^{II}$. It is helpful to differentiate between contributions caused by *mismatches* in stiffness between matrix and reinforcement (i.e., $\langle \underline{\underline{\sigma}} \rangle_M^{II,El}$), which disappear when any applied stress or macrostress is removed, and those arising from *misfits* between the matrix and ill-fitting reinforcement, including those resulting from differential thermal expansion, $\langle \underline{\underline{\sigma}} \rangle_M^{II,Th}$; plasticity, $\langle \underline{\underline{\sigma}} \rangle_M^{II,P}$; transformation, $\langle \underline{\underline{\sigma}} \rangle_M^{II,Tr}$; and so forth. Only the *misfits* give rise to residual stress.

It is possible to separate out the contribution arising directly as a result of the macrostress (i.e., the elastic *mismatch* microstress) from the terms arising from the shape *misfits* between stress-free shapes of the matrix and reinforcements, provided that a reliable predictive model of the relationship between macrostress and elastic mismatch stress [28] exists, such as the Eshelby model which relates them directly.

In the following sections, we consider each type of microstress in turn, looking at their origin, prediction, and measurement.

6.3.4 Elastic Mismatch Stresses

Many engineering composites achieve an increase in stiffness and strength over the matrix by including a stiffer and stronger reinforcing phase. As a result, applied loads are not distributed evenly between the phases. Instead, load is transferred from the more compliant matrix to the stiffer reinforcement. The extent of this load transfer is dependent on the mismatch in stiffnesses between the phases, phase geometry, and integrity of the interface between them. While the point-to-point variation in matrix stress is dependent on the local arrangement of the reinforcement, neutron diffraction samples the local average, or total, matrix and reinforcement stresses, $\bar{\underline{\underline{\sigma}}}_M$ and $\bar{\underline{\underline{\sigma}}}_R$. These are not so dependent on the local arrangement, and can be predicted with good reliability using finite element or Eshelby-type models [33]. The mismatch stresses are proportional to the local macrostress $\underline{\underline{\sigma}}^{Macro}$, with the components j given by:

$$\left\langle \sigma \right\rangle_{Mj}^{II,El} = \sum_k B_{Mjk}\, \sigma_k^{Macro}, \text{ and } \left\langle \sigma \right\rangle_{Rj}^{II,El} = \sum_k B_{Rjk}\, \sigma_k^{Macro} \tag{6.8}$$

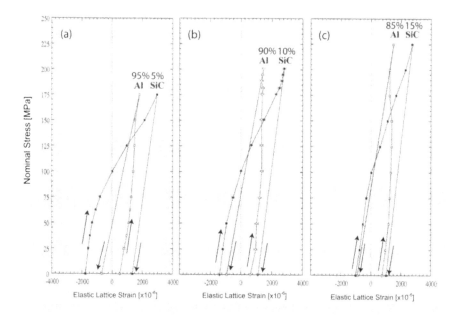

FIGURE 6.25
Response of the lattice strain component parallel to an applied uniaxial stress, of an Al matrix
(111) reflection (open circles), and (a) 5%, (b) 10%, and (c) 15% SiC whiskers (111) reflection
(solid squares), measured by neutron diffraction [50].

Here the tensor $\underline{\underline{B}}$ describes the strength of the proportional relationship.
Thus, in the case of a previously strain-free composite subjected to a uniform
applied stress $\underline{\sigma}^A$, the components of the internal mean phase stresses are

$$\left\langle \sigma \right\rangle_{Mj}^{II,El} = \sum_k B_{Mjk}\,\sigma_k^A,\ \text{and}\ \left\langle \sigma \right\rangle_{Rj}^{II,El} = \sum_k B_{Rjk}\,\sigma_k^A \qquad (6.9)$$

It is evident that in this case the mean phase stresses are not residual
stresses in that upon unloading the internal stresses disappear. Clearly, for
an effective composite, the diagonal components of B_{Mjk} should be large and
negative in sign since this reflects an effective transfer of load from the matrix
to the reinforcement. Neutron diffraction has been used extensively to
measure the extent of load transfer for particulate, whisker, and continuously
aligned composites. An example of the elastic and plastic loading response
is shown in Figure 6.25. Note that the SiC whiskers are initially in residual
compression and the matrix in tension due to thermal residual stresses. The
whisker thermal strains decrease and the matrix strains increase, as the
whisker content is increased from 5% to 15%, consistent with expectations
[51]. Upon loading, the two phases extend elastically at first, but soon the
matrix begins to yield, aided by the thermal stresses, causing the responses
to curve. Beyond the yield point for the matrix, the composite extends

plastically after which the measured elastic strain in the matrix remains approximately constant. As a consequence, the load is taken increasingly by the SiC whiskers. Note that while the composite is able to carry more load as the whisker fraction increases, the maximum strains in the whiskers and fibers are fairly similar for all three whisker contents. In other words, the maximum stress in the fibers, ~1250 MPa, is similar, but the total load borne by the fibers is three times larger in the 15% SiC composite relative to the 5% SiC composite. On unloading, it is seen that the residual phase strains have reversed sign due to a permanent retained plastic misfit strain [50].

Often it is helpful to compare the measured response with finite element or Eshelby model predictions in order to evaluate the efficacy of the matrix–reinforcement coupling. For elastic load transfer, the Eshelby model is quite sufficient [33]:

$$\langle \underline{\sigma} \rangle_M^{II,El} = f\, \underline{\underline{C}}_M\, (\underline{\underline{S}}^E - \underline{\underline{I}})\, \{(\underline{\underline{C}}_M - \underline{\underline{C}}_R)[\underline{\underline{S}}^E - f(\underline{\underline{S}}^E - \underline{\underline{I}})] - \underline{\underline{C}}_M\}^{-1} (\underline{\underline{C}}_M - \underline{\underline{C}}_R)\, \underline{\underline{C}}_M^{-1}\, \underline{\sigma}^{Macro} \tag{6.10}$$

where $\underline{\underline{S}}^E$, $\underline{\underline{I}}$, $\underline{\underline{C}}_M$, and $\underline{\underline{C}}_R$ are the Eshelby, identity, matrix and reinforcement stiffness, tensors respectively, represented in the contracted matrix notation (Section 5.3 [52]).

Fitzpatrick et al. [53] have used this equation to calculate the matrices B_{Mjk}, B_{Rjk} for the Al/SiC system, as given in Table 6.5. Each coefficient B_{jk} describes the matrix (M) or reinforcement (R) mean phase stress in the direction j caused by an applied stress of 1 Pa in the direction k. Note from Table 6.5 how the axial stress in the reinforcement increases with aspect ratio, but decreases with increasing volume fraction. This occurs because while more load is transferred from the matrix at higher reinforcement fractions, there are more reinforcements to carry the increased load.

6.3.5 Thermal Microstresses

Thermal residual stress is an inevitable consequence of processing composite materials at elevated temperatures. Residual stress measurements have been made on a very wide range of inorganic composite systems, as reviewed by Kupperman et al. [54]. The Type II mean phase thermal stresses are proportional to the effective thermal misfit between the phases and can be estimated using Eshelby's approach or by finite element analysis:

$$\langle \sigma \rangle_{Mj}^{II,Th} = \sum_k A_{Mjk}\Delta\alpha_k\Delta T, \text{ and } \langle \sigma \rangle_{Rj}^{II,Th} = \sum_k A_{Rjk}\Delta\alpha_k\Delta T \tag{6.11}$$

Here the subscripts j,k again denote the component directions. $\Delta\alpha = (\alpha_M - \alpha_R)$ is the difference in thermal expansivity between the matrix (α_M) and reinforcing inclusions (α_R), and ΔT is the temperature drop. In practice, $\Delta T =$

TABLE 6.5

Coefficients B_{Mjk} and B_{Rjk} Calculated for Al/SiC Composite for Three Reinforcement Volume Fractions and Reinforcement Aspect Ratios [53]

		Particles (AR = 1)		Short Fibers (AR = 5)		Long Fibers (AR = ∞)	
		Al	SiC	Al	SiC	Al	SiC
$f = 0.05$	B_{11}	−0.032	0.606	−0.125	2.379	−0.206	3.914
	$B_{22} = B_{33}$	−0.032	0.606	−0.019	0.368	−0.019	0.354
	$B_{21} = B_{31}$	0.0088	−0.168	0.0035	−0.0667	0.0051	−0.096
	$B_{12} = B_{13}$	0.0088	−0.168	0.0431	−0.819	0.0735	−1.397
	$B_{23} = B_{32}$	0.0088	−0.168	0.0057	−0.109	0.0045	−0.0864
	$B_{44} = B_{55}$	−0.041	0.775	−0.0403	0.767	−0.036	0.694
	B_{66}	−0.041	0.775	−0.025	0.472	−0.023	0.437
$f = 0.09$	B_{11}	−0.056	0.564	−0.204	2.068	−0.318	3.211
	$B_{22} = B_{33}$	−0.056	0.564	−0.034	0.344	−0.033	0.330
	$B_{21} = B_{31}$	0.0153	−0.154	0.0057	−0.057	0.0078	−0.079
	$B_{12} = B_{13}$	0.0153	−0.154	0.0698	−0.706	0.112	−1.136
	$B_{23} = B_{32}$	0.0153	−0.154	0.0102	−0.103	0.0084	−0.0849
	$B_{44} = B_{55}$	−0.071	0.719	−0.070	0.712	−0.0659	0.646
	B_{66}	−0.071	0.719	−0.044	0.443	−0.0406	0.411
$f = 0.17$	B_{11}	−0.10	−0.487	−0.326	1.592	−0.466	2.276
	$B_{22} = B_{33}$	−0.10	−0.487	−0.062	0.301	−0.059	0.286
	$B_{21} = B_{31}$	0.0265	−0.129	0.0089	−0.043	0.011	−0.055
	$B_{12} = B_{13}$	0.0265	−0.129	0.109	−0.533	0.162	−0.792
	$B_{23} = B_{32}$	0.0265	−0.129	0.0187	−0.091	0.0161	−0.0785
	$B_{44} = B_{55}$	−0.126	0.617	−0.125	0.611	−0.114	0.558
	B_{66}	−0.126	0.617	−0.079	0.389	−0.074	0.364

Note: The fibers are taken to be all aligned in one direction, 1. AR, aspect ratio; f, volume fraction.

$\Delta T_{esf} = T_{esf} - T$, where T_{esf} is effective stress-free temperature, that is, the temperature at which the composite would be free of residual stress, and T is the temperature at which the stress is to be calculated. As a result, $\Delta \alpha_j \Delta T_{esf}$ is the effective thermal expansion misfit that is accommodated elastically. $\Sigma A_{Mjk} \Delta \alpha_k$ and $\Sigma A_{Rjk} \Delta \alpha_k$ describe the level of mean stress in matrix and reinforcement in the direction j generated by a 1K change in temperature. They reflect the influences of reinforcement volume fraction, phase stiffnesses, and phase geometry on the level of internal stress generated.

The Eshelby model predicts that the stresses will have the form [33]

$$\langle \underline{\sigma} \rangle_M^{II,Th} = f\underline{C}_M \, (\underline{S}^E - \underline{I}) \, \{(\underline{C}_M - \underline{C}_R)[\underline{S}^E - f(\underline{S}^E - \underline{I})] - \underline{C}_M\}^{-1} \, \underline{C}_R \, \Delta \underline{\alpha} \Delta T_{esf} \quad (6.12)$$

where the matrices $\underline{S}^E$, $\underline{I}$, $\underline{C}_M$, and $\underline{C}_R$ are defined as in Equation (6.11). These equations have been used to calculate A_{Mjk}, A_{Rjk} for Al/SiC given in Table 6.6. Not surprisingly, the axial stress in both phases increases in magnitude with aspect ratio and reinforcement volume fraction, with opposite sign, being tensile in the Al matrix and compressive in the SiC inclusions. In some cases, the thermal expansivity is a strong function of temperature, in which case

TABLE 6.6

Calculated Stress in MPa Developed in Each Phase of Al/SiC Composite per Degree Drop in Temperature for Three Volume Fractions and Aspect Ratios [53]

		Particles (AR = 1)		Short Fibers (AR = 5)		Long Fibers (AR = ∞)	
		Al	SiC	Al	SiC	Al	SiC
$f = 0.05$	$\Delta\alpha A_{11}$	0.089	−1.685	0.238	−4.662	0.342	−6.603
	$\Delta\alpha A_{22} = \Delta\alpha A_{33}$	0.089	−1.685	0.063	−1.197	0.057	−1.091
$f = 0.09$	$\Delta\alpha A_{11}$	0.157	−1.597	0.400	−4.125	0.551	−5.543
	$\Delta\alpha A_{22} = \Delta\alpha A_{33}$	0.157	−1.597	0.113	−1.146	0.105	−1.054
$f = 0.17$	$\Delta\alpha A_{11}$	0.293	−1.423	0.670	−3.284	0.847	−4.175
	$\Delta\alpha A_{22} = \Delta\alpha A_{33}$	0.293	−1.423	0.213	−1.040	0.198	−0.967

Note: The fibers are taken to be all aligned in one direction, 1. *AR*, aspect ratio; *f*, volume fraction.

the accumulated misfit must be calculated by integration over the relevant temperature range.

By comparing neutron diffraction measurements of thermal stress with model predictions it is possible to identify the effective stress-free temperature, T_{esf}. This is not usually the temperature at which the composite was made, because relaxation effects such as creep at high temperatures, or plasticity when the misfit stresses become large, will tend to limit the build up of thermal stress. Nevertheless, the effective stress-free temperature is a useful concept for representing the magnitude of thermal stresses. It reflects the opportunities for relaxing the misfit by creep and yielding at high temperatures, and yielding, cracking, de-bonding, and so forth at low temperatures. Effective stress-free temperatures of around 200°C, 700°C, and 1300°C are typical of aluminum [55], titanium [56], and alumina [57] matrix systems, respectively, as measured by neutron diffraction. A related issue is the short lifetimes of uranium targets for neutron production in a spallation source that may be partly caused by thermal cycle intergranular stress-driven effects that result from large anisotropic thermal expansion stresses between the grains.

While the majority of studies made to date have focused on the measurement of the residual strain in one or both of the phases at room temperature, it is also possible to measure the thermal residual stresses at elevated temperatures provided that one allows for the stress-free changes in the reference lattice spacing due to thermal expansion. This is best done using single-phase reference samples, and requires accurate knowledge of temperature. For aluminum, an error in temperature of only 4°C between a strain-free reference and the composite would lead to a 100 με error. The difference in lattice plane spacing between the reference and the composite then allows the variation of thermal stress to be deduced, as shown in Figure 6.26. For materials of relatively low thermal expansion coefficient, it may be possible to calculate rather than measure the strain-free lattice parameter as a function of temperature from the room temperature value without serious error.

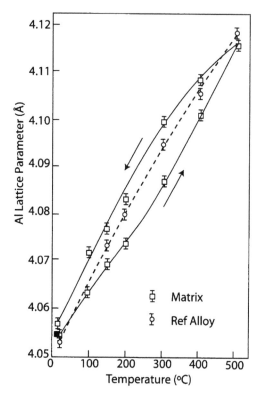

FIGURE 6.26

The variation in Al 111 lattice spacing (open squares) as a function of rising and descending temperature for the Al matrix of Al/5vol%SiC$_W$, compared with an unreinforced Al reference sample of the same alloy (open circles). The difference between the continuous and dashed curves is due to thermal residual stresses in the composite. The concomitant residual strains are shown in Figure 6.28.

6.3.5.1   Particulate Composites

Particulate metal matrix composites are economically attractive and a number of systems have received attention, most notably Al/SiC$_P$ [58]. Ceramic matrix composites, especially those containing particulate zirconia as inclusions, are of great interest because of the potential for increasing their fracture toughness through the exploitation of the structural phase transformation from tetragonal to monoclinic form. Wang et al. [59] have studied the thermal residual stresses in Al$_2$O$_3$/ZrO$_2$ composites as a function of the ZrO$_2$ volume fraction. In this case, the zirconia is in residual tension and the alumina in compression because the former has the larger coefficient of thermal expansion. Their results are shown in Figure 6.27, where it is seen that the macrostress calculated from Equation (6.6) is zero within the uncertainty, indicating that the thermal stresses self-equilibrate, as would be expected.

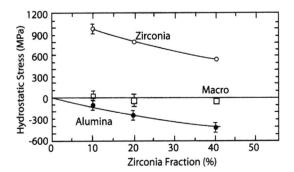

FIGURE 6.27
The variation in the residual hydrostatic thermal mean stresses in composites of an Al_2O_3 matrix containing ZrO_2 particulate, as a function of the ZrO_2 volume fraction. The macrostress is calculated from Equation (6.6) [59].

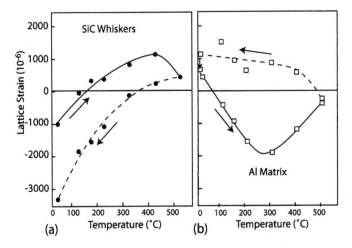

FIGURE 6.28
Neutron strain measurements of the mean Al and SiC axial 111 lattice strain components in (a) the whiskers, and (b) the matrix, for an Al/5%SiC$_W$ composite made over a thermal cycle. The arrows show the heating (solid) and cooling (dashed) curves [55]. The effectively stress-free temperature is approximately the point at which the heating curve crosses zero strain.

6.3.5.2 Whisker Composites

Perhaps the earliest neutron diffraction measurements of the variation of thermal residual strains with temperature in metal matrix composites were made on the Al/SiC$_W$ system. It was the high penetration of neutrons that made *in situ* studies within a furnace possible [55]. The evolution of internal strain within the two phases is found to be nonlinear, as shown in Figure 6.28. This is for two reasons. First, the matrix is plastically deformable. At temperatures in excess of 300°C, a sharp reduction in the Al yield stress accounts for the large decrease in the size of the residual strain in the matrix as the temperature rises. Second, the melting point of the Al is much lower than

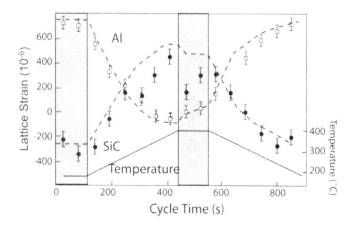

FIGURE 6.29
Variation in elastic axial strain component of the two phases in an Al/10vol%SiC$_w$ composite during a 175 to 400°C thermal cycle shown by the solid line. The data were accumulated using a strobe technique over many consecutive cycles. The dashed lines represent the predictions of finite element calculations incorporating both creep and plasticity. The thermal dwell periods are shaded [63].

is typical for a ceramic, giving rise to creep during the experiment. In view of the capacity for matrix creep, the effectively stress-free temperature for particulate and whisker-reinforced Al matrix systems tends to be between 100 and 200°C, as seen from the response shown in Figure 6.28. The presence of such thermal microstresses has been shown to influence mechanical behavior, but these microstresses become even more important when they are continually regenerated, such as when the temperature is thermally cycled. In these circumstances, composites have been observed to progressively distort [60,61], and, in the presence of small applied loads, to deform superplastically to large extensions in excess of 400% [62].

The strain changes occurring during thermal cycling can be studied by neutron diffraction. However, because the fastest measurement times for peaks in both phases of a composite are currently of the order of minutes, this places a limit on the shortest period of the thermal cycle that can be followed. This restriction can be overcome using stroboscopic techniques [63]. By accumulating diffracted signals acquired over many cycles into 16 time bins evenly distributed over each cycle, it is possible to accumulate sufficient diffracted intensity to obtain accurate strain measurements as illustrated in Figure 6.29. In this case, the variation in elastic lattice strain of the 111 reflections for the two phases of an Al/10vol%SiC$_w$ composite is shown at different times over a temperature cycle from 175 to 400°C. The temperature profile during the cycle period is plotted. The data were acquired over 175 cycles, with 50°C/min minimum ramp rate and a 100 s dwell time at the top and bottom of each cycle. The temperature variation of the reference strain-free lattice plane spacing has been subtracted to provide the plotted strain data. The data are compared with results of finite element predictions

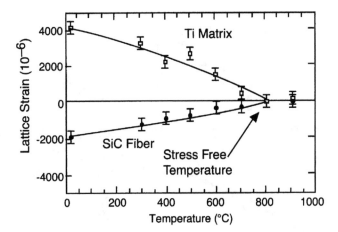

FIGURE 6.30
The evolution of internal strain components along the fiber direction in Ti-14Al-21Nb/35%SiC aligned fiber composite as a function of temperature [64].

incorporating creep, occurring mostly at high temperatures during the dwell, and plasticity, occurring mostly at the bottom of the cooling cycle.

6.3.5.3 Continuous Fiber Composites

The earliest internal stress studies using x-rays on aligned continuous fiber metal matrix composites focused on Al and Cu matrices [39]. More recent work, however, has focused on Ti- and Ni-based matrix systems, usually with SiC monofilamentary reinforcement, such as the textron SCS-6 fiber. As a result of the high melting point of their matrices and the difference in thermal expansion coefficients, these systems often have very large thermal residual stresses. These can severely limit their performance. The residual strains for the titanium aluminide matrix system, Ti-14Al-21Nb/35%SCS-6 SiC, shown in Figure 6.30 as a function of temperature, clearly point to a stress-free temperature of ~800°C [64]. Above this temperature, stress relaxation would seem to prevent the buildup of stress on the time scale of the neutron experiment. In a similar experiment on conventional titanium Ti-6Al-4V/35% SCS-6 SiC fiber composite, the stress-free temperature was found to be somewhat lower at around 630°C [65].

6.3.6 Plastic Misfit Stresses

Plastically generated misfit microstresses can occur when at least one of the phases is plastically deformable. The effect of plastically generated misfit stresses can be well characterized by neutron diffraction, since deformation studies can be made using the advantage of *in situ* measurements on the diffractometers (e.g., Figure 6.25). Plastic misfit stresses are rather difficult to predict *a priori*, partly because of difficulties in predicting the matrix yield

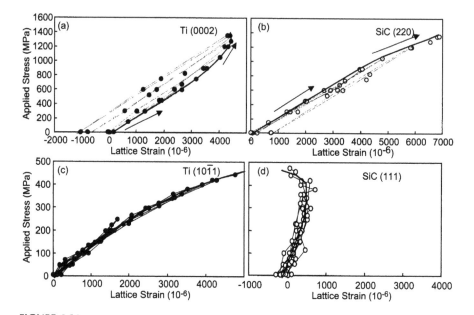

FIGURE 6.31
The response of the lattice strain component along the uniaxial loading direction, which is parallel to the fiber direction, of a Ti-6Al-4V/35%SiC aligned fiber composite in (a) the matrix, and (b) the fibers. The response of the lattice strain along the uniaxial loading direction, which is perpendicular to the fiber direction of the composite in (c) the matrix and (d) the fibers. The strains are shown relative to the unloaded state [56].

stress, which can be very different from the unreinforced alloy due to microstructural effects caused by the presence of the second phase. Nevertheless, finite element [66–68] and Eshelby models [69] have been used successfully to interpret experimental neutron data. As one might expect, it is evident from the following sections that the geometry of the reinforcement and the integrity of the interface are key parameters in determining the redistribution of load caused by plastically generated misfits.

6.3.6.1   Continuous Fiber Metal Matrix Systems

A good example of the use of neutron diffraction for investigating the plastic response of composite materials is given by a study of Ti/SiC continuous fiber composite carried out at Chalk River [56]. For loading parallel to the fiber direction, the elastic fiber strain component is equal to the composite strain. Significant matrix plasticity is observed at an applied stress, σ_A, of around 900 MPa. Following the approach of Section 6.3.4, the parallel stress component in the matrix arising from the applied load is predicted to be approximately $0.53\sigma^A$, and the original thermal residual stresses were found to be about +450 MPa, giving a total load of around 930 MPa. This is in good agreement with the unreinforced Ti yield stress of around 850 MPa. Above 900 MPa, matrix plasticity causes a transfer of load from matrix to fiber

(Figure 6.31). This causes the matrix lattice strain component along the loading direction to curve upward (Figure 6.31a), reducing its rate of straining, and the corresponding SiC lattice strain response to curve toward greater elastic strain to compensate (Figure 6.31b). As a consequence of this plastic extension, upon unloading the matrix resides in compression and the fibers in tension.

The situation is quite different for loading perpendicular to the fiber direction. In this case, the curves deviate from a linear response at an applied stress of only 300 MPa. Because the fibers are less effective when loaded perpendicular to their length, the fraction of the applied stress borne by the matrix in the elastic regime is higher, at approximately $0.92\sigma^A$, and the thermal residual stress component in the matrix is around 155 MPa, so that the matrix stress component along the loading direction is around 430 MPa, which is significantly below the matrix yield stress. It is clear that the curvature in strain response is not due to macroscopic matrix yielding, but rather is due to matrix–fiber interface de-bonding. This can also be deduced from the shape of the curves. The sense of load transfer in Figure 6.31c and d is contrary to that in Figure 6.31a and b. In the case of loading perpendicular to the fiber direction, the load is transferred from fiber to matrix as de-bonding progresses. Recent work using the ENGIN-X instrument has extended this study from test pieces to aeroengine components [70]. ENGIN-X is particularly well suited to such studies because it has excellent flux and diffraction peak resolution, allowing large penetration depths even in titanium and the simultaneous measurement of diffraction peaks from both Ti and SiC (see Sections 6.3.9 and 6.6.1).

6.3.7 Phase Transformation Stresses

Steels are perhaps the best developed and exploited family of multiphase metal composites of all. Key to their excellent performance is an understanding about the role of transforming phases. Martensitic transformations, or more generally displacive transformations, have important effects on the mechanical properties of many types of materials. Unlike diffusive transformations, they involve the ordered reorganization of a structure and they are accompanied by significant residual stresses. As a result, the extent of transformation, morphology, and texture of the transformed phase all depend on internal stress. How such a transformation can affect the state of residual stress during welding was discussed in Section 6.2.3. An understanding of such transformations is also important for understanding the behavior of superelastic, shape memory, and ferroelectric materials, all of which are currently being investigated as active components for smart structures.

A good example of the use of neutron diffraction is provided by a study on NiTi, which is superelastic [71]. Superelastic materials have the ability to absorb large amounts of strain energy and release it as the applied strain is removed, as shown schematically in the Figure 6.32. For example, the

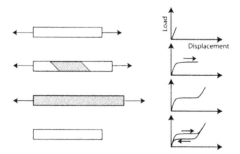

FIGURE 6.32
Schematic illustration of superelastic behaviour, by which a large recoverable strain can be introduced by "elastic" loading. The gray region has undergone a reversible displacive transformation.

elasticity of NiTi is approximately ten times that of steel. The use of neutron diffraction to monitor the extent of transformation, texture, and internal stresses caused by a phase transformation is illustrated by the results shown in Figure 6.33. Figure 6.33a shows the bulk superelastic strain behavior of the response to loading stress. In Figure 6.33b, the extent of the austenite to martensite transformation is plotted as deduced from their relative diffraction peak intensities. The transformation is linear with strain and shows no apparent hysteresis as a function of superelastic strain. In Figure 6.33c, the preferred orientation of the martensite can be seen in terms of the variation in diffraction peak intensity as a function of sample orientation. This is due to the fact that the most favorable austenite orientations transform first. Note that values below zero are an artifact of the normalization used. In Figure 6.33d, the elastic strain in the residual austenite, as measured by diffraction peak shifts, remains essentially linear even after significant transformation has occurred. This is thought to be because the martensite that forms initially comprises favorable variants with low internal stress. A small amount of load transfer to martensite occurs at large strains when the remainder transforms.

It is anticipated that diffraction techniques will play an increasingly important role in the development of smart materials containing transforming materials because of their capability to reveal the extent and orientation of the transformed products, as well as the internal stresses that play a role in governing the transformation.

6.3.8   Microstress-Related Peak Broadening

As described in Section 5.7, broadening of the intrinsic Bragg diffraction peak width can arise for many reasons. These can be divided into three groups:

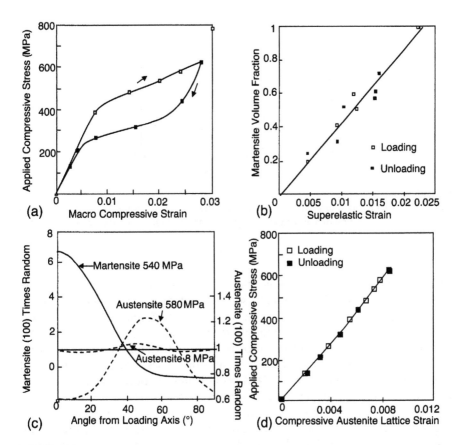

FIGURE 6.33
(a) Stress–strain curve of polycrystalline NiTi showing characteristic superelastic response. (b) Variation in volume fraction of martensite during superelastic loading and unloading. (c) Martensite and austenite (100) peak intensity as a function of angle to the loading axis normalized by the intensity expected for a random texture. (d) The variation in (100) austenite lattice strain with applied compressive stress [71].

- Type II stresses, arising from microscale mismatches, such as are common in composites
- Type III stresses, arising from defects such as dislocations and twins
- Size effects such as grain size broadening that may arise from dislocations due to plasticity

A unique interpretation of peak broadening effects is therefore very difficult, and while a number of theories exist — for example, Warren and Averbach [72] — they are difficult to apply. They require high quality, high resolution, data over a number of reflections, and rely heavily on information in the tails of peaks for which good counting statistics are difficult to obtain. In addition, due to the poorer instrumental resolution, neutron diffraction

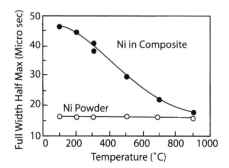

FIGURE 6.34

Full-width at half-maximum variation for 311 Ni reflection from a 6 μm WC/26wt%Ni composite, and from free powder, as a function of temperature. Widths are given in terms of time of flight [74].

peaks are typically much broader than x-ray peaks, further complicating the task. Nevertheless, it is possible to see that reflections are much broader in composites than are observed from for the separate phases, and that this broadening can provide the materials scientist with important information.

Todd and Derby [73] interpreted peak asymmetries for alumina/SiC composites (i.e., a compressive tail for the alumina phase) directly in terms of the strain distribution. Krawitz [74] reported large changes in full-width half-maximum with increasing temperature for cermets. These changes, which are shown for the 311 Ni reflection in a 6 μm WC/26wt%Ni composite in Figure 6.34, are reversible upon heating and cooling. This is consistent with the evidence from analytical methods of separation of the strain and size effect contributions, which point to the broadening being almost entirely due to elastic strain [74].

6.3.9   Future Directions for Neutron Diffraction Measurements of Composites

Neutron diffraction measurements have provided key information on residual stress in metal and ceramic matrix composites, and thus have been instrumental to understanding their development. To date, this work has focused primarily on the characterization of thermal residual stresses, and on individual phase responses to tensile loading. As composite materials find more widespread applications, greater emphasis will be placed on their in-service performance. Neutron diffraction can provide information both about the stress distribution during in-service loading, fatigue, thermal fatigue, creep, and so on, and postmortem after-service loading. Although much work has concentrated on the study of test samples, in future neutron diffraction will become more focused on studies of real composite components as the materials find new applications.

FIGURE 6.35
A hoop-wound Ti-6Al-4V/SiC fiber integrally bladed ring, or bling, prototype aeroengine component. Neutron strain mapping is able to examine the strains in matrix and fibers even when the fibers are well below the surface. The inset shows the microstructure of the reinforced region. (Photograph courtesy of P. Doorbar.)

One example of this is the drive to develop integrally bladed ring structures, called *blings* (Figure 6.35), to replace the traditional blade and aeroengine turbine disk designs. Weight savings as great as 75% are possible, but this requires the application of Ti/SiC composites in order to obtain sufficient high-temperature strength and creep resistance [33]. In this application, the fibers can be as deep as 10 mm from the surface, and in such cases only neutron diffraction has sufficient penetration to measure their stresses. The move away from test samples to real components also involves a step jump in geometric complexity. Simple manipulation of the sample and the need for reduced setup times could well lead to the use of coordinate measurement technology, or the next generation of laser location devices, to ensure accurate and repeatable placement of the sample on the instrument's stage (see Section 3.6.3).

6.4   Plastically Deformed Components and Materials

While residual stresses are often detrimental to component life, in certain circumstances it is possible to introduce residual stresses that can prolong life. In many cases, these stresses are introduced by plastic deformation. In this section, we consider the measurement of stress in plastically deformed samples.

6.4.1   Origin and Nature of Plastic Misfit Stresses

As demonstrated in Figure 1.3, residual stresses usually arise because of misfits between regions within a component or material. These can be generated plastically at the macroscale by, for example, the plastic bending of a bar, or at the microscale due to the inhomogeneous grain-to-grain plasticity of a polycrystalline metal. In this section, we first consider the challenges posed by microstresses termed *intergranular stresses*. These were discussed at length in Section 5.6 because they are almost always an important consideration when trying to determine macrostress. We then examine some cases where residual stresses have been introduced by plastic deformation.

6.4.2   Challenges to Accurate Neutron Diffraction Measurement of Stress in Plastically Deformed Components

The main challenges to the accurate application of neutron diffraction to the measurement of residual stresses in plastically deformed structures lie in the accurate analysis of intergranular stresses. In Chapter 5, it was seen that these can have two origins, namely, elastic mismatch strains due to the anisotropy of single crystal stiffness with crystal direction, and plastically generated misfit stresses because of the anisotropy of plastic behavior. The former is often termed *elastic anisotropy*, and the latter *plastic anisotropy*.

Elastic anisotropy is usually accounted for relatively easily simply by selecting a diffraction elastic constant appropriate for the chosen reflection, as seen in Sections 4.2.2, and 5.5. Plastic anisotropy is much more difficult to deal with. The effects of plastic anisotropy can be accounted for in a number of ways as discussed in Section 5.6, and much further work is required to elucidate the optimum method of analysis for different types of material. The following strategies have been used with some success:

- If the extent of plastic flow is known *a priori*, then it is sometimes possible to correct the measured strains for a given reflection for plastic anisotropy. For example, see the measurements on a bent tube described below.
- It is possible to carry out calibration measurements of the lattice strain response to applied uniaxial loading, and to choose reflections

which give relatively linear stress–lattice strain response for subsequent measurements on engineering samples (i.e., to choose reflections that are relatively unaffected by plastic anisotropy).

- On a time-of-flight instrument it is possible to use a Rietveld refinement as mentioned in Section 4.5 in order to fit the lattice parameters using all the available reflections. It has been shown that in many cases the refined lattice parameter response is essentially linear even in the plastic regime, having an elastic modulus similar to the bulk material [75]. This is because the refined value takes into account the behavior of many reflections to provide an ensemble average as described in Section 5.6.6.

- On reactor continuous beam instruments, it may be possible to study more than one reflection to minimize the likelihood of obtaining unrepresentative results.

In most cases, the aim is to remove the effects of plastic anisotropy in order to obtain a reliable measure of the macrostrains and stresses. The ultimate challenge is to be able to infer, from the measurement of the residual strain from many individual reflections, the past history of plastic straining as well as the resulting residual macrostress in a component.

6.4.3 Examples of Stress Measurement in Plastically Deformed Materials

Examples of cases where neutron diffraction has provided important information about plastic deformation can be divided into two categories: those for which intergranular stresses provide fundamental insight into the micromechanics of polycrystalline slip in materials, and those in which the long-range residual macrostress caused by plastic deformation is important to engineers for understanding component performance, such as the fatigue performance of a cold expanded hole. Some basic examples are described in Chapter 5; here, we focus more on the application of the understanding developed in that chapter.

6.4.3.1 Studies of Fundamental Aspects of Material Deformation

As described in Section 5.6, there have been major advances in the theoretical description of the response of polycrystalline aggregates of grains to applied loads based on elastoplastic self-consistent (EPSC) models, such as that implemented by Lebensohn and Tomé [76]. The elastic and plastic deformations of each of many thousands of grains of known crystallite orientation are tracked and averaged to find the macroscopic deformation in the aggregate. These models represent accurate constitutive models of material behavior because they incorporate the correct slip behavior as well as the elastic anisotropy at grain level. Lattice strain measurement by diffraction gives the elastic deformation of the grains, which is calculated in the model, and so provides a detailed check on the model at the grain level, particularly with regard to the strain in different crystallographic orientations of grains as seen in Figure 5.18.

Idealized experiments such as those described in Section 5.6 have now been carried out on uniaxial tensile loading of many materials used in industry. These include austenitic [77] and ferritic steels, Al alloys [78], and hexagonal close-packed alloys of Zr, Mg, Ti, and Be. Although good agreement between the experimental results and model predictions has been achieved for the axial strain components under uniaxial loading where the main deformation mode is simple slip, other deformation geometries and deformation micromechanisms are much less well understood. Furthermore, the majority of work on the elastic response of individual sets of lattice planes to plastic deformation has been carried out on randomly oriented textures with no preferred grain orientations. However, as Tomé pointed out [79], an important consideration is that plastic anisotropy is texture dependent, as is the effective elastic constants of textured polycrystalline samples in the elastic regime. If the material under examination is strongly textured, this must be taken into consideration, and plastic anisotropies calculated or measured for random aggregates will generally not be appropriate.

6.4.3.2 Practical Example of Intergranular Stresses: Bent Monel Tube

The stress distribution in a bent bar was one of the early examples of strain measurement by neutron diffraction [80,81]. The bending of a tube, while slightly more complex, generates similar stresses around the diameter of the tube from the extrados or point of least curvature (i.e., the point of maximum tensile stress during bending) through the flank to the intrados, or point of most curvature (i.e., the point of maximum compressive stress during bending). A study by Holden et al. [82] on a bent *fcc* Ni/Cu Monel-400 tube used in steam generators has highlighted large differences between the axial component of residual lattice strains found after bending as recorded by the 002 and 111 reflections shown in Figure 6.36a. In fact, direct interpretation of the data from the 002 reflection seemed to indicate that the extrados was under tension, contrary to expectations. At a practical level, the very large differences between the strains measured from the two reflections present difficulties of interpretation. Indeed, if the strain from the 002 were used uncorrected, then a tensile stress of around 55 MPa at the extrados would be inferred (Figure 6.36c). The reason for the difference is due to the generation of intergranular strains between the differently oriented grain subsets in the plastically deformed regions near the extrados and intrados regions. Given that the point-to-point variation of plastic deformation is easily calculated for a bent tube, it is possible to predict the extent of the intergranular stresses and thus to correct for them. This is done either by using a calibration experiment in which a Monel-400 sample is subjected to known uniaxial tensile loading into the plastic regime, or by calculation directly from models of the type mentioned above and described in detail in Chapter 5.

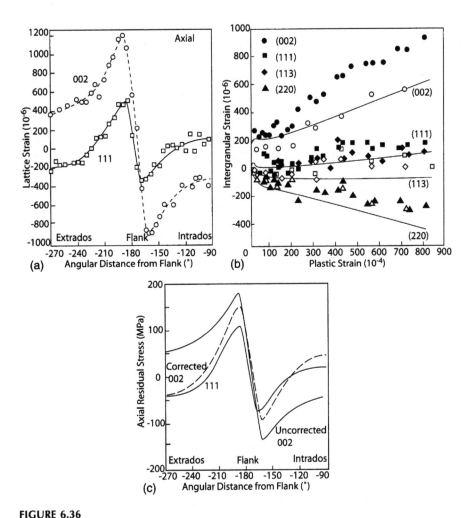

FIGURE 6.36
(a) The residual axial component of lattice strains measured at various positions around a bent Monel-400 tube using the 002 and 111 reflections. (b) The intergranular strain component parallel to the stress axis, with load applied (solid) and upon unloading (open symbols) generated in a Monel-400 test piece as a function of tensile plastic strain. Solid curves are results of self-consistent elastoplastic calculations with no adjustable parameters. (c) The inferred macrostress variation around the pipe based on the 111 and 002 reflections using their diffraction elastic constants. The 002 is shown with (dashed) and without (continuous) a correction for intergranular strains [82].

Figure 6.36b shows the results of such a calibration experiment in which the intergranular strain, determined as the deviation from the linear part of the lattice strain–macroscopic strain curve, is plotted against the measured macroscopic plastic strain for uniaxial loading. This has been done with the load applied (filled symbols) by drawing a return curve parallel to the elastic gradient, and more simply by measuring the residual lattice strain after unloading (open symbols). The difference between these values, which is especially marked for the 002, is noteworthy, and is due to some redistribution of the plastic misfit upon unloading. In other words, the unloading response is not completely elastic. More generally, note the variation in intergranular strain, from the large net tensile response of 002 lattice planes to the net compressive response of 220 lattice planes. For most reflections, the agreement between the EPSC model [83] and experiment is as close as the experimental uncertainty. However, for some reflections, such as (002), there are systematic errors of as much as 30%, indicating that improvements to the model are necessary. Using this experimental intergranular strain–plastic strain response for Monel it is possible to deduce the correct macroscopic stress field in the bent tube. First the contribution from intergranular effects strain to the measured strain from each reflection at each point around the tube is determined. Assuming that the deformation behavior is the same for the bent tube as in the calibration uniaxial tensile test, from the measured macroscopic strain around the tube the intergranular contribution is found. This can then be subtracted from the total strain recorded for the reflection, leaving just the elastic contribution. This elastic contribution can then be used to make an estimate of the macrostress using the appropriate plane-specific diffraction elastic constant. Using the measured strains for the 002 reflection in Figure 6.36a and the data in Figure 6.36b, the corrected macrostresses are shown in Figure 6.36c. In this case, the corrections are up to 180%, including a change of sign in the vicinity of the extrados of the stress inferred from the 002, reflection from +55 MPa to −35 MPa. To within the experimental uncertainties, the corrected 002 data and the 111 data (which do not require correction because the intergranular strains were small) are in agreement. Both reflections give a balance of stress and moment. This example illustrates how essential it can be to correct for intergranular stress effects in plastically deformed samples.

6.4.3.3   Stress in a Highly Textured Sample

The existence of texture in a sample may indicate prior plastic deformation since this can preferentially orient the grains. Because plastic deformation occurs by crystallographic slip, it is normally heterogeneous at the grain level. Unless the sample has undergone a stress-relieving heat treatment, residual intergranular strains will exist, being positive in some grain families corresponding to certain *hkl*, and negative in other grain orientations. The total intergranular stresses balance for a given sample orientation when integrated over a small volume of sample containing a sufficient number of differently oriented grains, but will not normally be zero when evaluated using a single reflection as discussed above.

A different approach to carrying out and interpreting measurements aimed at determining the macrostrain must therefore be taken for highly textured samples as illustrated by the case of rolled plate [83,84]. Although a quantitative account of the effect of texture on strain is beyond the scope of this section, the following qualitative ideas are useful. Figure 2.13 shows the

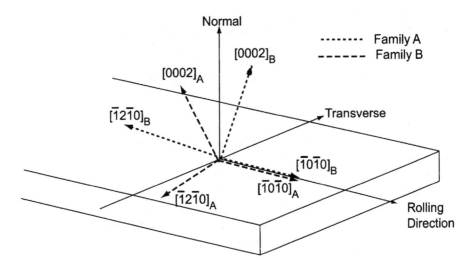

FIGURE 6.37
A schematic representation of the two "families" of grains in a Zircaloy-2 rolled plate with a common [10$\bar{1}$0] crystallographic orientation along the rolling direction, and the two [0002] orientations at ±40° to the plate normal in the normal-transverse plane [84].

(0002) and (10$\bar{1}$0) pole figures of a strongly textured Zircaloy-2 plate. The interpretation of the pole figures of the textured plate is illustrated schematically in Figure 6.37, which indicates that there are two principal "families" of grain orientations in the rolled plate. Both families have [10$\bar{1}$0] orientations along the rolling direction. The first "family" has [0002] orientations concentrated around +40° from the normal direction in the normal–transverse plane, while the second is concentrated around −40° from the normal direction. The third orthogonal axis, [$\bar{1}$2$\bar{1}$0], of the first "family" is not so far removed in angle from the [0002] direction for the second "family" and vice versa. In sample directions near 40° from the sample normal, there are therefore grains with both [0002] and [$\bar{1}$2$\bar{1}$0] crystal orientations, and these have coefficients of thermal expansion that differ by a factor of 2. Because of the difference in thermal expansion, grains with these orientations constrain each other as the sample cools from the annealing temperature, and the [0002] grains end up in tension at room temperature with the [$\bar{1}$2$\bar{1}$0] grains in balancing compression. However, in the rolling direction, where there is a preponderance of grains with [10$\bar{1}$0] directions, only a few grains have different thermal expansion, so there are few constraints to generate residual intergranular strains. Measurements show very small intergranular strains in this direction in [10$\bar{1}$0] grains [84]. It is thus clear that the nature of the texture can modify the values of strain generated in subsequent thermomechanical operations. More generally, however, intergranular stresses are generated by the difference in the plastic and elastic response to applied load for different crystal orientations. This situation occurs in cooled Zircaloy plate when it is further rolled.

It is worthwhile noting that the lattice strains measured — for example, from the 0002 reflection, which is a minority orientation in the *rolling* direction for the textured plate — will not be characteristic of most of the grains making up the plate. The 0002 reflection will be difficult to measure since the intensity will be low. In fact, the 0002 reflection may even indicate a strain that balances the strains exhibited by the majority of grains, and therefore have the opposite sign. However, knowing the orientations of the grains, it was possible to map out the complete strain tensor by examining the strain components along the principal *crystal* coordinates rather than along the obvious *geometric* sample coordinates [85]. In this case, the strains are actually rather small along the principal directions in the plate so measurements in these directions alone would give a misleading picture of the stress state. However, in the principal *crystal* directions, for example at 40° to the normal direction, the strains are very large since the strain tensor is, in a sense, tied to the crystal orientations.

If the strain tensor describing the macroscopic state of strain in, say, a strongly textured Zircaloy tube were required, the measurements would have to be corrected for the intergranular effects. These effects could be obtained by examining a small coupon, or coupons as necessary, cut from the tube in order to obtain the correct reference d^0 values, since cutting does not destroy the intergranular stresses and strains on the spatial scale of the grains, but does destroy the macroscopic stress field. For low-symmetry materials such as hexagonal close-packed structures, there is no reflection, analogous to the 113 reflection for nickel-based alloys, which can be identified as generally having very small intergranular effects.

6.4.3.4   *Intergranular Lattice Strain as an Indicator of Macroscopic Plastic Strain*

As discussed in Section 5.6.6, if the deviation from the linear response of lattice strain to uniaxial loading beyond yield can be predicted, then it opens the possibility of interpreting these deviations in terms of the plastic strain history of a sample component. A number of methods of analysis for this are currently being explored. Two simple approaches have been investigated by Korsunsky et al. [86]. They have considered the deformation of a face-centered cubic Al composite bar that had been plastically bent. Following Equation (5.46), they have defined two parameters; the anisotropy strain parameter $\varepsilon_{An} = \gamma/C_0$ and the difference strain parameter ε_D:

$$\varepsilon^{hkl} = \varepsilon^{h00} - \varepsilon_{An}\, A_{hkl}$$

and

$$\varepsilon_D = \varepsilon^{111} - \varepsilon^{h00} \tag{6.13}$$

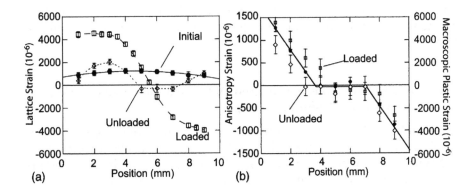

FIGURE 6.38
(a) Variation in the average matrix lattice strain component along the bar with position across the thickness of an Al/17%SiC particle composite bar in the initial, loaded, and unloaded conditions under four-point bending. (b) Comparison between the measured plastic strain (filled circles) and anisotropy strain in the loaded (squares) and unloaded (triangles) conditions. The solid line is a guide to the eye [86].

where ε^{hkl} is the measured lattice strain for the reflection hkl, and A_{hkl} is the anisotropy factor defined by Equation (5.24). In effect, the anisotropy strain is an extra parameter to be fitted in the Rietveld analysis that allows for the hkl variation peak shift arising from plastically generated misfits. As discussed in Section 5.6.6, it has been shown to be a good descriptor for the shifts for untextured polycrystals under elastic loading [75].

By way of an example case, residual plastic and elastic strain were introduced into a 10 mm thick Al/17%SiC composite bar by four-point bending. The average matrix elastic lattice strain as determined by a Rietveld fit is shown in Figure 6.38a. It is clear that prior to four-point bending, the matrix was under residual tension due to thermal residual misfit strains (Section 6.3.5), along with a macroscopic approximately parabolic quench stress. The difference between the strains with the bending load applied and after unloading is linear across the thickness, which confirms that no reverse plasticity occurred on unloading. The most appropriate value of the anisotropy strain was deduced from the Rietveld fitting routine for the time-of-flight diffraction spectrum for each sampled gauge volume position. Figure 6.38b shows a comparison between the plastic strain variation introduced by bending and the anisotropy strain derived for each time-of-flight spectrum. The correlation between the plastic strain and the anisotropy strain is clear. This suggests that the anisotropy strain is related to the overall inelastic deformation, and that it is fairly insensitive to elastic strain or residual strain. This is to be expected for aluminum because of its low level of elastic anisotropy (Table 5.1). Good agreement was also found between the difference strain parameter, ε_D, and the plastic anisotropy.

Clearly, the anisotropy strain may not be the best method of analysis for materials that show a large elastic anisotropy. However, this simple example illustrates the principle. Extensive research is ongoing at the current time, and it is possible that in the future more sophisticated analyses may allow the deduction of plastic strain directly from measurements of intergranular strain.

6.4.4 Future Directions for Neutron Diffraction Measurement of Stress in Plastically Deformed Components

It is clear that a thorough understanding of the intergranular stresses induced in plastically deformed materials, which will enable detailed predictions of behavior to be made, will require much further work on both samples of typical materials under known stress loading, and on practical cases of residual stress in engineering components. To date, research has focused on uniaxial deformation, whereas many manufacturing and in-service environments are multiaxial, both in terms of applied stresses and plastic strain. In addition, little attention has yet been focused on complex strain paths, such as those encountered in forging, rolling, and so on.

Results will lead to further important contributions to materials science in understanding of the behavior of polycrystalline materials under stress, and to a more reliable means of interpreting strain data from engineering component samples in terms of macrostress. It is anticipated that measurements made on both continuous source instruments, where a few chosen reflections can be studied in detail, and on pulsed source instruments, where very many reflections can be studied, will both contribute to this work. With more intense pulsed sources being planned, the latter promise to provide increasingly detailed information on the simultaneous response of many grain orientations to plastic loading.

6.5 Near-Surface Stresses

One tends to think that the principal advantage of neutron diffraction is its use for measurement at positions deep within samples, compared with x-ray diffraction, which is mainly useful for near-surface stress measurement. However, neutrons are providing important information even in the region within less than 2 mm of the surface. Normally, in strain mapping measurements the component completely fills the instrumental gauge volume. If, however, the instrumental gauge volume is only partly filled, which can be the situation when measuring strains near surfaces resulting from surface stress modification by processes such as shot-peening or carburizing, a number of systematic instrumental errors can occur as discussed extensively in Section 3.6. These must be avoided by careful design of the experimental procedure or by subsequent correction of the data. In general, very high spatial resolution is needed, calling for beam aperture dimensions to be 0.5 mm or less to define the instrumental gauge volume.

6.5.1 Origin and Nature of Near-Surface Stresses

The near-surface state of stress can have very important consequences for the behavior of the component. For example, the failure strength of glass can be increased many times by engineering a compressive state of stress into the surface. This is because the surface is often a key site for the creation of defects and flaws, which then often propagate into the interior at fairly low levels of applied stress. For window glass, this is usually done by cooling the glass from above the glass transition temperature — the temperature above which it behaves like a liquid — very rapidly using air jets. A similar effect can also be introduced by doping the surface with ions such as potassium.

Peening, quenching, carburizing, nitriding, and coating are just some of the ways in which the engineer can manipulate the near-surface residual stresses, and hence component behavior. In most cases, the aim is to place the surface in a biaxial state of compressive stress.

6.5.2 Challenges to Accurate Neutron Diffraction Measurement of Stress Near Surfaces

Measurement of near-surface stress has presented a challenge for many years. The surface sensitivity and depth capability of various stress measurement methods are illustrated in Figure 6.39. In general, diffraction techniques can only be used on crystalline materials, Raman techniques on those that produce Raman excitations within the appropriate part of the electromagnetic spectrum, and Barkhausen and other magnetic techniques are suitable only for ferromagnetic materials. Glancing incidence x-ray diffraction is capable of providing submicron surface information. Traditional x-ray diffraction can provide measurement of the stress state within the first 5 to 30 μm according to the atomic number of the material; layer removal methods are required to penetrate deeper. Hole drilling can provide important data, but the surface stresses can often be so high that yielding can occur during hole drilling, invalidating the results. Furthermore, not all materials are susceptible to hole drilling. The use of synchrotron x-rays is beginning to have an impact in this area, with over 50% of synchrotron x-ray strain measurements to date made on coatings and surface treatments, as summarized in Table 6.1. The main advantage of the neutron method over other methods is that it can provide information from within 50 μm of the surface to a depth centimeters or even tens of centimeters with a spatial resolution of millimeters.

As seen in Section 3.6, there are at least two important hurdles to be overcome if the accurate measurement of stress is to be achieved using

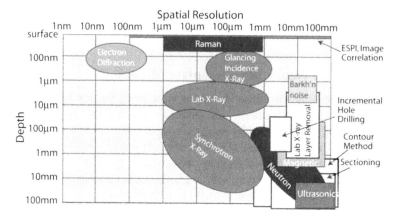

FIGURE 6.39

Schematic illustrating the approximate spatial resolution and depth capabilities of various stress measurement techniques. Destructive techniques are represented by white boxes. Image correlation and electronic speckle pattern interferometry involve the analysis of surface images, while the depth of Raman spectroscopy is limited by the transparency of the object.

neutron diffraction in the near-surface region: accurate location of the effective position of measurement in the sample, and correction of instrumental aberration effects that give rise to anomalous shifts in diffraction angle and hence spurious strains. Accurate corrections for the aberrations in turn depend on knowledge of exact positioning. It is not uncommon for very steep strain and stress gradients to exist in the near-surface region. Gradients in excess of 2000 MPa/mm are not exceptional for machined/ground or peened surfaces. In such cases, an error in the sample position of just 100 μm represents an error in stress measurement of ~200 MPa.

One way of accurately locating the surface of the sample is to fit the measured diffracted intensity profile to the expected variation as the surface is scanned into the instrumental gauge volume (IGV) in reflection geometry, as illustrated in Figure 6.40. The form of the intensity variation is given in Appendix Section A.2.2, and can be broken down into three regions in which the IGV is less than half filled, more than half filled, and completely filled by the sample. If there is zero attenuation in the sample, then the surface is at the reference point at the centroid of the IGV when the intensity is just half of the maximum, as expected. However, if there is attenuation, then the intensity peaks earlier and the half-maximum intensity occurs at a position shifted from the centroid. By comparing the observed variation with that calculated from the model, it is possible to accurately determine the surface position — that is, the translator reading, T — when the surface is at the reference point to better than ±25 μm, and this provides a useful "origin" for translator readings. The reading is set to zero in Figure 6.40.

When scanning the surface through the IGV, the position recorded by the translator must be accurately related to the effective centroid of the sampled gauge volume (SGV) representative of the material from which neutrons are

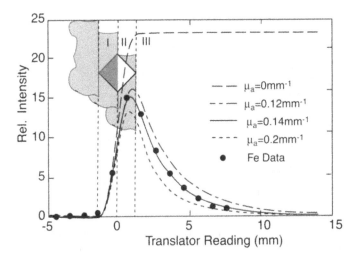

FIGURE 6.40
Comparison between the variation in measured integrated intensity for the 110 reflection at ϕ^s = 90°, when scanning a steel surface, in the reflection configuration, into an instrumental gauge volume (IGV) defined by 2 mm incoming and outgoing apertures, and the calculated response for various linear attenuation coefficients, μ_a. For steel, $\mu_a = 0.12$ mm^{-1}, but the best fit is achieved for $\mu_a = 0.14$ mm^{-1}. The translator reading, T, is zero when the surface is at the reference point at the center of the IGV. The intensity in the regions I, II, and III is described by the relations given in Equations (A.2.2).

diffracted. As described in Sections 3.6.1 and 3.6.2, this is the weighted centroid calculated by taking into account both the fraction of the IGV occupied by scattering material, and the point-to-point variation in intensity of the diffracted signal over the SGV that account for attenuation. It is the effective position in the sample at which the average strain is measured. The situation is illustrated in Figure 3.24, and an example of the way the centroid varies with translator position appears in in Figure 3.25.

Having located the effective position of strain measurement relative to the surface as given by the translator position, it is then necessary to correct each measurement of angle or strain at this position for any anomalies resulting from the off-center of this centroid of the SGV from the reference point. From the details of the instrumental angular and wavelength spreads on a continuous source instrument, and from the geometry of the detectors on a pulsed source instrument, the anomalous angular or flight time shifts arising from a partially filled IGV can be corrected for by simulation calculations, as described in Section 3.6.4.

An example showing measured and calculated spurious strain as a stress-free sample of Al or Fe scanned into the IGV in the reflection configuration is shown in Figure 6.41. There is good agreement between the modeled and observed spurious strains, so these may be subtracted from the strains measured in a sample with surface stresses to give the true surface strain. An accurate knowledge of the position of the surface relative to the center of the IGV, determined from the experimental intensity variation as described

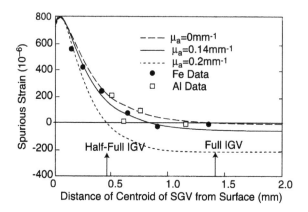

FIGURE 6.41
The observed spurious surface strains (points) determined using a Rietveld refinement of many peaks measured on ENGIN at ISIS for strain-free steel and aluminum powder samples scanned into the ~2 × 2 mm² section of the instrumental gauge volume (IGV) in the reflection configuration. They are compared with those calculated from a model for various linear attenuation coefficients, μ_a (lines). The value for aluminum is indistinguishable from the line for $\mu_a = 0$ mm⁻¹. The abscissa is the distance of the effective centroid of the sampled gauge volume from the surface, $|d_{cs}|$ of Figure 3.24. This distance varies nonlinearly from 0 to $\sqrt{2}$ as the sample fills the IGV depending on the attenuation coefficient, as described for this size of gauge volume in Figure 3.25.

above, is essential. Note that, even when the sample completely fills the IGV, for strong attenuation there will be a shift in position of the centroid of the SGV from the reference point at the center of the IGV as seen in Figure 3.25, with spurious strain given by the short broken line in Figure 6.41. It can be shown that for ENGIN at the ISIS pulsed source, the positional shifts are relatively small, at least until less than half the IGV is filled. Even in the extreme case when the IGV is less than 10% filled the spurious strain is less than 600 με. This is partly because the radial collimator on ENGIN closely defines the neutron trajectories. However, at reactor source instruments using 2 mm apertures with an otherwise unobstructed PSD, the spurious strains can be as large as 10,000 με. As discussed in Section 3.6.4, there are a number of strategies to correct for, or to suppress, the effects of spurious strains.

In the following sections, practical aspects of near-surface stress measurement using neutron diffraction are described by considering a series of examples with stress fields extending to increasing depths in the sample.

6.5.3 Examples of Measurement of Stress Near Surfaces

6.5.3.1 Machining Stresses

In many cases, machining and grinding stresses have a detrimental effect on component life. Normally, machining stresses extend only 100 to 200 μm below the surface. Such a shallow range presents significant technical difficulties for accurate measurement by neutron diffraction. Nevertheless, in

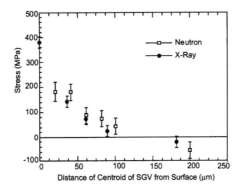

FIGURE 6.42
In-plane stress component parallel to the grinding direction in a ground EN9 steel plate measured by neutron diffraction using the vertical Z-scanning method and a Rietveld fit on time-
of-flight data from ENGIN at ISIS, and by Cr x-ray diffraction with layer removal using the 211
reflection [87].

cases where a nondestructive method is required, it is possible to map the
depth profile reliably. A good example shown in Figure 6.42 is the work by
Balart et al. [87] who measured the tensile stresses arising from the grinding
of EN9, a plain medium C steel. They used the Z-scan method, as described
in Section 3.6.4, which is not sensitive to geometric shifts. Strain-free lattice
spacing was measured at a depth of 4 mm. If one assumes that the out-of-
plane stress is zero, the in-plane stresses can be calculated simply from the
in-plane strains.

6.5.3.2 Shot-Peening
Shot-peening, conventionally carried out by repeatedly impacting the surface with a stream of small shot, introduces a compressive residual strain
and stress in the surface region of components. It is usually carried out to
improve their resistance to fatigue, but can also be used to introduce stresses
that form the shape of components. Measurement of the resulting residual
stress distribution is desirable both to monitor the shot peening procedure
and to predict the fatigue life of components. One example of such a study
is that carried out on the α-β titanium alloy IMI-318, Ti-6Al-4V, which is
widely used in gas turbine engines. This alloy shows a high neutron attenuation, twice that of iron, a substantial incoherent scattering component that
increases the general background, and a low diffraction intensity, one quarter
that of iron, which makes it a difficult material for neutron diffraction strain
measurement, as mentioned in Section 2.3.6.

A $23 \times 20 \times 50$ mm^3 IMI 318 (Ti-6Al-4V) alloy plate was shot-peened using
230R shot at an intensity of 20 Almen. The Almen is a measure of shot-
peening intensity. For the measurements made using the ENGIN instrument
at ISIS, it was assumed that a biaxial in-plane stress exists in directions in
the shot-peened surface. The strain component normal to the surface was

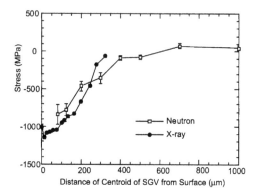

FIGURE 6.43
The in-plane stress component, measured by neutron diffraction at a pulsed source, near the shot-peened surface of a Ti-6Al-4V alloy sample. The in-plane strains have been measured by the z-scan method, while the out-of-plane strain has been measured in reflection and corrected for spurious strain. The results are compared with measurements made using the $(2\bar{1}33)$ reflection of x-rays, removing layers to obtain data at depth [88].

measured in reflection geometry by conventionally translating the sample horizontally through the IGV [88]. The low diffraction intensity of this alloy and relatively high attenuation results in strong geometric anomalous shifts, and the steep strain gradient necessitated accurate positioning to avoid errors. The Z-scan method described in Section 3.6.4 was then used to measure the in-plane strains, using an IGV of $1.4 \times 20 \times 0.5$ mm^3. The strain-free reference lattice parameter was inferred from a point 2 mm below the shot-peened surface. These strain data were combined to give the stress distribution shown in Figure 6.43, plotted against the position most representative of measurement, that is, the distance of the effective centroid of the SGV from the surface. The stress variation is compared with results of destructive x-ray measurements made using layer removal. The two distributions, measured by neutron and x-ray diffraction, agree well. Note that in this case the in-plane and out-of-plane strains were combined to calculate the in-plane stress and the neutron measurements provide data up to 100 µm from the surface. The fact that when combining the three components of the strain, the out-of plane stresses were found to be small at all depths suggests that the correction for the geometric shifts in the out-of-plane strain is appropriate. Another example of the measurement of shot-peening stresses determined in this way for IN 718 is given in Figure 3.34.

6.5.3.3   *Carburized Layers*

Carburizing is often used to surface harden steels, with case depths varying depending on the application, but 0.4 mm is typical. It is achieved by the diffusion of carbon into the steel from a carbon-rich environment, by gas carburizing using a gas rich in CO, or pack carburizing using charcoal or coke with alkali carbonates bound together by oil or tar. Carburizing presents

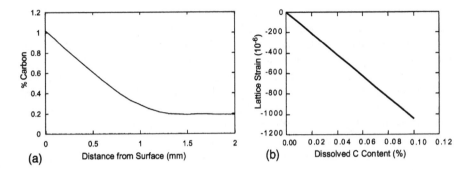

FIGURE 6.44
(a) The variation in carbon content with depth through a carburized zone [89]. (b) The compositionally induced spurious lattice strain as a function of dissolved C content (%) in C steel. If unaccounted for, this change in lattice parameter would be interpreted as elastic strain.

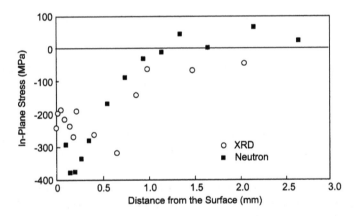

FIGURE 6.45
Variation of the in-plane component of residual stress through a carburized layer on a steel disk as measured by neutron diffraction nondestructively, and by x-ray diffraction with layer removal [90].

a challenge to neutron diffraction measurement of the residual strains since it changes the chemical composition of the material near the surface. An example of such variation is shown in Figure 6.44a. Furthermore, the lattice parameter of steel is particularly sensitive to changes in C concentration, as shown in Figure 6.44b. Carburizing can also introduce Type II strains and stresses [89]. Great care therefore needs to be exercised in analyzing measured strain data from carburized layers.

An example is the measurement of strain in a carburized layer on a steel disk [90]. In this case, in order to interpret the results of neutron measurements of strain in terms of stress, two reasonable assumptions were made: that the stress normal to the surface was zero at the surface, and that the gradient of the stress normal to the surface was also zero. By adopting these assumptions, the problem of finding the strain-free lattice parameter of the steel in the carburized layer as a function of depth was circumvented. Figure 6.45 shows the residual stress profile thus determined by neutron diffraction, and this is compared with that found from x-ray diffraction using the layer removal technique. The latter profile was obtained by electropolishing the surface, and subsequently correcting for layer removal. In the region very close to the surface, 0 to 50 μm, the x-ray measurements have a unique advantage, but at greater depths the neutron results are more informative as well as being nondestructive. Indeed, in this case the x-ray results failed to show tensile stress at the depth that must be present in order to balance the compression near the surface. This failure was partly because of compounded errors in the necessary corrections for layer removal with increasing depth.

6.5.4 Future Directions for Neutron Diffraction Measurement of Stress Near Surfaces

Neutron diffraction techniques have special advantages for strain and stress measurement near surfaces when:

- Surface stresses or their effects extend beyond 1 mm into the material, and the depth profile is required to these depths, which are often up to 15 to 20 mm.
- Nondestructive information is required, especially on complex/large components. In these cases, the geometric restraints of working at high x-ray energies preclude synchrotron x-ray diffraction.
- *In situ* behavior is required such as the behavior of coatings in high temperature environments corresponding to in-service conditions, and the effect of fatigue cycling.

In each of these cases, improved flux, precise IGV definition, and better positional accuracy will extend the power of the technique. All of these will be addressed by the next generation of neutron instrumentation. The increasing understanding of corrections for near-surface anomalous effects will lead to more reliable data interpretation.

Standard x-ray diffraction will continue to be the method of choice for determining near-surface stress at depths less than 50 to 100 μm, and its portable capabilities make it especially attractive. Layer removal can extend the depths it can probe, albeit partially destructively and with necessary data corrections. The use of synchrotron x-rays with much smaller gauge volumes will increasingly challenge that of neutrons for the cases where it is most applicable, although the full advantage of high x-ray energies in terms of increased penetration distance is not carried over into near-surface penetration. This can be understood if the maximum feasible path lengths calculated on the basis of a maximum acceptable count time or minimum acceptable

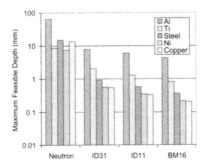

FIGURE 6.46
The approximate maximum feasible depth of measurement in several materials for neutron and synchrotron x-ray diffraction in the reflection geometry configuration [91].

peak to background ratio in Section 4.4.2 are used to calculate penetration distances. The maximum feasible path lengths for thermal neutrons were summarized in Table 4.2. In Figure 6.46, this approach has been used to calculate maximum penetration distances for neutrons and high-energy synchrotron x-rays in reflection geometry [91]. The neutron estimate is for the L3 instrument at Chalk River, with an SGV of area 40 mm² in the scattering plane and a count time of 1 h. The three x-ray estimates are for instruments at the ESRF with an SGV of 1 mm³ and peak to a background ratio of unity. Note that upon increasing the x-ray energy from 40 keV typical of BM16 (assuming $\phi^s = 2\theta = 20°$), to 50 keV typical of ID11 (assuming $\phi^s = 16°$), and to 60 keV typical of ID31 (assuming $\phi^s = 16°$), little advantage in penetration depth is achieved due to the low scattering angles. Indeed, angles as low as $\phi^s = 5$ to $10°$ are typically used on ID15 at 150 keV. In this respect, the $\phi^s = 90°$ arrangement characteristic of neutron measurements is advantageous.

While hole drilling will remain the standard technique for *in situ* measurement on large industrial component samples, the recent development of magnetic techniques that are portable and nondestructive in their examination of magnetic materials, will enable them to play an increasingly important role [92].

6.6 *In Situ* and Through-Process or Life Studies

One area in which neutron diffraction is playing an increasingly important role is in following the evolution of residual or applied stress either through successive manufacturing stages, or as a function of in-service life. This is because the large working area available at dedicated neutron strain instruments, combined with the penetrating power of thermal neutrons, make neutron diffraction one of the few tools that can probe real components to provide in-depth nondestructive stress measurement in three dimensions.

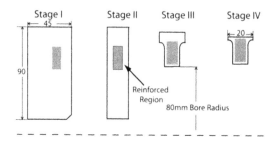

FIGURE 6.47
A schematic of the machining sequence required to machine the bling component shown in Figure 6.35. The hoop-wound composite region is shown in gray. Stage IV corresponds to the breaking of contiguity in the blade region.

The high penetration can also be utilized for furnaces and stress rigs to simulate environmental conditions.

6.6.1 Manufacturing Process Studies

An example of a manufacturing study is provided by the progressive machining of an integrally bladed ring, or bling, of the type shown in Figure 6.35 [70]. This was made by a proprietary route from Ti-6Al-4V-coated SiC fibers, placed within Ti-6Al-4V cladding. The approximately 250-mm diameter prototype ring comprised a hoop of continuous SiC fiber-reinforced Ti-6Al-4V composite embedded in monolithic Ti-6Al-4V. It was consolidated by hot isostatic pressing (HIP) at a temperature above 1173 K. The ring geometry after consolidation, Stage I, and subsequent machining, Stages II, III and IV, are shown schematically in Figure 6.47. The microstructure of the composite reinforced region is shown in Figure 6.35. The overall fiber fractions are approximately 1.5%, 3%, 12%, and 24% through Stage I to IV, respectively. Since large amounts of monolithic material were removed in machining to Stage II, a significant change in the residual stress distribution may be expected. Separate samples of coated fibers, uncoated fibers, and forged Ti-6Al-4V from parts of the cladding removed after Stage II, were used to provide stress-free reference lattice parameters.

The neutron diffraction measurements were undertaken using the ENGIN instrument at ISIS. Diffraction patterns ranging over lattice spacings from 0.5 to 3.0 Å were collected at each subsurface measurement location, from which the average lattice parameters for each phase, Ti-Al-4V and SiC, were determined by Rietveld refinement. The residual elastic strain, ε, in the titanium alloy phase was calculated as described in Section 4.5 using the equation

$$\varepsilon = \frac{2(a-a^0)}{3c^0} + \frac{(c-c^0)}{3c^0} \tag{6.14}$$

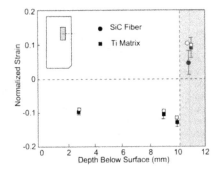

FIGURE 6.48
The measured axial strain components with depth in Stage I (full symbols), compared with those calculated from a finite element model (open symbols). The inset shows the positions of measurement [70].

where a and c are the measured average hexagonal lattice parameters, and a^0 and c^0, the stress-free lattice parameters. For reasons of texture the strain in the SiC phase was calculated only from the lattice parameters a and a^0. The principal strains were assumed to be in the radial, axial and hoop directions with respect to the ring axis. In interpreting the measured strains, the stiffnesses of matrix and reinforcement, around 115 and 450 GPa, respectively, must be kept in mind.

On cooling from the HIP temperature, large compressive thermal misfit hoop strains are set up in the fibers because their volume fraction is low, while lower balancing tensile hoop strains are introduced into the matrix. As discussed in Section 2.3.6, Ti-6Al-4V alloy is a difficult material for neutron diffraction experiments due to the large incoherent scattering of Ti and V. Nevertheless, it was possible to probe the fiber-reinforced region in Stage I, as well as to investigate the stress field within the peripheral cladding. Excellent agreement with a finite element model was achieved throughout the cladding as shown in Figure 6.48. As expected, strain magnitudes within the cladding were small, as the locations measured were remote from the reinforcement and the overall reinforcement fraction at this stage was low.

On machining to Stage II, the ring cross-sectional area was reduced by more than 50%. Extensive residual strain redistribution was expected in Stages III and IV, especially in the region close to the inner diameter of the ring where the volume of the cladding was drastically reduced. While FE calculations of strain were in good agreement with neutron diffraction measurements in the main region of cladding, poor agreement was found in the 1 mm walls due to the difficulty of completely filling the IGV. Synchrotron radiation was used to determine the residual strains in this region. Only a small number of the locations that were measured in Stage I remain at Stage IV. In Figure 6.49, the variation in strain in the reinforced region is shown across the four stages. Note that because the relative volume fraction of the fibers increases through machining, the thermal residual strains are increasingly accommodated in the matrix and cladding through the progressive

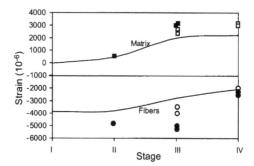

FIGURE 6.49
Neutron diffraction (full symbols) and synchrotron x-ray diffraction (open symbols) measurements of the evolution of hoop strain components of the matrix and fibers in the composite across the four stages, compared with finite element predictions (lines).

stages, causing the matrix hoop strain to rise and the fiber strain to fall. Overall, it would appear that the finite element model underestimates the fiber and matrix stressing in the composite region.

6.6.2   Heat Treatment

It is often the case that manufacturing processes generate unacceptable residual stresses in a component that must be relieved prior to putting it into service. Welding, for example, often introduces large residual stresses, as discussed in Section 6.2. Neutron diffraction can be an invaluable tool in the development of stress relief procedures for welded joints. A good example where the ability to follow the evolution of weld residual stresses through heat treatment has led to better heat treatment procedures, is for the inertia welding of new-generation RR1000 Ni-base γ/γ' superalloys [93]. These alloys achieve their high-temperature strength primarily by having a very high γ' content. Unfortunately, this makes them difficult to weld by traditional fusion welding techniques, so friction-based inertia welding was used to join two 50 mm diameter tubes made from this alloy. As a result of the higher-temperature capability of these alloys relative to that of existing alloys, new postweld heat treatments needed to be developed. Neutron diffraction is the only tool able to measure the residual strain components along the axial, radial, and hoop directions across the 8 mm tube thickness of a weldment between Ni superalloy tubes [27]. As is evident from Figure 6.50a, the microstructure varies markedly across the weld line. In the parent, there is a large fraction of secondary ($\sim$200 nm) and tertiary ($\sim$20 nm) ordered γ' precipitates. These have been dissolved at the welding temperatures such that fine ($\sim$8 nm) tertiary γ' has reprecipitated. As a result, the overall γ' content, evaluated in terms of the strength of the superlattice (001) γ' reflection relative to the (002) reflection for which both γ and γ' contribute, varies significantly in the as-welded condition within 2 mm of the weld, as shown

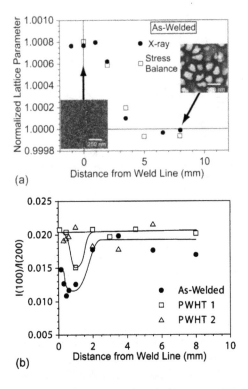

(a)

(b)

FIGURE 6.50

(a) Variation of stress-free lattice parameter, normalized by that for the parent material, with axial distance from the weld line in a Ni-based superalloy inertia friction weldment. The points are from laboratory x-ray measurements (full circles), and from calculations based on imposing stress balance (squares) on the axial stress component measured in the as-welded condition [27]. Scanning electron microscope images representative of the weld line and parent material microstructures are shown in inset. (b) Variation in γ' content with distance from the weld line, given in terms of the relative strength of the γ' (100) superlattice reflection, in the as-welded condition and after two post-weld heat treatments [93].

in Figure 6.50b. These data were obtained using synchrotron x-rays at ESRF, Grenoble. As a consequence, one might expect that the stress-free lattice parameter would vary as a function of position across the weld for the reasons of compositional changes discussed in Sections 4.6.2 and 6.2.3. These data were obtained using synchrotron x-rays at ESRF, Grenoble. The actual variation with position of the stress-free lattice spacing for the as-welded sample was determined by x-ray methods as described in Section 4.6.3. As is clear from Figure 6.50a, a marked variation, equivalent to 1000×10^{-6} strain, was observed with distance from the weld, with the largest lattice spacing on the weld line. On the other hand, little variation was found across the tube radius [27]. As a result, stress balance applied to the results of the neutron diffraction measurements could also be used to infer the variation of stress-free lattice parameter, as described in Section 4.6.3. Figure 6.50a shows that the two methods are equivalent in this case, the latter being nondestructive and thus easier to apply. The application of stress balance

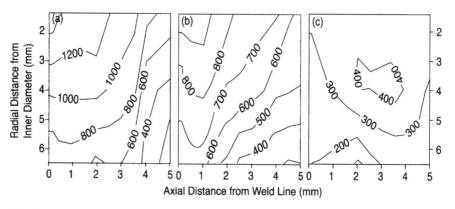

FIGURE 6.51

Hoop stress components in MPa mapped across the tube cross-section of the Ni superalloy inertia weldment for (a) the as-inertia welded tube, (b) after conventional post-weld heat treatment, (c) after post-weld heat treatment at +50°C above the conventional treatment temperature. The strains were determined from Rietveld refinements of the tof neutron diffraction spectra using the axially variant stress-free lattice parameter from Figure 6.50a in (a), and a constant value in (b) and (c) [27].

revealed that for the postweld heat-treated components, there was no variation in lattice parameters. This is consistent with the observation that due to the uniform exposure to elevated temperatures, the γ' content becomes more uniform across the weld line for the PWHT tubes. It is self-evident that failure to account for changes in stress-free lattice spacing would lead to incorrect measurement of residual stress.

The resulting residual stress maps obtained from the neutron diffraction measurements are shown in Figure 6.51. The yield stress of the material is ~1050 MPa, and the hoop residual stresses near the weld line caused by inertia welding, especially toward the inner diameter, are very large. Not unexpectedly, the residual stresses in the sample after the conventional heat treatment (PWHT1) for Ni-base superalloys indicate that this treatment is not sufficient to relax the residual stresses to safe values. However, a heat treatment of 50°C above the conventional temperature (PWHT2) is sufficient to relax the stresses to within design limits, while retaining acceptable weld microstructures.

6.6.3   Service Life Studies

There are many instances where it is advantageous to be able to follow the development of residual stresses as a function of in-service life. One such instance may be when examining the extent to which residual stresses deliberately introduced to extend life have faded, or quantifying the extent to which damaging residual stresses have been introduced by the in-service

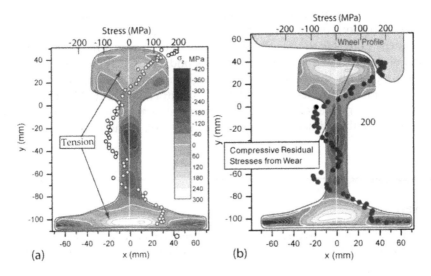

FIGURE 6.52
Longitudinal residual stress component measured by neutron diffraction down the centerline of a 10 mm thick slice cut longitudinally from (a) a new rail (open circles), and (b) a used rail (solid circles) [95]. The superimposed contour maps of longitudinal stress are from more recent contour method measurements on a similar set of new and used British rails. The white contour represents zero stress [96].

history. Fatigue, fretting or rolling contact fatigue, thermal cycling, and creep fatigue are just some of the loading types experienced in service that can have a strong effect on residual stresses.

One of the earliest uses of neutron diffraction to examine the extent to which residual stresses change as a function of life was the study of residual stress field in rails [94,95]. Residual stresses are generated in the rail during manufacture due to differential cooling after hot rolling, and these are modified by roller straightening prior to placement. The longitudinal residual stress component in the head is important from the point of view of fatigue cracking due to railhead contact with the train wheels. The variation in longitudinal stress component has been measured down the center line of a 10 mm thick slice taken from a new rail, using the D20 instrument at the Institut Laue-Langevin. The results are shown as the points in Figure 6.52a, together with those measured on a corresponding slice taken from a used rail in Figure 6.52b. The rails have tensile longitudinal stresses in the head and foot balanced by compressive stresses in the web. Consistent with expectations, the used rail has a very similar profile except near the contact face on the head, where the rolling contact has plastically deformed the near-surface region, setting up a compressive stress that is balanced by an increased level of tension in the center of the railhead. More recent measurements using the contour method (Section 6.2.2) have been made on similar rails nearly 15 years after the original investigation, and the results are plotted as contour diagrams (Figures 6.52a and 6.52b). There is general good agreement between the results of the two sets of measurements, both in terms of compressive and tensile regions and stress magnitudes, which is reassuring. With the new dedicated high-intensity instruments such as ENGIN-X at ISIS, Rutherford Appleton Laboratory, UK, and SMARTS at Los Alamos, USA, now available, it is now possible to map longitudinal stresses over much of the railhead of larger sections of rail, without needing to take thin slices.

References

1. K. Masubuchi, Residual stresses and distortion in welds, in *Encyclopedia of Materials: Science and Technology*, K.H.J. Buschow, et al., Eds., Elsevier, Oxford, 2003, 8121–8126.
2. J. Goldak, Modeling thermal stresses and distortions in welds, in *Recent Trends in Welding Research*, S.A. David and J.M. Vitek, Eds., ASM International, Materials Park, OH, 1989, 71–82.
3. H.J. Stone, S.M. Roberts, T. Holden, P.J. Withers, and R.C. Reed, The development and validation of a model for the electron beam welding of aeroengine components, in Proceedings, 5th International Conference on Trends in Welding Research, ASM International, 1998, 955.
4. R.V. Preston, S.D. Smith, H.R. Shercliff, and P.J. Withers, finite element modelling of TIG welding aluminum alloy 2024, *Sci. Technol. Weld. Joining*, 8, 10–18, 2003.
5. K. Masubuchi, *Analysis of Welded Structures: Residual Stresses, Distortion, and Their Consequence*, Pergamon, Oxford, 1980.
6. D. Rosenthal and T. Norton, A method of measuring triaxial residual stress in plates, *Weld. J.*, 24, 295–307, 1945.
7. G. Schajer, Residual stresses: destructive measurements, in *Encyclopedia of Materials Science and Technology*, K.H.J. Buschow, et al., Eds., Elsevier, Oxford, 2001, 8152–8158.
8. K. Masubuchi and D.C. Martin, Investigation of residual stresses by use of hydrogen cracking, part I, *Weld. J.*, 40, 553–563, 1961.
9. M.B. Prime, D.J. Hughes, and P.J. Webster, Weld application of a new method for cross-sectional residual stress mapping, Proceedings SEM Annual Conf. on Experimental and Analytical Methods, Portland, Oregon, 2001, 608–611.
10. D.J. Buttle and C. Scruby, Residual stresses: measurement using magnetoelastic effects, in *Encyclopedia of Materials: Science and Technology*, K.H.J. Buschow, et al., Eds., Elsevier, Oxford, 2001, 8173–8180.
11. R.A. Owen, R.V. Preston, P.J. Withers, H.R. Shercliff, and P.J. Webster, Neutron and synchrotron measurements of residual strain in an aluminum alloy TIG weld, *Mater. Sci. Eng.*, A346, 159–167, 2003.
12. H.J. Stone, *The Characterisation and Modelling of Electron Beam Welding*, Ph.D. thesis, Cambridge University, 1999.

13. H.J. Stone, P.J. Withers, T. Holden, S.M. Roberts, and R.C. Reed, Comparison of three different techniques for measuring the residual stresses in an electron beam-welded plate of WASPALOY, *Metall. Trans.*, 30A, 1797–1808, 1999.

14. A. Mohr and H.G. Priesmeyer, Experimental correlation between the elemental composition and the lattice dimensions in alloys, using Al-Cu as a model substance, Proceedings, International Conference on Neutrons and Their Applications, Crete, *Soc. of Photo-optical Instrumentation Engineers*, vol. 2339, 350–355, 1994.

15. W.K.C. Jones and P.J. Alberry, *Ferritic Steels for Fast Reactor Steam Generators*, British Nuclear Engineering Society, London, 1977.

16. W.K.C. Jones and P.J. Alberry, A model for stress accumulation in steels during welding, in Proceedings, Residual Stresses in Welded Construction and their Effects, London, 1977, Welding Institute, Cambridge, Paper No. 2, 1977.

17. H.K.D.H. Bhadeshia, Possible effects of stress on steel weld microstructures, in *Mathematical Modelling of Weld Phenomena*, vol. 2, H. Cerjak and H.K.D.H. Bhadeshia, Eds., Institute of Materials, London, 71–118, 1995.

18. J. Yamamoto, Y. Muramatsu, S. Zenitani, N. Hayakawa, and K. Hiraoka, Control of welding residual stress by low-temperature transformation welding consumables, influence of restraint of joint and transformation-temperature on residual stress, in Proceedings of 6th International Conference on Trends in Welding Research, Pine Mountain, Georgia, 2002, ASM International, 2002, 902–905.

19. P.J. Webster, G. Mills, X.D. Wang, W.P. Kang, and T.M. Holden, Residual stresses in alumino-thermic welded rails, *J. Strain Anal.*, 32, 389–400, 1997.

20. L. Edwards, P.J. Bouchard, M. Dutta, and M.E. Fitzpatrick, Direct measurement of residual stresses in a repair weld in an austenitic steel tube, in Proc. International Conference on Integrity of High-Temperature Welds, Inst. Mech. Eng., London, 1999, 181–182.

21. H.J. Stone, T.M. Holden, and R.C. Reed, Determination of the plane specific elastic constants of waspaloy using neutron diffraction, *Scripta Mater.*, 40, 353–358, 1999.

22. A.N. Ezeilo, G.A. Webster, P.J. Webster, and X. Wang, Characterisation of elastic and plastic deformation in a Ni superalloy using pulsed neutrons, *Physica B*, 180, 1044–1046, 1992.

23. S.M. Roberts, H.J. Stone, J.M. Robinson, P.J. Withers, R.C. Reed, D.R. Crooke, B.J. Glassey, and D.J. Horwood, Chracterisation and modelling of the electron beam welding of waspaloy, in *Mathematical Modelling of Weld Phenomena IV*, H. Cerjak, Ed., Institute of Materials, London, 1998, 631–648.

24. A. Steuwer, M. Peel, and P.J. Withers, Measurement of the unstrained lattice parameter using a through thickness $\sin^2 \Psi$ method in AA5083 and AA7010 aluminum friction stir welds, (in preparation, Univ. Manchester, UK) 2004.

25. A.N. Ezelio, and G.A. Webster, Linear friction welding of Ni-superalloys, VAMAS Report, 2000.

26. R.V. Preston, S.D. Smith, H.R. Shercliff, and P.J. Withers, Finite element modelling of TIG welded aluminum alloy 2024, *Sci. and Tech. of Welding and Joining*, 8, 10–18, 2003.

27. M. Preuss, J.W.L. Pang, P.J. Withers, and G.J. Baxter, Inertia welding nickel-based superalloy, Part II: Residual stress characterization, *Metall. Mater. Trans.*, 33A, 3227–3234, 2002.

28. I.C. Noyan and J.B. Cohen, An x-ray diffraction study of the residual stress-strain distributions in shotpeened two-phase brass, *Mater. Sci. Eng.*, 75, 179–193, 1985.

29. X. Yang and R.J. Young, Determination of residual strains in ceramic-fibre reinforced composites using fluorescence spectroscopy, *Acta Metal. Mater.*, 43, 2407–2416, 1995.

30. M.A. Qing and D.R. Clarke, Measurement of residual stresses in sapphire fibre composites using optical flourescence, *Acta Metall. Mater.*, 41, 1817–1823, 1993.

31. L.S. Schadler and C. Galiotis, Fundamentals and applications of micro-raman spectroscopy to strain–measurements in fiber-reinforced composites, *Int. Mater. Rev.*, 40, 116–134, 1995.

32. Y. Huang and R.J. Young, Analysis of the fragmentation test for carbon-fibre epoxy model composites by means of raman spectroscopy, *Composites Sci. Technol.*, 52, 505–517, 1994.

33. T.W. Clyne and P.J. Withers, *An Introduction to Metal Matrix Composites*, Cambridge Solid State Series, Cambridge University Press, Cambridge, 1993.

34. R.C. Ecob, R.A. Ricks, and A.J. Porter, The measurement of precipitate matrix lattice mismatch in Ni-base super-alloys, *Scripta Metall.*, 31, 1085–1088, 1982.

35. V. Randle and B. Ralph, The assessment of compositional changes in nimonic PE16 by convergent beam electron diffraction, *J. Microsc.*, 14, 305–312, 1987.

36. H. Feng, H. Biermann and H. Mughrabi, Microstructure based 3D FEM of lattice misfit and long range internal stresses in creep deformed Ni base superalloy single crystals, *Mater. Sci. Eng. A*, 214, 1–16, 1996.

37. M. Fahrmann, J.G. Wolf, and T.M. Pollock, The influence of microstructure on the measurement of gamma-gamma' lattice mismatch in single-crystal Ni-base superalloys, *Mater. Sci. Eng. A*, A210, 8–15, 1996.

38. D.V. Wilson and Y.A. Konnan, Work hardening of a steel containing a coarse dispersion of cementite particles, *Acta Metall.*, 12, 617–628, 1964.

39. H.P. Cheskis and R.W. Heckel, In-situ measurements of deformation behaviour of individual phases in composites by x-ray diffraction, in *Metal Matrix Composites. 1968*, ASTM STP 438, American Society for Testing and Materials, Philadelphia, PA, 1968, 76–91.

40. J.P. Bonnafe and J.L. Lebrun, Residual stresses in matrix and reinforcement phases of a SiC short fibre reinforced Al matrix composite: application of x-ray diffraction to their determination, in Proceedings, 12th International Risø Conference, 1991, 229–234.

41. H.M. Leadbetter and M.W. Austin, Internal strain (stress) in a Al-SiC$_p$ reinforced composite: an x-ray diffraction study, *Mater. Sci. Eng.*, 89, 53–61, 1987.

42. S.D. Tsai, D. Mahulikar, and H.L. Marcus, Residual stress measurements on Al-graphite composites using x-ray diffraction, *Mat. Sci. Eng.*, 47, 145–149, 1981.

43. T. Hanabusa, K. Nishioka, and H. Fujiwara, Criterion for the triaxial x-ray residual stress analysis, *Z. Metallkd.*, 74, 307–313, 1983.

44. M.R. Watts and P.J. Withers, The interpretation of x–ray stress measurements of metal matrix composites using computer modelling techniques, in Proceedings, 11th International Conference on Composite Materials, M.L. Scott, P.A. Lagace, and A. Poursartip, Eds., Woodhead Pubkishing, 1997, 1821–31.

45. B.D. Cullity, *Elements of X-ray Diffraction*, 2nd ed., Addison-Wesley, Reading, MA, 1978.

46. I.C. Noyan and J.B. Cohen, *Residual Stress — Measurement by Diffraction and Interpretation*, in *Materials Research and Engineering*, B. Ilschner and N.J. Grant, Eds., Springer-Verlag, New York, 1987.

47. R. Sinclair, M. Preuss, E. Maire, J.-Y. Buffiere, P. Bowen, and P.J. Withers, The effect of fibre fractures in the bridging zone of fatigue cracked Ti-6Al-4V/SiC fibre composites, *Acta Mater.*, 52, 1423–1438, 2004.

48. P.J. Cotterill and P. Bowen, Fatigue crack growth in a Ti-15-3 matrix/SCS-6 fibre material matrix composite under tension-tension loading, *Mater. Sci. Technol.*, 12, 523–529, 1996.

49. R. Sinclair, M. Preuss, and P.J. Withers, Imaging and stain mapping in the vicinity of a fatigue crack in a Ti/SiC fibre composite, *Mat. Sci. Tech.*, 2004 (accepted for publication).

50. T. Lorentzen, N.J. Sorensen, and Y.L.Liu, A comparison of numerical predictions and in situ neutron diffraction measurements of MMC phase response, in Proceedings, 15th Risø International Symposium on Numerical Predictions of Deformation Processes and the Behaviour of Real Materials, 1994, 405–412.

51. P.J. Withers, W.M. Stobbs, and O.B. Pedersen, The application of the Eshelby method of internal stress determination for short fibre metal matrix composites, *Acta Metall.*, 37, 3061–3084, 1989.

52. J.F. Nye, *Physical Properties of Crystals — Their Representation by Tensors and Matrices*, Clarendon, Oxford, 1985.

53. M.E. Fitzpatrick, M.T. Hutchings, and P.J. Withers, Separation of macroscopic, elastic mismatch and thermal expansion misfit stresses in metal matrix composite quenched plates from neutron diffraction measurements, *Acta Mater.*, 45, 4867–4876, 1997.

54. D.S. Kupperman, S. Majumdar, J.P. Singh, and A. Saigal, Application of neutron diffraction time-of-flight measurements to the study of strain in composites, in *Measurement of Residual and Applied Stress Using Neutron Diffraction*, M.T. Hutchings and A.D. Krawitz, Eds., NATO ASI Series E: Applied Sciences vol. 216, Kluwer: Dordrecht, 1992, 439–450.

55. P.J. Withers, D.J. Jensen, H. Lilholt, and W.M. Stobbs, The evaluation of internal stresses in a short fibre metal matrix composite by neutron diffraction, in Proceedings, Intl. Conf. on Composite Materials VI/European Conf. on Composite Materials, 1987, 255–264.

56. P.J. Withers and A.P. Clarke, A neutron diffraction study of load partitioning in continuous Ti/SiC composites, *Acta Mater.*, 46, 6585–6598, 1998.

57. R.I. Todd, M.A.M. Bourke, C.E. Borsa, and R.J. Brook, Neutron diffraction measurements of residual stresses in alumina/SiC nanocomposites, *Acta Metal. Mater.*, 45, 1791–1800, 1997.

58. P.J. Withers, H. Lilholt, D.J. Jensen, and W.M. Stobbs, An examination of diffusional stress relief in MMCs, in Proceedings, 9th Risø International Symposium on Metals and Materials Science: Mechanical and Physical Behavior of Metal and Ceramic Composites, 1988, 503–510.

59. X.L. Wang, C.R. Hubbard, K.B. Alexander, P.F. Becher, J.A. Fernandez–Baca, and S. Spooner, Neutron diffraction measurements of the residual stresses in Al_2O_3-ZrO_2 (CeO_2) ceramic composites, *J. Am. Ceram. Soc.*, 77, 1569–75, 1994.

60. B. Derby, Internal stress superplasticity in metal matrix composites, *Scripta Metal.*, 19, 703–707, 1985.

61. G.S. Daehn and T. Oyama, The mechanism of thermal cycling enhanced deformation in whisker-reinforced composites, *Scripta Metal.*, 22, 1097–1102, 1988.

62. M.Y. Wu and O.D. Sherby, Superplasticity in a silicon carbide whisker reinforced aluminum alloy, *Scripta Metal.*, 18, 773–776, 1984.

63. M.R. Daymond and P.J. Withers, Neutron diffraction and laser extensometry in situ monitoring of thermal cycled metal matrix composites, *J. Appl. Composites*, 34, 621–628, 1997.

64. A. Saigal, D.S. Kupperman, and S. Majumdar, Residual strains in titanium matrix high temperature composites, *Mater. Sci. Eng.*, A150, 59–66, 1992.

65. H. Choo, P. Rangaswamy, and M.A.M. Bourke, Internal strain evolution during heating of Ti-6Al–4V/SCS-6 composite, *Scripta Mater.*, 42, 175–181, 2000.

66. G.L. Povirk, M.G. Stout, M. Bourke, J.A. Goldstone, A.C. Lawson, M. Lovato, S.R. MacEwen, S.R. Nutt, and A. Needleman, Mechanically induced residual stresses in Al-SiC composites, *Scripta Metall.*, 25, 1883–1888, 1991.

67. A. Levy and J.M. Papazian, Elastoplastic finite element analysis of short-fiber-reinforced SiC/Al composites: effects of thermal treatment, *Acta Metall. Mater.*, 39, 2255–2266, 1991.

68. M.R. Daymond and P.J. Withers, An examination of tensile/compressive loading asymmetries in aluminum-based MMCs using the finite element method, *Mater. Sci. Tech.*, 11, 228–235, 1995.

69. A.J. Allen, M. Bourke, S. Dawes, M.T. Hutchings, and P.J. Withers, The analysis of internal strains measured by neutron diffraction in Al/SiC MMCs, *Acta. Metall.*, 40, 2361–2373, 1992.

70. J.W.L. Pang, G. Rauchs, P.J. Withers, N.W. Bonner, and E.S. Twigg, Measurement and prediction of residual stresses in Ti MMC rings, *J. Neutron Res.*, 9, 373–380, 2002.

71. R. Vaidyanathan, M.A.M. Bourke, and D.C. Dunand, Texture and strain evolution in superelastic NiTi and NiTi-TiC composites, *Acta Mater.*, 47, 3353–3366, 1999.

72. B.E. Warren and B.L. Averbach, The effect of cold work distortion on x-ray patterns, *J. Appl. Phys.*, 21, 595–599, 1950.

73. R.I. Todd and B. Derby, Deconvolution of measured internal strain distribution functions for Al_2O_3/SiC_P composites from neutron diffraction studies, in Proceedings, Residual Stresses in Composites, The Metallurgical Society-American Institute of Mining, Metallurgical, and Petroleum Engineers, 1993, 147–160.

74. A.D. Krawitz, Stress measurements in composites using neutron diffraction, in *Measurement of Residual and Applied Stress Using Neutron Diffraction*, M.T. Hutchings and A.D. Krawitz, Eds., NATO ASI Series E: Applied Sciences, vol. 216, Kluwer, Dordrecht, 1992, 405–420.

75. M.R. Daymond, M.A.M. Bourke, R.B. Von Dreele, B. Clausen, and T. Lorentzen, Use of Rietveld refinement for elastic macrostrain determination and for the evaluation of plastic strain history from different spectra, *J. Appl. Phys.*, 82, 1554–1556, 1997.

76. R.A. Lebensohn and C.N. Tomé, A self-consistent anisotropic approach for the simulation of plastic deformation and texture development of polycrystals: application to zirconium alloys, *Acta Metall. Mater.*, 41, 2611–2624, 1993.

77. B. Clausen, T. Lorentzen, M.A.M. Bourke, and M.R. Daymond, Lattice strain evolution during tensile loading of stainless steel, *Mater. Sci. Eng.*, A259, 17–24, 1999.

78. B. Clausen and T. Lorentzen, Polycrystal deformation: experimental evaluation by neutron diffraction, *Metall. Mater. Trans.*, 28A, 2537–2541, 1997.

79. C.N. Tomé, Elastic behaviour of polycrystals, in *Encyclopedia of Materials: Science and Technology*, K.H.J. Buschow, et al., Eds., Elsevier, Oxford, 2001, 2409–2415.
80. L. Pintschovius and V. Jung, Residual stress measurements by means of neutron diffraction, *Mat. Sci. Eng.*, 61, 43–50, 1983.
81. A.J. Allen, M.T. Hutchings, C.G. Windsor, and C. Andreani, Neutron diffraction methods for the study of residual stress fields, *Adv. Phys.*, 34, 445–473, 1985.
82. T.M. Holden, A.P. Clarke, and R.A. Holt, Neutron diffraction measurements of intergranular strains in Monel 400, *Metall. Mater. Trans.*, 28A, 2565–2576, 1997.
83. C.N. Tomé and P.A. Turner, A study of residual stresses in Zircaloy-2 with rod texture, *Acta Metall. Mater.*, 42, 4143–53, 1994.
84. T.M. Holden, B.M. Powel, G. Dolling, and S.R. MacEwen. Internal strain measurements in over rolled cold-worked Zr 2.5 wt% Nb pressure tubes by neutron diffraction, in Proceedings, 5th Risø International Symposium on Materials Science, Microstructural Characterization of Materials by Non-Destructive Techniques, 1984, 291–294.
85. T.M. Holden, J.H. Root, V. Fidleris, R.A. Holt, and G. Roy, Applications of neutron diffraction to engineering problems, *Mater. Sci. Forum*, 359, 27–28, 1988.
86. A.M. Korsunsky, M.R. Daymond, and K.E. James, The correlation between plastic strain and anisotropy strain in aluminum alloys polycrystals, *Mater. Sci. Eng.*, A334, 41–48, 2002.
87. M.J. Balart, A. Bouzina, L. Edwards, and M.E. Fitzpatrick, The onset of tensile residual stresses in grinding of hardened steels, *Mater. Sci. Eng.*, 367A, 132–142, 2004.
88. D. Wang, I.B. Harris, P.J. Withers, and L. Edwards, Near-surface strain measurement using neutron diffraction, in 4th European Conference on Residual Stress, Societé Francaise de Metallurgie et de Materials, 1996, 69–78.
89. C.A. Stickels and C.M. Mack, Heat treatment of some temperature-resistant carburizing steels, *J. Heat Treatment*, 4, 223–236, 1986.
90. M.A.M. Bourke, P. Rangaswamy, T.M. Holden, and R. Leachman, Complementary x-ray and neutron strain measurements of a carburized surface, *Mater. Sci. Eng.*, A257, 333–340, 1998.
91. P.J. Withers, Depth capabilities of neutron and synchrotron diffraction strain measurement instruments: Part II — Practical implications, *J. Appl. Crystallogr.*, 37, 596–606, 2004.
92. D. Buttle, P.M. Mummery, and P.J. Withers, Depth profiling of residual stress by magnetic analysis, (in preparation, Univ. Manchester, UK) 2004.
93. M. Preuss, J. Pang, P.J. Withers, and G.J. Baxter, Inertia welding nickel-based superalloy, Part I: Metallurgical characterization, *Metall. Mater. Trans.*, 33A, 3215–25, 2002.
94. G.A. Webster, M.A.M. Bourke, H.J. MacGillivray, P.J. Webster, K.S. Low, D.F. Cannon, and R.J. Allen, Measurement of residual stresses in railway rails, in Proceedings, 2nd International Conference on Residual Stresses, C. Beck, S. Denis, and A. Simon, Eds., Elsevier Applied Science, London, 1989, 203–208.
95. P.J. Webster, Neutron strain scanning, *Neutron News*, 2, 19–22, 1991.
96. J. Kelleher, M.B. Prime, D. Buttle, P.M. Mummery, P.J. Webster, J. Shackelton, and P.J. Withers, The measurement of residual stress in railway rails by diffraction and other methods, *J. Neutron Res.*, 11, 187–193, 2004.

7

The Future

This book has attempted to provide an introduction to the use of neutron diffraction for strain and stress measurement in samples of crystalline materials. It has emphasized the progress made over the past 20 years or so, and pointed to where the development and understanding of concepts is not complete. There are indeed many important areas of future work and development for both the experimentalist and theoretician alike. In particular, a thorough understanding of intergranular stresses in the plastic regime is required, as is understanding of many aspects of the details of stress in multiphase materials, understanding of the contributions to the anomalous effects due to partially filled instrumental gauge volumes, and measurement of strain-free lattice constants. These areas fall between the traditional lines of the physicist and materials scientist, and require an important input from the engineer, both as to his/her requirements and his/her ultimate use of the technique in order to improve the manufacture and life expectancy of industrial components.

The applications described in previous chapters were chosen to illustrate the capabilities of the neutron strain mapping technique, and the precautions that must be taken in various circumstances likely to ensure that accurate and reliable measurements and interpretation are made. Guides to future development were given for a number of application areas. There is little doubt that in forthcoming years the number and breadth of studies are set to increase sharply.

Two recent developments in particular point to further growth in the application of neutron strain mapping to materials development and engineering design.

- The development of a new generation of both pulsed and constant-flux strain mapping instruments. These will see an increase in the size and weight of samples that can be studied in more complex environments, more accurate sample location procedures using laser positioning systems or co-ordinate measurement machines, and better gauge definition for improved measurement accuracy. Software developments will accelerate analysis procedures and simplify instrument operation and allow virtual experimentation to steer data acquisition strategies. Furthermore, they promise better strain resolution and reduced data acquisition times relative to existing instruments. These aspects will no doubt benefit from parallel development of synchrotron x-ray techniques, where instrument positioning times are often so much longer than actual counting times that it is imperative to reduce the former. Engineering applications require that improving the cost-effectiveness of measurements be of a high priority.

- The definition of measurement *standards* for the repeatable and accurate measurement of strain by neutron diffraction. This will accelerate the uptake of neutron strain measurements into engineering usage.

The new generation of strain measurement instruments will open up new avenues of research and development. Most notably, through submillimeter spatial resolution, faster, more economic area scanning, and the capability to study more realistic engineering conditions. Many phenomena occur at time scales too short to study using today's instruments, such as phase transformations, thermal and mechanical fatigue effects, production processes, and so on. With accredited standards for measurement, area mapping of strain and stress promises to play an increasing role in the validation and definition of finite element models of the behavior of engineering components and of engineering processes, such as joining, milling, and sintering, among others. Another field that as yet has received little attention but will no doubt be important in the future, is the extension to studies of physical phenomena coupled to stress, such as in the behavior of ferroelectrics, piezoelectrics, and shape memory materials. Furthermore, increased interaction with other complementary measurement techniques promises to provide extra insight into, for example, simultaneous measurement of texture and stress as a function of position, and of imaging stress (by diffraction) and structure (by tomography).

Symbols and Abbreviations

The following table provides a summary of the symbols and abbreviations used in this book. The section in which each first appears is given. Vectors are in boldface type, and tensors and matrices in double-underlined boldface. Symbols for variables only used once and defined in the text are not included here.

Symbol	Meaning	Section
α_1	source–monochromator collimation angle	3.2.1
α_2	monochromator–sample collimation angle	3.2.1
α_3	sample–detector collimation angle	3.2.1
α, β, γ	angles between unit cell axes	2.3.1
γ	anisotropy strain parameter	5.6.6
γ^{el}	elastic anisotropy parameter	5.6.6
γ^{pl}	plastic anisotropy parameter	5.6.6
γ_{ij}	ij component engineering shear strain	5.1.1
$\delta(x)$	Dirac delta function	2.3.3
$\Delta C(x_i)$	uncertainty in count $C(x_i)$ at point x_i	4.3.2
$\Delta_u(x_0)$	uncertainty in measurement of x_0	4.4.1
Δx	increment in x_i in profile scan	4.4.1
$\varepsilon(l,m,n)$	continuum strain component in direction with cosines l,m,n	5.1.1
$\underline{\underline{\varepsilon}}$	continuum strain tensor	5.1.1
ε_{ij}^{hkl}	ij component of lattice strain tensor for hkl reflection	1.4.2, 5.1.2
$\underline{\underline{\varepsilon}}_c^P$	plastic strain in each grain of EPSC model	5.6.2
$\varepsilon_{ij}^{Dev.}$	deviatory strain tensor component	5.1.1
$\varepsilon_{i_D i_D}$	principal strain component	5.1.1
η_f	fraction of epithermal flux	3.1.1
η_M	mosaic spread of the monochromator	3.2.1
θ_c	critical angle for neutron total external reflection	3.1.1
θ_D	Debye temperature	2.5
θ_{hkl}^B	Bragg angle (from sample)	1.2, 2.3.2
θ^M	Bragg angle at monochromator	3.2.1

Symbol	Meaning	Section
$\phi^s = 2\theta$	scattering angle	2.3.2
λ	neutron wavelength	1.2, 2.1
λ_T	neutron wavelength corresponding to energy at peak of flux distribution as function of energy	3.1.1, A.1.3
λ_c	wavelength below which there is low transmission down a guide tube	3.1.1
λ_{hkl}	wavelength corresponding to hkl reflection in tof pattern	3.2.2
λ_e^{hkl}	wavelength corresponding to Bragg edge of hkl planes	2.3.3
μ	continuum shear modulus	5.1.1
μ^X	bulk shear modulus calculated from theory $X = V, R,$ or K	5.4
μ_a	linear attenuation coefficient	2.4
$\mu\varepsilon$	unit of strain = 10^{-6} (pronounced microstrain)	4.6.2
ν	wave frequency	2.1
ν	continuum Poisson's ratio	5.1.1
ν^X	bulk Poisson's ratio calculated from theory $X = V, R,$ or K	5.4
ν_{hkl}	plane-specific Poisson's ratio	5.1.2
ν_{hkl}^X	Poisson's ratio of polycrystalline hkl reflection calculated from theory $X = V, R, K$	5.5
ρ_j	density of element j	2.3.7
ρ_n	number density of atoms	3.1.1
σ^{coh}	atomic coherent scattering cross-section	2.2.1
σ^{incoh}	atomic incoherent scattering cross-section	2.2.1
σ_λ^{abs}	atomic absorption cross-section for neutrons of wavelength λ	2.2.1
$\sigma_{i_D i_D}$	principal stress component	5.1.1
$\sigma_{ij}^{Dev.}$	deviatory stress tensor component	5.1.1
$\langle \underline{\underline{\sigma}} \rangle_i^{II}$	mean Type II stress for phase i (excluding applied macrostress)	1.4.2
$\bar{\underline{\underline{\sigma}}}_i$	average, or total, stress for phase i (including macrostress)	1.4.2
$\underline{\underline{\sigma}}_i$	continuum stress tensor	1.4.1, 5.1.1
$\underline{\underline{\sigma}}^i$	Type i stress tensor, $i = I, II, III$	1.4.2
$\underline{\underline{\sigma}}^{Macro}$	macrostress (Type I stress) tensor	6.3.3
$\underline{\underline{\sigma}}_{hkl}$	stress tensor from hkl lattice planes	1.4.2
σ^A	uniaxial applied stress	5.5.1
$\sigma_{0.2}$	0.2% macroscopic yield stress	5.6.4
σ	atomic total scattering cross section	2.2.1
$d\sigma/d\Omega$	differential atomic scattering cross-section	2.2.1
Σ^{coh}	macroscopic coherent scattering cross-section per unit volume	2.3.7

Symbol	Meaning	Section
Σ^{incoh}	macroscopic incoherent scattering cross-section per unit volume	2.3.7
Σ^{abs}	macroscopic absorption cross-section per unit volume	2.3.7
$d\Sigma/d\sigma$	macroscopic differential scattering cross-section	2.2.1
τ^i	critical resolved shear stress	5.6.2
τ_{ij}	shear stress component	5.1.1
ϕ, ϕ'	scattering angle (from sample)	2.3.2
ϕ	azimuthal angle between diffraction plane normal and the x in-plane principal axis	5.1.3
ϕ^M	monochromator scattering angle	3.2.2
$\phi(\theta_D/T)$	Debye function	2.6
Φ_0	incident neutron flux	2.4
Φ_l	flux after path length l	2.4
$\Phi(x)$	flux distribution as function of $x = v$, E, or λ	3.1.1, A.1.2
Φ_F	total flux in neutron beam	A.1.2
χ^2	goodness-of-fit parameter	4.3.2
ψ	polar angle between diffraction plane normal and plate normal	5.1.3
ψ^*	point of invariance	4.6.3
Ω	sample rotation axis normal to scattering plane	3.6.3
$d\Omega$	solid angle	2.2.1
Ω	crystallite, or grain, identity	5.4.2
A_{hkl}^{abs}	angle-dependent absorption factor	2.3.5
A_{hkl}	orientation dependency of elastic anisotropy	5.3.1
A, B, C	reciprocal lattice vectors	2.3.3
a_j	atomic fraction of atom j	2.3.6
a, b, c	unit cell parameters also termed lattice constants or parameters	2.3.1
a, b, c	lattice vectors	2.3.1
a_i^I	relative abundance of isotope i	2.2.1
a^0, b^0, c^0	strain-free lattice parameters	4.6.1
$\bar{b}$	average nuclear scattering length	2.2.1
B_0	background count	4.4.1
B_0	background count rate	4.4.2
c	speed of light in vacuo (2.9979×10^8 m/s)	2.1
$C(x_i)$	number of counts recorded for diffraction peak at point x_i	4.3.2
$\underline{C}$	stiffness tensor	5.1.1
C_{ijkl}	component of stiffness tensor	5.1.1
C_{mn}	component of stiffness in contracted matrix notation	5.3

Symbol	Meaning	Section
d_{hkl}	lattice spacing for *hkl* planes	1.2, 2.3.1
d^0_{hkl}	strain-free *hkl* lattice spacing	1.2, 3.2.1
$\Delta d/d$	lattice strain; resolution	3.2.2, 3.4.1
d_{CS}	depth of SGV centroid from sample surface	3.6.2
d_{CR}	distance from reference point to centroid of SGV	3.6.2
d_{SR}	distance of reference point from sample surface	3.6.2
E	continuum Young's modulus	5.1.1
E^X	bulk Young's modulus calculated from theory $X = V, R,$ or K	5.4
E_{hkl}	plane-specific Young's modulus	5.1.2
E^x_{hkl}	Young's modulus for polycrystalline *hkl* reflection calculated from theory $X = V, R,$ or K	5.5
E	neutron energy	2.1
f_i	volume fraction of composite phase i	5.1.3
f	volume fraction of reinforcing phase in two-phase composite.	6.3.1
$F(hkl)$	structure factor	2.3.3
$F(x_i)$	theory (fitted) value of diffraction peak profile at point x_i	4.3.1
G_{hkl}	reciprocal lattice vector	2.3.3, 3.2.1
h	Planck's constant (6.6261×10^{-34} Js)	2.1
$\hbar$	$h/2\pi$ (1.0546×10^{-34} JS)	2.1
h_d	detector height	2.3.5
h_γ	hardening parameter	5.6.2
hkl	Miller indices	2.3.1
H_0	theory (fitted) value of peak height	4.3.1
H_0	rate of acquisition of peak height	4.4.2
I_C	total number of signal counts in measured peak	4.4.1
I_c	measured peak summed count rate	4.4.2
I	theory (fitted) integrated, diffraction peak intensity	4.3.1
$I(t)$	intensity profile on a *tof* instrument	3.2.2
j	multiplicity of crystal plane	2.3.4
k_i	incident neutron wave vector	2.3
k_f	scattered neutron wave vector	2.3
k_B	Boltzmann's constant (1.3807×10^{-23}J$/K$)	3.1.1
K^X	bulk modulus calculated from theory $X = V, R,$ or K	5.4
K	stress intensity factor	
ΔK	stress intensity factor range	1.5.2
K_{min}, K_{max}	minimum and maximum stress-intensity factor	1.5.2

Symbol	Meaning	Section
l_μ	1/e attenuation length, or penetration depth	2.4
l	neutron path length through sample	2.4
l_p	interparticle spacing in composites	6.3.2
l	lattice translation vector	2.3.1
l,m,n	direction cosines	5.1.1
l_{1j}	$j = 1,2,3$ direction cosines between direction 1 and crystal axes	5.3.1
L_i	moderator–sample distance	3.2.2
L_f	sample–detector distance	2.3.5, 3.2.2
L	total neutron flight path	2.3.2
l_0^j	characteristic size of misfitting region over which the residual stress Type j = I, II, III varies rapidly	1.4
L_0^j	characteristic distance over which Type j = I, II, II residual stress self equilibrates	1.4
m_n	neutron rest mass (1.6749×10^{-27}kg)	2.1
M	atomic mass	2.5
n	number of points	4.3.2
N	number of planes	2.3.2
N_A	Avagadro's number (6.0221×10^{23} mole^{-1}).	2.3.7
N_c	number of unit cells in crystal.	2.3.2
N_a	number of atoms per unit volume	2.2.1
N_f	number of cycles to failure	1.5.2
$\hat{n}_A$	unit vector normal to area A	1.4.1
n_{hkl}	number of neutrons scattered by *hkl* planes per unit of time falling on detector	2.3.5
$n(\lambda)$	neutron refractive index of surface material	3.1.1
$n_d(v)$	neutron velocity distribution function	A2.1
n_D	number of neutrons/unit volume in source	A2.1
p	neutron momentum	2.1
P_1,P_2,P_3	sample coordinate axes	5.1.3
Q	scattering vector	2.3
r_u	position of atom in basis unit cell	2.3.3
R_i	position of atom i in real space	2.3.3
S_1, S_2	x-ray or diffraction elastic constants	5.1.3
$\underline{\underline{S}}$	compliance tensor	5.1.1
S_{ijkl}	component of compliance tensor	5.1.1
$\underline{\underline{S}}'_{ijkl}$	component of transformed compliance tensor	5.3.1

Symbol	Meaning	Section
S_{mn}	component of compliance in contracted matrix notation	5.3
$\underline{\underline{S}}^E$	Eshelby tensor	5.6.2
t_c	thickness of perfect crystal	2.3.2
t_s	thickness of plate sample	2.3.5
t_M	measurement time for n points	3.4.1, 4.4.1
t	time of flight (tof)	2.3.2
t_{hkl}	time of flight corresponding to *hkl* peak in tof pattern	3.2.2
T	temperature	3.1.1
T	sample translator position	3.6.2
u_x	standard deviation of Gaussian function	4.3.1
$<u^2>$	mean square amplitude of vibrating atom	2.5
v_0	volume of unit cell	2.3.3
v	neutron velocity vector	2.1
v_T	neutron velocity at peak of velocity distribution	A.1.1
V_0^j	characteristic volume over which the Type j = I, II, III residual stresses self-equilibrate	1.4
V_v	sampled gauge volume	1.4.2, 3.6.1
V_g	instrumental gauge volume	2.3.5, 3.5.2
V	total volume of sample	1.4.1
w_j	weight fraction of atom *j*	2.3.7
W_j	atomic weight of atom *j*	2.3.7
$e^{-2W(Q)}$	Debye–Waller factor	2.5
x_i	angle, or time, point on scan of peak profile	4.3.1
x,y,z	coordinate axes, relative to sample	5.1.1
Z	atomic number	2.2.2

Abbreviation	Meaning	Section
ALARA	as low as reasonably achievable	3.1.3
CPFEM	crystal plasticity finite element modeling	5.6
CVD	controlled vapor deposition	2.6.2
DEC	diffraction elastic constant	5.1.2, 5.1.3
DSW	diffracted-beam slit width	3.2.1
EBW	electron beam weld	6.2.2
EDM	electrodischarge machining	4.6.3
EPSC	elastoplastic self-consistent	5.6
FE	finite element	4.6.3

Abbreviation	Meaning	Section
FoM	figure of merit	3.4.1, 4.4.2
FSS	Fourier strain spectrometer	3.7.3
FSW	friction stir welding	6.2.4
FWHM	full-width half-maximum	3.4
HAZ	heat-affected zone	4.3.6.2
HCF	high-cycle fatigue	1.5.2
HEM	homogeneous effective medium	5.4.2
HFR	high flux reactor	3.1.1
HIP	hot isostatic pressing	6.6.1
HRPD	high-resolution powder diffractometer	3.4.2
IGV	instrumental gauge volume	3.5.2
ISW	incident-beam slit width	3.2.1
LCF	low-cycle fatigue	1.5.2
LINAC	linear accelerator	3.1.2
LSND	low stress, no distortion	6.2.5
MMC	metal matrix composite	1.3
MTR	materials testing reactor	3.1.1
NGV	nominal gauge volume	3.5.2
NTP	normal temperature and pressure	2.1
ODF	orientation distribution function	2.6.2
PDE	positional discrimination effect	3.6.4
PG	pyrolytic graphite	3.2.1
PSD	positional sensitive detector	3.2.1
PWHT	post-weld heat treatment	6.6.2
QA	quality assurance	4.7.1
rms	root mean square	5.7
RTOF	reverse time of flight	3.7.3
SGV	sampled gauge volume	3.6.1
SNS	spallation neutron source	1.3
TDS	thermal diffuse scattering	2.5
TIG	tungsten inert gas	4.6.3
TMAZ	thermomechanically affected zone	6.2.4
tof	time of flight	3.2.2
UTS	ultimate tensile strength	1.5.2
WE	wavelength effect	3.6.4
XEC	x-ray elastic constant	5.1.3

Glossary

The meanings of certain terms used in this book are not yet universally accepted. To avoid confusion, the definitions of such terms as used in this book are summarized here, along with the section in which each is first used (in parentheses following term).

Shift (1.2)   Difference between the measured Bragg scattering angle ϕ^s, or diffraction angle θ^B, from the stressed sample, and that measured from the reference sample, or the corresponding difference in time of flight, wavelength, or lattice spacing.

Sample (1.2.1)   Test component under study for strain measurement by neutron diffraction.

Engineering materials (1.2.1)   Materials used in industrial plant or components.

Lattice strain (1.2)   Elastic strain measured by diffraction from defined lattice planes.

Reference axis (3.5.1)   Axis of instrument at the intersection of incident and scattered neutron beams, and about which the instrument components in these beams rotate.

Reference point (3.5.1)   Point on the reference axis that is at the center of the instrumental gauge volume, and to which the position of the sample may be referred.

Nominal gauge volume (NGV) (3.5.2)   Geometric instrumental gauge volume calculated from the aperture dimensions, assuming non-divergent beams.

Instrumental gauge volume (IGV) (3.5.2)   Effective measurement gauge volume, measured or calculated to include all divergent beams. Due to the penumbra the IGV is larger than the NGV. Conventionally, the IGV is defined in terms of a contour map of intensity contribution.

Sampled gauge volume (SGV) (3.6.1)   Part of the instrumental gauge volume from which a measurement of the sample diffraction peak is obtained. The SGV is the volume of sample over which the strain measurement is averaged. The measured strain is affected by partial filling of the IGV, wavelength distribution of incident neutrons, attenuation of neutrons within the sample, and texture gradient in the sample.

Effective centroid of SGV (3.6.1 and 3.6.2) Calculated or modeled weighted average position in the SGV giving rise to the diffraction peak, including effects of wavelength, texture, and attenuation distribution. It is the most representative position of strain measurement in the sample.

Surface effects (3.6.4) Used as a generic term for all possible errors introduced when measuring near surfaces, either in the value of strain measured or the location of measurement, or both.

Anomalous shifts (3.6.4 and 4.2.2) Apparent shifts in angle, time of flight, and so on due to uncorrected geometric effects, such as those from partially filled gauge volumes, the effects of an incorrect reference value of d^0, or contaminating Bragg edges.

Wavelength effect (WE) (3.6.4) Contribution to an anomalous shift in diffraction angle (etc.) due to the variation in wavelength over the SGV, when a sample partially fills the IGV. Even if the IGV is filled by the sample, a WE anomalous shift may occur due to attenuation effects.

Positional discrimination effect (PDE) (3.6.4) Contribution to an anomalous shift in diffraction angle (etc.) due to the mean position of the diffracting grains in a sample, that is, the position of the centroid of the SGV relative to the instrument reference point, in a partially filled IGV. Even if the IGV is filled by the sample, a PDE may occur.

Spurious strain (3.6.4) The apparent lattice strain that may be deduced from anomalous shifts, but which are not physically present.

Spurious location (3.6.4) An incorrect position of measurement in the sample, which typically does not take into account the effects of partially filled IGV, wavelength, texture, or attenuation distribution.

Pseudo strain (5.6.5) Used only for a residual strain that may be measured correctly, that is, after correction for spurious strains, but differs from the true macroscopic residual strain solely because the effect of intergranular strains has not been taken into account.

Macroscopic plastic strain (5.6.5) Deviation from elastic linearity of the bulk macroscopic strain on loading.

Intergranular strain (5.2) Elastic strain arising from misfits between grains. In effect, it is the difference in strain between the grain and the polycrystalline aggregate.

Elastic anisotropy (4.1) Anisotropy in the linear lattice strain response of differently oriented crystals, or crystallites in a polycrystalline sample, to an applied load in the elastic regime due to anisotropic crystalline elastic properties.

Plastic anisotropy (4.1)   Anisotropy in the lattice response of differently oriented crystallites in a polycrystalline sample to an applied load in the plastic regime, due to effects of intergranular strains and the anisotropic nature of crystallite slip planes.

Average (or total) stress $\bar{\sigma}_i$ (1.4.2 and 6.3.3)   Stress averaged over the SGV as deduced from the average strain, as measured by diffraction from phase i. It is the sum of the macrostress and mean phase i stress.

Mean phase stress $\langle \underline{\underline{\sigma}} \rangle_i^{II}$ (1.4.2 and 6.3.3)   Mean stress in phase i over the SGV, excluding any macrostress.

Economically feasible path length (4.4.2)   Maximum total path length of incident and diffracted beams in a sample that will allow sufficient peak intensity to give a strain measurement to an accepted accuracy. It takes account of the scattering cross-section, background counts, and attenuation of the beam.

Figure of merit (3.4.1 and 4.4.2)   For an instrument, the inverse of the time to taken to measure a lattice spacing to a given accuracy.

Appendix 1

Note on Reactor Flux Spectrum

The distribution of thermalized neutron flux can be expressed in terms of the neutron velocity, energy, or wavelength. The form of the distribution is slightly different according to which form is chosen; the basic formulas given in this Appendix follow the approach of Lomer and Low [1].

A.1.1 Velocity Distribution

The neutron velocity distribution $n_d(v)$ in equilibrium with a moderator at temperature T is given by the Maxwell–Boltzmann distribution. $n_d(v)dv$ is the number of neutrons with velocities between v and v + dv.

$$n_d\left(v\right) = \frac{4n_D}{\pi^{1/2}}\left[\frac{v^2}{v_T^3}e^{-\frac{v^2}{v_T^2}}\right] \qquad \text{(A.1.1)}$$

Here, n_D is the number of neutrons/unit volume. v_T is the equilibrium velocity at which this distribution peaks:

$$v_T = (2k_B T/m_n)^{1/2} \qquad \text{(A.1.2)}$$

A.1.2 Flux Distribution

The distribution of flux Φ in terms of velocity, is given by:

$$\Phi(v) = v.n_d(v) \qquad \text{(A.1.3)}$$

It can be expressed in terms of energy, E, or wavelength, λ, as variable from the relations:

$$\Phi(v)dv = \Phi(E)dE = \Phi(\lambda)d\lambda \qquad (A.1.4)$$

where the variables v, E, and λ are related by:

$$E = m_n v^2/2, \quad \lambda = h/m_n v \qquad (A.1.5)$$

A.1.2.1 Flux in Terms of Neutron Velocities

The reactor flux spectrum may be given as a function of v, $\Phi(v)$, where $\Phi(v)$ dv is flux of neutrons with velocity between v and $v + dv$. From $\Phi(v) = v.n_d(v)$, it follows that:

$$\Phi(v) = 2\Phi_F \left[\frac{v^3}{v_T^4} e^{-\frac{v^2}{v_T^2}} \right] \qquad (A.1.6)$$

Here, Φ_F is the total flux, $\Phi_F = \dfrac{2n_D}{\pi^{1/2}} v_T$.

A.1.2.2 Flux in Terms of Neutron Energies

The flux spectrum can be given as a function of E, $\Phi(E)$, where $\Phi(E)dE$ is the flux of neutrons with energies between E and $E + dE$:

$$\Phi(E) = \Phi_F \left[\frac{E}{(k_B T)^2} e^{-\frac{E}{k_B T}} \right] \qquad (A.1.7)$$

A.1.2.3 Flux in Terms of Neutron Wavelength

The flux spectrum can be expressed as a function of λ, $\Phi(\lambda)$, where $\Phi(\lambda)d\lambda$ is the flux of neutrons with wavelengths between λ and $\lambda + d\lambda$:

$$\Phi(\lambda) = 2\Phi_F \frac{\lambda_T^4}{\lambda^5} e^{-\frac{\lambda_T^2}{\lambda^2}} \qquad (A.1.8)$$

where $\lambda_T = h/m_n v_T$.

A.1.3 Peak in Flux Distributions

The maximum flux per velocity interval is given by a maximum of $\Phi(v)$ at

$$v = v(maxflux) = (3\ k_B T/m_n)^{1/2} \tag{A.1.9}$$

The maximum flux per energy interval given by the maximum of $\Phi(E)$ at

$$E = E(maxflux) = E_T = k_B T \tag{A.1.10}$$

with corresponding $\lambda_T = h/(2m_n k_B T)^{1/2}$, and $v_T = (2\ k_B T/m_n)^{1/2}$.
The maximum flux per wavelength interval is given by a maximum of $\Phi(\lambda)$ at

$$\lambda = \lambda(maxflux) = h/(5m_n k_B T)^{1/2} \tag{A.1.11}$$

So we have for a moderator at temperature T:

$$\lambda(maxflux) = \lambda_T\ (2/5)^{1/2} = 0.632\ \lambda_T \tag{A.1.2}$$

where λ_T is the wavelength at which the flux spectrum versus energy peaks.

The peak values of the variables are summarized in Table A.1.1. They may be changed a little by the collimation. The function used depends on the application. For example, when considering a crystal monochromator, the distribution $\Phi(\lambda)$ should be used, since a slice $d\lambda = 2d\cos\theta^M.d\theta^M$ of this flux spectrum is taken, with $d\theta^M$ being the effective collimation angle in the collimated beam.

TABLE A.1.1

Velocities, Energies, and Wavelengths at Peak of Flux from Maxwell–Boltzmann Velocity Distribution, Expressed as a Function of Each Variable

Variable	v at Peak	E at Peak	λ at Peak
v	$(3\ k_B T/m_n)^{1/2}$	$3\ k_B T/2$	$h/(3m_n\ k_B T)^{1/2} = 0.816\ \lambda_T$
E	$v_T = (2\ k_B T/m_n)^{1/2}$	$E_T = k_B T$	$\lambda_T = h/(2m_n\ k_B T)^{1/2}$
λ	$(5\ k_B T/m_n)^{1/2}$	$5\ k_B T/2$	$h/(5m_n\ k_B T)^{1/2} = 0.632\ \lambda_T$

Reference

1. W.M. Lomer and G.G. Low, in *Thermal Neutron Scattering*, P.A. Egelstaff, Ed., Academic Press, New York, 1965, chapter 1.

Appendix 2

Relation Between the Centroid of Sampled Gauge Volume and Translator Reading

In this Appendix the expressions used to calculate the curves in Figures 3.25 and 6.40 are given. The relevant distances are defined in Figure 3.24. Figure 3.25 shows the relationship between the position of the centroid of the sampled gauge volume relative to the surface and the translator encoder reading, T, giving the depth, $|d_{CS}|$, of the effective measurement position below the surface. The geometry is for a sample scanned into the instrument nominal gauge volume in the reflection configuration. The nominal gauge volume (NGV) is defined by two slits of width ISW = DSW = w. T is set equal to zero when the surface is at the instrument reference point at the center of the NGV, that is $T_R = 0$, so that $T = d_{SR}$. The half-diagonal of the NGV normal to the surface, D, is given by $D = w/2\cos\theta$. d_{CS} is by definition always negative in Figure 3.24.

The total neutron path length through the material for diffraction from an element of the sample at depth y is $2y/\sin\theta$. The linear attenuation coefficient is μ_a, and we define $a = 2\mu_a/\sin\theta$, and $z = D + d_{SR}$. The relationship between the variables is shown in Figure A2.1.

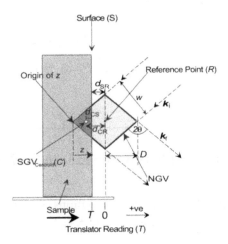

FIGURE A2.1
Schematic illustration showing the definition of the variables used in the expressions below. The translator reading T = d_{SR} and is negative as shown. d_{CS} is always negative in the convention used.

A.2.1   Depth of Centroid of Sampled Gauge Volume from the Surface

1. If the sample is completely outside the nominal gauge volume, that is, $d_{SR} \leq -D$, then clearly no measurement can be made. For $d_{SR} \leq -D$, or equivalently $z \leq 0$, then

$$d_{CS} = 0 \tag{A.2.1a}$$

2. If the sample is less than half inside the nominal gauge volume, that is, $-D \leq d_{SR} \leq 0$, or equivalently $0 \leq z \leq D$, then

$$|d_{CS}| = \frac{\left[(2+az)e^{-az} - (2-az)\right]}{a\left[e^{-az} + az - 1\right]} \tag{A.2.1b}$$

3. If the sample is more than half inside the nominal gauge, that is, $0 \leq d_{SR} \leq D$ or equivalently $D \leq z \leq 2D$, then

$$|d_{CS}| = \frac{\left[(2+az)\cdot e^{-az} + (2-az+2aD) + (2aD-2az-4)\cdot e^{-ad_{SR}}\right]}{a\left[\left(e^{-az} - az + 1\right) + 2\left(a\cdot D - e^{-ad_{SR}}\right)\right]} \tag{A.2.1c}$$

4. If the sample is completely inside the nominal gauge, that is, $d_{SR} \geq D$, or equivalently $z \geq 2D$, then

$$|d_{CS}| = d_{SR} + \frac{\left[(2+aD)e^{-2aD} + (2-aD-4e^{-aD})\right]}{a\left[1+e^{-2aD} - 2e^{-aD}\right]} \tag{A.2.1d}$$

A.2.2   Intensity as Sample is Scanned into the Nominal Gauge Volume

The relative intensity, $I(z)$ for the four situations is given by the following expressions, where $B = 2\cot\theta/a^2$, and $z = D + d_{SR}$:

$d_{SR} \leq -D$, or equivalently $z \leq 0$,

$$I(z) = 0 \tag{A.2.2a}$$

$-D \le d_{SR} \le 0$, or equivalently $0 \le z \le D$,

$$I(z) = Be^{-az}\left[e^{az}(az-1)+1\right]$$

(A.2.2b)

$0 \le d_{SR} \le D$, or equivalently $D \le z \le 2D$,

$$I(z) = Be^{-az}\left[1-2e^{aD}+e^{az}(2aD-az+1)\right]$$

(A.2.2c)

$d_{SR} \ge D$, or equivalently $z \ge 2D$,

$$I(z) = Be^{-az}\left[1-2e^{aD}+e^{2aD}\right]$$

(A.2.2d)

Appendix 3

Points for Consideration When Making a Neutron Diffraction Stress Measurement

A.3.1 Preliminary Considerations

In this section, a summary is provided of points that need careful consideration before and during a set of measurements made using neutron diffraction to determine the stress field in a sample of an engineering component. Most of these points have been discussed in Chapters 3 and 4, and are listed here to provide a more formal approach to addressing a strain/stress measurement problem.

A.3.2 Residual Stress Measurement Issues

The following basic questions to be addressed when embarking on a series of stress measurements are pertinent regardless of the measurement technique.

1. *For the sample under consideration, what measurement of stress is required and to what accuracy?* Neutron diffraction tends to measure Type I macrostress, and phase- or grain-averaged Type II microstress. It is usually routine to measure strain to 100×10^{-6} accuracy, and with millimeter spatial resolution. A spatial resolution of a third of a millimeter is about the current technological limit. Sampled volumes are normally considerably greater than 1 mm^3 (Section 1.4).

2. *Of what material is the sample made?* Neutron powder diffraction is confined to crystalline materials with grain sizes of less than 100 microns. Diffuse scattering from H may make the measurement of strain in crystalline polymers impractical, but most other materials

are suitable unless they have high-volume fractions of absorbing atoms such as B, Cd, and Gd (Section 2.2).

3. *At exactly which positions in the sample is the stress to be determined?* A good understanding of the sample geometry and key locations of measurement is needed before undertaking measurements (Section 3.6).

4. *What components of stress, and therefore strain, should be measured?* Keep in mind that even if the directions of principal axes are assumed, at least three components of strain are required to determine the triaxial stress at each position, and that the errors in each strain measurement accumulate when converting to stress. If the object of measurement is to compare with model predictions, it is better to convert the modeled stresses to strains for comparison, rather than the measured strains to stress (Section 5.1).

5. *Are the principal axes known? If not, should they be determined?* In some cases the principal stresses can be deduced from the symmetry, but if they are unknown at least six directions of strain measurement are required at each position (Section 5.1).

6. *What sampled gauge volume is required?* Small gauge volumes increase the measurement time. If possible, estimate the likely stress gradients and enlarge the gauge dimensions in directions over which only gradual changes in strain are likely (Section 3.6).

A.3.3   Suitability of Neutron Diffraction

You must determine whether neutron diffraction is the best technique to facilitate obtaining the required information. You may find that it is possible to obtain only some, but not all, of the required information. It may be too expensive or take too long to collect all the data required.

1. *Is the sample capable of being taken to a neutron source? Can it physically be accommodated on a diffractometer?* As yet, there are no portable neutron diffraction stress measurement facilities. The size, weight, and complexity envelope of samples that can be investigated are constantly being extended with new dedicated instruments. Many can accommodate samples whose longest dimension is less than a meter, although it may be difficult to measure stress in such a sample at all positions. Weight limits in many cases range from 250 to 1000 kg.

2. *Can the measurements be justified?* Often an underlying consideration is the cost of the measurements, and whether this is justified by the

anticipated results. The number of measurement positions in a day can vary from hundreds to just one or two, depending on the depth of the required position of measurement. Most measurements take tens of minutes. At the present time, commercial daily beam costs range from about $4,000 to $15,000 for confidential work. For openly publishable research, the question is rather one of whether the science and engineering case is strong enough to win beam time under peer review competition. Careful planning can optimize time allocated, so it is necessary to find out as much as possible about the instruments before starting the experiments. Carry out as much testing as possible "off-line". The accommodation, interface, and control of any environmental or loading rigs can be prepared ahead of time during periods when the source is off and the instrument idle.

3. Several questions about the sample material under examination must be asked:

a. *What is the material's density?* This will affect the weight to be accommodated and ease of orientation.

b. *What is the neutron absorption and overall attenuation length in the material?* Besides absorption, incoherent scattering and strong coherent scattering can cause high attenuation of the beam and thus make deep measurements difficult.

c. *Is the material's coherent scattering length high, medium or low?* It should be kept in mind that some elements have negative scattering lengths, such as Ti and Mn. When alloyed with most other elements with positive scattering lengths the average scattering length can be reduced to give weak coherent Bragg peaks and thus long count times. This occurs for Ti-6Al-4V alloys, for example.

d. *What is the crystal structure?* For some structures, those with a basis of several atoms at each primitive unit cell point, the Bragg intensities will vary through the structure factor. An intense Bragg peak must be used for measurements, consistent with a knowledge of the elastic constants.

e. *Are the lattice elastic constants of the material known?* There are now a considerable number of measurements of lattice elastic constants for common material, but if these are not known a calibration experiment may be necessary.

f. *What is the grain size of the material?* A conservative estimate would say that grains should typically be a few tens of microns at maximum size in order give a fully random orientation distribution in a sampled 1 mm cube gauge volume. Plastic working of a sample may break up individual grains into smaller mosaics. As a test, a fine scan through 1 to 2° of the sample angle Ω about an axis normal to the scattering plane should give a smooth variation of a Bragg peak intensity. Alternatively, the sample can be scanned in small steps through the IGV. A sign that the grain size is too small is given by large fluctuations in integrated diffraction peak intensity from measurement to measurement (Section 4.2.1).

g. *Is the material textured?* A scan of the sample angle Ω over a wide range of angles can reveal the extent of texture. Texture may vary with position in a sample, especially near welds. It will affect the directions in which strain can be measured using one reflection, and can affect the elastic constants used to convert strain to stress (Section 4.2.1).

h. *How is the stress-free reference lattice spacing d^0 determined?* A good strain-free reference is needed for absolute strain measurements. The optimum method of determining d^0 will depend on the particular application under consideration. Of particular importance is its sensitivity to chemical composition and temperature. Care should be taken with the preparation of any "stress-free" material to avoid the introduction of residual stresses during manufacture (Section 4.6).

i. *Does the material exhibit any hazard?* For example, is it radioactive requiring additional shielding, or will it become active in the neutron beam? The answers to these questions are best provided by the instrument scientists and health physicist at the respective facility (Section 3.1.3).

j. *In light of the attenuation in the material, is the neutron path length sufficiently short to allow data to be collected in a reasonable time for each of the required measurement positions and orientations, and for the required sampled gauge volume?* A number of attenuation lengths are given in Tables 2.3 and A.4.1. Figure 4.14 shows a more useful measure of the strength of the diffracted signal with depth in a sample of common materials. Correspondingly, a rough guide to practical path lengths for observable diffraction peak intensity is given by the maximum economic path lengths, as listed in Table 4.2 (Section 4.4).

A.3.4   Identifying the Most Suitable Neutron Source and Diffractometer Instrument

Once it is established that it is appropriate to use neutron diffraction to determine the required stress field, the next elements to determine are the type of stress measurement diffractometer required and the most suitable type of neutron source. Availability and convenience are typically important variables. Ideally, however, you can ask if a steady source is best, and if a two- or three-axis instrument is most suitable, or if a time-pulsed source would be most suitable and, if so, in which configuration (standard, modulated intensity, or transmission using Bragg edges).

1. *Is information required from a single reflection or a number of reflections? Are significant elastic or plastic anisotropy and consequent intergranular stresses expected? Is the time structure of a pulsed source an advantage?* Generally, single-peak measurements for mapping strain at many positions is best undertaken at a steady-state source. However, new dedicated instruments currently being built at spallation sources, may change this balance. If there is concern about the effect of intergranular stresses, or more than one reflection is required, such as for study of composites, then it may be advisable to use a time-of-flight instrument. Finally, stroboscopic experiments can exploit the time structure of pulsed source instruments.

2. *Can the instrument accommodate the sample, and position it to the required accuracy in all necessary orientations?* In other words, are the available x-y-z movement and Ω rotations sufficient? This information is usually available from instrument websites, or from the responsible instrument scientist.

3. *How is the sample to be mounted on the sample table in each orientation?* The use of standard mounting plates assists rapid mounting of small samples. However, large or irregular samples inevitably require individual mounting. If possible, test the mounting procedure off-line.

4. *Can the measurements be completed without disturbing the instrument hardware, such as removing collimators or detectors? If these have to be moved will the gauge volume be disturbed?* On the whole, aperture based systems on steady state sources offer more freedom. Unless high-quality accurate positioning is provided, the realignment of radial collimators after removal or replacement can be time consuming.

5. *What type of detector should be used, a single detector or position-sensitive detector (PSD)?* Single detectors give data more directly, but PSDs can offer faster rates of data acquisition albeit sometimes with accompanying higher background levels. Definition of the gauge volume, and the avoidance of anomalous surface effects can be more difficult with PSDs, and a single detector is preferable.

A.3.4 Details of Instrument Setup

Generally, there are more options to consider on a continuous beam instrument than on a time-of-flight instrument. The optimum monochromator and sample diffraction angles to be used must be chosen after consideration of several factors, often a compromise having to be made between optimizing different aspects of the measurement.

1. *What is the optimum sample reflection to use?*

   a. Bearing in mind the available wavelengths, typically you should choose an *hkl* that will give a scattering angle of $\phi^s \sim 90°$ in order to obtain a cuboidal instrument gauge volume, although in certain cases one might choose other scattering angles in order for the beam to pass through apertures in the sample in the equipment or to minimize path lengths. The chosen wavelength must be well away from any Bragg edge in the transmission spectrum of the sample.

   b. Ideally, the reflection *hkl* should have a high multiplicity and structure factor to give optimum intensity.

   c. Preferably, the diffraction elastic constants should be known, and be close to the bulk moduli to reduce the effect of elastic anisotropy. Although higher-order reflections pose no problems, "accidental" coincidences of peaks such as (333) and (511) for cubic structures should be avoided, since these may have different elastic constants.

   d. If the sample might have been plastically deformed, the reflection chosen should be known to give low plastic pseudostrains and intergranular stresses. In other words, the lattice strain versus applied uniaxial load relationship should be essentially linear in both elastic and plastic regimes. See Table 4.1.

   e. The texture should be such that the chosen *hkl* reflection will give good intensity in all the required measurement directions.

2. *On a continuous beam source, what are the best choices for monochromator and wavelength?* For each of the available monochromators, find out what range of wavelengths/scattering angles are available. It is important to be able to measure all desired reflections *hkl* without having to change monochromator. Are these consistent with the required resolution?

   a. Check the expected flux for the various monochromator options at the desired wavelengths, and that the sample diffraction angles will not be too far from $\phi^s = 90°$. In addition, determine the monochromator has the size and focusing to optimize the required intensity or resolution required. One further consideration — if the beam is shared with another instrument closer to the source — is to ensure that the monochromator on that instrument does not vary in a manner that might affect the wavelength composition of the beam on the strain measurement instrument.

3. *What is the optimum position for apertures necessary to define the gauge volume? Can these be moved automatically?* The answer is usually as close as possible to the sample, but this may risk collisions when translating and rotating the sample. Except for automated apertures it is best not to alter settings during a series of measurements.

4. *What are the optimum collimation angles, and are these best achieved by soller collimators or by slit-apertures?* Over-collimation will result in loss of intensity, and under-collimation in poor resolution. Ideally, the resolution should be matched to other elements in the instrument setup.

5. *How can background shielding be optimized?* Background increases the time taken to measure a peak to a required uncertainty, it also decreases the sensitivity to weak reflections. Diffuse scattering from the sample cannot be avoided. Stray background peaks from sample mounts, slits, rogue beams, and so on, can be disastrous. Always check the background with no sample present to make sure that such peaks are absent, and investigate the origin of any unexpected peaks.

A.3.5 Steps in the Measurement Procedure

The steps in carrying out an experiment are listed here as a general guide only. One must always consider what action to take in the light of data already obtained. A standard two-axis reactor based instrument is used as an example.

A.3.5.1 General Points

A good rule is to ensure that all the data taken are secure, in the sense that the order of measurements is such that, if the instrument or source fails, key calibrations and checks are not outstanding. Once an adjustment, angle or translation, of a component of the instrument has been made to give its optimum setting for a series of measurements, it is sensible to switch off all motors that might affect its position if accidentally activated. It is important to set up repeat measurement scans of peaks at standard positions in the sample, in order to monitor if anything has accidentally changed during measurements.

Photographs of the neutron beam, using a scintillator camera, at various points along the beam can be very useful in ensuring that the form of the beam is the expected one.

A careful complete record of all the measurement parameters and measurement sequences must be made as described in Section 4.7.1.

A.3.5.2 Procedure

1. Set up the overall configuration of the instrument, such as optical benches, sample table, and so on. Ensure that their alignment is accurate.

2. Set up the monochromator-sample (M-S) and sample-detector (S-D) collimation required. Place aperture mounts in position.

3. Determine the reference axis and reference point relative to the sample table.

4. Set the monochromator take-off angle $2\theta^M$. Align the monochromator angle θ^M to give maximum counts in a monitor counter after the M-S collimator. Make initial adjustment of the monochromator tilt curvature to maximize the flux at the sample. This can be done, for example, by placing an aperture of approximately the size of the gauge volume in front of a low-efficiency monitor counter placed on the table, and maximizing the count.

5. Center a standard powder sample on the table, such as Al_2O_3, and scan ϕ^s to measure a number of Bragg peak angles to give the wavelength and zero angle of ϕ^s. This will also enable measured data to be used on other instruments.

6. Center the apertures so that instrumental gauge volume center lies at the reference point. A neutron photograph can be taken at the gauge volume center to verify that the volume is the correct order of magnitude.

7. Check the background, blocking off the beam at stages along its path. Add shielding as required. With no sample, scan ϕ^s through the peak position to ensure that only the background level is recorded.

8. Mount test sample. Using lasers, theodolites, and so on, check alignment. Measure distance from reference point on sample to instrument reference point at gauge volume center. Check settings by scanning sample edges through the gauge volume.

9. It is important to test for grain size and texture effects by performing an Ω-scan of the sample, or by monitoring the variation in integrated intensity from point to point.

10. Set the instrumental gauge volume to the required size. This is assuming that adjustment of apertures is possible about an accurate center; otherwise this should be done at Step 6.

11. Block beam after last aperture before sample and check that only background is observed on scanning ϕ^s through the Bragg reflection.

12. If a series of automatic settings are to be made, check extreme movements of sample do not involve any collisions with instrument hardware. Simulation software may be available to assist these checks.

13. Carry out desired measurements of strain. Include reproducibility checks at a convenient high-intensity "standard" position.

14. Remove sample.

15. Measure stress-free reference sample under identical conditions.

16. Repeat standard powder over one peak to ensure that no movement has occurred.

17. Remount sample in new orientation and repeat Steps 8 through 16.

18. Once the measurements are concluded, check the sample for any induced radioactivity. If any is found, transfer the sample to a safe radioactive storage area to allow it to decay before removal from the facility.

Appendix 4

Macroscopic Scattering Cross-Sections of All Elements

Table A.4.1 below lists the macroscopic cross sections, penetration lengths as defined in Section 2.3.7, and scattering length densities. The table is based on material provided by Hutchings and Windsor [1], but has been updated using the scattering lengths and atomic cross-sections for 1.8Å (25.2 meV) neutrons given by Sears [2]. The atomic cross-sections for Pu are derived from Dianoux and Lander [3], and densities from Lide [4]. The densities of gaseous elements at normal temperature and pressure are in the liquid state at the boiling point, so the cross-sections apply to this state. The macroscopic coherent cross-sections for thermal neutrons fluctuate rapidly at long wavelengths, and those given for 25 meV neutrons should be taken as typical mean values. In order to calculate the penetration depths for long wavelength neutrons 4.0Å, with 5 meV energies, the coherent macroscopic cross-section is taken to be zero, and the macroscopic absorption cross-section is increased in proportion to the wavelength. The uncertainty in the numbers in the six right-hand columns is governed by the uncertainty in the density and atomic cross sections from which they are calculated.

TABLE A.4.1

Macroscopic Cross Sections and Scattering Length Densities for the Elements

Atom	Atomic Number Z	Atomic Weight W	Density ρ (g/cm³)	Σ^{coh} (cm⁻¹)	Σ^{incoh} (cm⁻¹)	Σ^{abs} (25 meV) (cm⁻¹)	I_μ (Cold) 5 meV (cm)	I_μ (Thermal) 25 meV (cm)	$\rho\bar{b}10^{12}$ (cm⁻²)
H	1	1.00794	0.0708	0.07431	3.39507	0.01407	0.292	0.287	-0.01582
D	1	2.01500	0.01	0.01671	0.00613	0.00000	163.126	43.781	0.00199
He	2	4.002602	0.122	0.02460	0.00000	0.00014	3261.565	40.431	0.00598
Li	3	6.941	0.534	0.02103	0.04262	3.26632	0.136	0.300	-0.00880
Be	4	9.012182	1.848	0.94221	0.00022	0.00094	430.879	1.060	0.09620
B	5	10.811	2.34	0.46143	0.22159	99.97614	0.004	0.010	0.06908
C	6	12.0107	2.266	0.63057	0.00011	0.00040	997.199	1.585	0.07551
N	7	14.00674	0.808	0.38248	0.01737	0.06601	6.062	2.147	0.03252
O	8	15.9994	1.14	0.18159	0.00000	0.00001	54854.120	5.507	0.02490
F	9	18.9984	1.5	0.19100	0.00004	0.00046	944.556	5.222	0.02688
Ne	10	20.1797	1.207	0.09437	0.00029	0.00140	291.601	10.410	0.01645
Na	11	22.9898	0.971	0.04222	0.04121	0.01348	14.016	10.319	0.00923
Mg	12	24.3050	1.738	0.15636	0.00345	0.00271	105.137	6.153	0.02315
Al	13	26.9815	2.6989	0.09006	0.00049	0.01392	31.637	9.573	0.02078
Si	14	28.0855	2.33	0.10808	0.00020	0.00854	51.805	8.560	0.02073
P	15	30.9738	1.82	0.11702	0.00018	0.00609	72.535	8.111	0.01815
S	16	32.0660	2.07	0.03960	0.00027	0.02060	21.578	16.536	0.01107
Cl	17	35.4527	1.56	0.30543	0.14044	0.88771	0.470	0.750	0.02538
Ar	18	39.948	1.4	0.00967	0.00475	0.01425	27.320	34.891	0.00403
K	19	39.0983	0.862	0.02244	0.00358	0.02788	15.168	18.551	0.00487
Ca	20	40.078	1.55	0.06475	0.00116	0.01001	42.448	13.171	0.01095
Sc	21	44.9559	2.989	0.76075	0.18018	1.10109	0.378	0.490	0.04921

Z	El									
22	Ti	47.867	4.54	0.08482	0.16393	0.34785	1.062	1.676	-0.01964
23	V	50.9415	6.11	0.00133	0.36693	0.36693	0.842	1.360	-0.00276
24	Cr	51.9961	7.19	0.13823	0.15239	0.25399	1.388	1.836	0.03027
25	Mn	54.9380	7.3	0.14004	0.03201	1.06427	0.415	0.809	-0.02985
26	Fe	55.8450	7.874	0.95270	0.03396	0.21737	1.923	0.831	0.08024
27	Co	58.9332	8.9	0.07085	0.43654	3.38135	0.125	0.257	0.02265
28	Ni	58.6934	8.902	1.21479	0.47496	0.41011	0.718	0.476	0.09408
29	Cu	63.546	8.96	0.63557	0.04670	0.32097	1.308	0.997	0.06554
30	Zn	65.39	7.133	0.26631	0.00506	0.07292	5.949	2.905	0.03731
31	Ga	69.723	5.904	0.34039	0.00816	0.14023	3.108	2.046	0.03716
32	Ge	72.61	5.323	0.37173	0.00795	0.09713	4.442	2.097	0.03614
33	As	74.9216	5.73	0.25055	0.00276	0.20726	2.145	2.171	0.03031
34	Se	78.96	4.79	0.29153	0.01169	0.42743	1.034	1.369	0.02912
35	Br	79.904	3.12	0.13638	0.00235	0.16225	2.739	3.322	0.01598
36	Kr	83.80	2.16	0.11906	0.00016	0.38806	1.152	1.971	0.01212
37	Rb	85.4678	1.532	0.06822	0.00540	0.00410	68.636	12.866	0.00765
38	Sr	87.62	2.54	0.10806	0.00105	0.02235	19.603	7.607	0.01226
39	Y	88.9059	4.469	0.22855	0.00454	0.03875	10.967	3.679	0.02346
40	Zr	91.224	6.506	0.27659	0.00086	0.00795	53.689	3.504	0.03075
41	Nb	92.9064	8.57	0.34736	0.00013	0.06388	6.994	2.431	0.03919
42	Mo	95.94	10.22	0.36374	0.00257	0.15909	2.791	1.903	0.04308
43	Tc	98	11.5	0.40987	0.03533	1.41336	0.313	0.538	0.04805
44	Ru	101.07	12.41	0.45919	0.02958	0.18930	2.208	1.475	0.05198
45	Rh	102.9055	12.41	0.31519	0.02179	10.51605	0.042	0.092	0.04270
46	Pd	106.42	12.02	0.29860	0.00633	0.46933	0.947	1.292	0.04020
47	Ag	107.8682	10.5	0.25834	0.03400	3.71065	0.120	0.250	0.03471
48	Cd	112.411	8.65	0.14087	0.16034	116.77741	0.004	0.009	0.02257

TABLE A.4.1 (CONTINUED)

Atom	Atomic Number Z	Atomic Weight W	Density ρ (g/cm³)	Σ^{coh} (cm⁻¹)	Σ^{incoh} (cm⁻¹)	Σ^{abs} (25 meV) (cm⁻¹)	l_μ (Cold) 5 meV (cm)	l_μ (Thermal) 25 meV (cm)	$\rho b 10^{12}$ (cm⁻²)
In	49	114.818	7.31	0.07975	0.02070	7.43040	0.060	0.133	0.01559
Sn	50	118.710	7.31	0.18060	0.00082	0.02321	18.966	4.887	0.02308
Sb	51	121.760	6.691	0.12906	0.00000	0.16249	2.752	3.430	0.01843
Te	52	127.60	6.24	0.12457	0.00265	0.13841	3.204	3.765	0.01708
I	53	126.9045	4.93	0.08188	0.00725	0.14388	3.040	4.292	0.01235
Xe	54	131.29	3.52	0.04908	0.00000	0.38589	1.159	2.299	0.00794
Cs	55	132.9055	1.873	0.03132	0.00178	0.24612	1.811	3.581	0.00460
Ba	56	137.327	3.5	0.04958	0.00230	0.01688	24.966	14.543	0.00778
La	57	138.9055	6.145	0.22725	0.03010	0.23897	1.772	2.015	0.02195
Ce	58	140.116	6.77	0.08555	0.00000	0.01833	24.396	9.627	0.01408
Pr	59	140.9077	6.773	0.07642	0.00043	0.33289	1.343	2.441	0.01326
Nd	60	144.24	7.008	0.21739	0.26918	1.47758	0.280	0.509	0.02250
Pm	61	145	7.264	0.60338	0.03922	5.08044	0.088	0.175	0.03801
Sm	62	150.36	7.52	0.01271	1.17463	178.36306	0.002	0.006	0.00241
Eu	63	151.964	5.244	0.14027	0.05195	94.13934	0.005	0.011	0.01500
Gd	64	157.25	7.901	0.88656	4.56898	1503.83015	0.000	0.001	0.01967
Tb	65	158.9253	8.23	0.21331	0.00012	0.72975	0.613	1.060	0.02302
Dy	66	162.50	8.551	1.13765	1.72391	31.49930	0.014	0.029	0.05356
Ho	67	164.9303	8.795	0.25883	0.01156	2.07774	0.215	0.426	0.02572
Er	68	167.26	9.066	0.24906	0.03591	5.19005	0.086	0.183	0.02543

Tm	69	168.9342	9.321	0.20867	0.00332	3.32274	0.135	0.283	0.02349
Yb	70	173.04	6.966	0.47080	0.09697	0.84366	0.504	0.708	0.03013
Lu	71	174.967	9.841	0.22118	0.02371	2.50649	0.178	0.363	0.02442
Hf	72	178.49	13.31	0.34129	0.11676	4.67483	0.095	0.195	0.03489
Ta	73	180.9479	16.654	0.33256	0.00055	1.14178	0.392	0.678	0.03830
W	74	183.84	19.3	0.18777	0.10305	1.15696	0.372	0.691	0.03073
Re	75	186.207	21.02	0.72060	0.06118	6.09790	0.073	0.145	0.06254
Os	76	190.23	22.57	1.02888	0.02144	1.14320	0.388	0.456	0.07645
Ir	77	192.217	22.42	0.99041	0.00000	29.85271	0.015	0.032	0.07446
Pt	78	195.078	21.45	0.76679	0.00861	0.68204	0.652	0.686	0.06357
Au	79	196.9666	19.3	0.43194	0.02537	5.82121	0.077	0.159	0.04502
Hg	80	200.59	13.546	0.82312	0.26841	15.14070	0.029	0.062	0.05162
Tl	81	204.3833	11.85	0.33792	0.00733	0.11976	3.635	2.150	0.03064
Pb	82	207.2	11.35	0.36666	0.00010	0.00564	78.662	2.685	0.03103
Bi	83	208.9804	9.747	0.25695	0.00024	0.00095	423.948	3.874	0.02396
Ra	88	226	5.0	0.17320	0.00000	0.17054	2.622	2.909	0.01332
Th	90	232.0381	11.72	0.40637	0.00000	0.22417	1.995	1.586	0.03136
Pa	91	231.0359	15.37	0.41666	0.00401	8.03667	0.056	0.118	0.03646
U	92	238.0289	18.95	0.42684	0.00024	0.36293	1.232	1.266	0.04035
Np	93	237	20.25	0.72037	0.02573	9.05094	0.049	0.102	0.05429
Pu	94	239	19.84	1.24978	0.00000	27.89517	0.016	0.034	0.07049
Am	95	243	13.67	0.29474	0.01016	2.55099	0.175	0.350	0.02812

References

1. M.T. Hutchings and C.G. Windsor, Industrial applications, in *Methods of Experimental Physics*, vol. 23C, *Neutron Scattering*, K. Sköld and D.L. Price, Eds., Academic Press, New York, 1987, 405–482.
2. V.F. Sears, Neutron scattering lengths and cross-sections, *Neutron News*, 3, 26–36, 1992.
3. A.-J. Dianoux and G. Lander, *Neutron Data Booklet*, Institut Laue-Langevin, Grenoble, 2002.
4. D.R. Lide, Ed., *CRC Handbook of Chemistry and Physics*, 84th ed., CRC Press, Boca Raton, FL, 2003–2004.

Author Index

A

Aborn, R.H., 3
Ageron, P., 66
Alberry, P. J., 276
Albertini, G., 88
Alexander, K. B., 306
Allen, A. J., 5, 7, 39, 78, 105, 211, 310, 318
Allen, B. L., 6
Allen, R. J., 339
Allenspach, P., 75
Allman, B. E., 73
Anderson, I. S., 73
Andreani, C., 5, 39, 78, 105, 318
Armstrong, W. D., 8
Arsenault, R. J., 232
Asano, H., 182
Ashby, M. F., 18, 20
Ashcroft, N. W., 35
Austin, M. W., 292
Averbach, B. L., 62, 256, 313
Avis, N. J., 125
Axe, J. D., 87

B

Bacon, G. E., 26, 30, 51, 59, 62, 182
Balart, M. J., 329
Barrett, C. S., 185, 192
Barton, N., 233
Bar-Ziv, S., 87
Bate, P., 233
Baxter, G. J., 197, 288, 336
Beaudoin, A. J., 233
Becher, P. F., 306
Becker, R., 233
Berliner, R. R., 102
Berveiller, M., 233
Beyerlein, R. A., 162, 182
Bhadeshia, H. K. D. H., 15, 276
Biermann, H., 292
Bishop, J. F. W., 232
Blech, I. A., 62
Bollenrath, F., 210

Bonnafe, J. P., 292
Bonner, N. W., 246, 311, 334, 335
Borsa, C. E., 305
Bouchard, P. J., 12, 18, 126, 190, 280
Bourke, M. A. M., , 7, 8, 96, 99, 150, 183, 211, 244, 246, 251, 305, 309, 310, 311, 317, 318, 323, 331, 339
Bouzina, A., 329
Bowen, P., 295
Boyce, D., 233, 244, 246
Bragg, W. H., 2
Brand, P. C., 89, 109, 174
Breiting, B., 87
Brockhouse, B. N., 5
Brøndsted, P., 7, 8
Brook, R. J., 305
Browne, P. A., 130, 131
Brun, T. O., 162, 182
Brune, J. E., 5
Bruno, G., 123
Budai, J. D., 34
Budiansky, B., 232
Buerger, M. J., 35
Buffiere, J.-Y., 295
Buttle, D. J., 272, 333

C

Caglioti, G., 105
Cannon, D. F., 339
Carlile, C., 123
Carpenter, J. M., 9, 66, 73, 97, 162, 182
Chakera, A., 51
Chastel, Y., 245
Chen, H., 73
Cheskis, H. P., 292, 294, 309
Cheung, S., 87
Choo, H., 309
Christodoulou, N., 233, 247
Cimmino, A., 73
Clarke, A. P., 8, 246, 305, 310, 318, 320
Clarke, D. R., 291

Index

O

One-site model, 232
Optical bench, 83, 88, 90
    alignment, 376
Orientation distribution function,60
Overlapping peak fitting, 178

P

Paris regime, 20
Particulate composites
    thermal stresses, 306
Pawley-Rietveld refinement. See Rietveld
        refinement
    Bragg edge, 142, 192-6
Peak broadening, 254-7
    analysis, 256, 313
    dislocations, 255
    instrumental, 103-8
    microstress, 13, 16, 97, 312
    particle size, 38, 255, 313
    stacking faults, 255
Peak fitting, 164–82
    correlated fit parameters, 175
    multiple peaks, *see* Pawley-Rietveld
        overlapping peaks, 178
    range of scan, 174
    sensitivity analysis, 181
    single peak, 159, 162
    strategy, 181
    uncertainty in intensity, 167
    uncertainty in position, 164, 166
    uncertainty in width, 167
    VAMAS recommendations, 178
    weighting, 177
Pearson Type VII function, 160
Peening stresses, 10, 329
Penalty factor of background, 166
Penetration of neutrons, 55
    table of, 33, 379
Penetration of x-rays, 3–4, 30, 33, 294
Phase analysis, 311
    across welds, 288
Phase selectivity, 7, 16, 17, 39, 210, 227, 264
    composites, 292, 297
Phase transformations
    residual stress, 10, 276, 311
    strain-free lattice spacing, 185
    Type-II stress, 276
    weldments, 288
Piezoelectrics, 348
Plastic anisotropy, 151, 191, 211, 230-54
    appropiate instrument, 373

correction by experiment, 249, 318
    less affected reflections, 154, 191.242,254,
        316
Plastic deformation
    effects of residual stress, 17
    lattice strain response, 239–54, 316–24
    multiaxial, 324
Plastic misfit stresses, 309
    continuous fibre composites, 310
    polycrystal, 230–54, 239, 316
Plastic macroscopic strain
    inferred, 211-2, 242-4, 251, 322
Poisson's ratio
    hkl specific, 227
    polycrystalline bulk, 223
POLDI, 108, 142
Pole figure, 60
Polycrystalline deformation, 210–54
Position sensitive detectors, 101
    collimation, 88
    radial collimators, 96
Positioner, 97, 124
    effect of uncertainty, 122
    specification, 121
    Stewart platform, 123
Postweld heat treatment, 338
Powders
    calibration, 82
    strain-free lattice spacing, 188
Precipitation, 282
    effect on strain-free lattice spacing, 184,
        275
Pre-stressed concrete, 17
Primary extinction, 58
Principal strains, 205
Principal stresses, 205
Pseudo strains, 127, 191, 242
Pulsed beam source, 73–76
    asymmetric peaks, 161
    instrument, 92–100
    spallation, 74

R

Radial collimators. See Collimation
Radiation exposure, 77
Radioactive samples, 136
Radiography
    of phase, 144
    of strain, 142
Railway rails
    through life study, 339
    welded, 278
Raman spectroscopy